SOIL PHYSICS

SOIL PHYSICS

FIFTH EDITION

William A. Jury
University of California, Riverside

Wilford R. Gardner
University of California, Berkeley

Walter H. Gardner
Washington State University, Pullman

JOHN WILEY & SONS, INC.
New York • Chichester • Brisbane • Toronto • Singapore

Library of Congress Cataloging in Publication Data:
Jury, William A., 1946–
  Soil physics / William A. Jury, Wilford R. Gardner, Walter H.
 Gardner.—5th ed.
  Includes bibliographical references and index.
  1. Soil physics. I. Gardner, Wilford R. II. Gardner, Walter H.
 (Hale), 1917– . III. Title.
 S592.3.J86 1991
 631.4′3—dc20   90-24351
 ISBN 0-471-83108-5   CIP

Printed and bound in the United States of America

10 9 8 7 6

To the memory of Champ B. Tanner:
Scientist, teacher, and friend

PREFACE

The discipline of soil physics has changed considerably since the previous edition of this book was published in 1972. Research on chemical pollution of the environment is now of the highest priority worldwide. As a result, the once minor field of solute transport through soil is now a mainstream research area of soil physics.

Another obvious change in the focus of soil physics over the past 20 years has been the shift in orientation from the laboratory to the field regime. This transition has given rise to new areas of research characterizing the effects of spatial and temporal variations in soil properties on the mean behavior of large-scale soil systems.

The final and perhaps most significant evolution in the soil physics discipline over the past two decades has been the increased use of mathematics and statistics in research publications. Transport equations, once solvable only by approximate analytical or iterative methods in one dimension, are now routinely solved by high-speed computers, even in two and three dimensions. The variability of the field regime has prompted the development of new methods of statistical analysis based on the theory of stochastic processes.

This edition of *Soil Physics* has been altered significantly to reflect the changes in the discipline. The book focuses heavily on transport processes and problem solving, teaching the reader to simplify the general theory for specific applications. This approach is developed systematically using physical principles rather than empirical laws and is illustrated throughout the text with over 70 completely worked examples that provide the reader with a model for using the theory in a practical manner.

The book should be suitable as an advanced undergraduate or graduate level instruction text in soil physics as well as a reference text for professional scientists. Chapter 1 is used to describe the important physical and chemical properties of the soil solid phase, emphasizing those characteristics that are most relevant to the transport, retention, or transformation of water, heat, gases, and solutes as well as emphasizing the properties of agricultural significance.

Chapter 2 characterizes water in soil, first describing the molecular and fluid properties of water and then developing the thermodynamic description of water potential energy. The relationship of the potential components to soil water measurement devices is established quantitatively by equilibrium analysis.

Chapters 3–4 introduce the theory of water transport through saturated and unsaturated soil and provide numerous approximate models of water flow. Chapters 5–7 deal with the transport of heat, gases, and dissolved solutes in soil, with emphasis on practical problems encountered in the field. Chapter 7, covering solute movement, offers a general treatment of dispersion not tied to any single set of assumptions and provides many examples of environmental pollution calculations. Chapter 8 deals with methods of assessing the properties of spatially variable soil, including a description of the most modern methods.

The material in this book has formed the basis of a soil physics course taught for many years by the senior author at the University of California, Riverside. Numerous students have contributed to the text by making helpful suggestions for improvement of draft sections used in the course. Special thanks are owed to Professors Robert Graham and Chris Amrhein of the University of California at Riverside for their critical comments on Chapter 1 and to Dr. Jeremy Dyson for his review of and suggestions on Chapter 7. Appreciation is also expressed to Mrs. Dorothy McAtee for her careful typing of the many drafts of the manuscript.

Riverside, California   WILLIAM A. JURY

Berkeley, California   WILFORD R. GARDNER

Pullman, Washington   WALTER H. GARDNER

May 1991

CONTENTS

SOIL PHYSICS

1 The Soil Solid Phase

The soil is a very complex system, made up of a heterogeneous mixture of solid, liquid, and gaseous material. The solid phase is composed of a mineral portion, containing particles of varying sizes, shapes, and chemical composition, and an organic fraction. The organic fraction is very heterogeneous, containing a diverse population of live, active organisms as well as plant and animal residues in different stages of decomposition. The liquid phase consists of soil water, which fills part or all of the open spaces between the solid particles. The water phase contains solutes that may have been dissolved from the soil mineral phase or may have entered through the soil surface. The water phase is held by forces in the soil matrix and varies significantly in mobility depending on its location. The gaseous or vapor phase occupies the part of the pore space between the soil particles that is not filled with water; its composition can differ considerably from that of the air above soil surface and may change dramatically in a short time period. The chemical and physical relationships among the constituents of the solid, liquid, and gaseous phases are affected not only by the properties of each component but also by temperature, pressure, and light.

The soil solid phase has a dominant influence on many water, heat, and solute transport and retention processes. Therefore, characterizing the physical and chemical properties of the soil solid phase is essential to an understanding of many of the practical agricultural and environmental problems within the purview of modern soil physics research. The solid phase consists of particles of vastly different shapes and sizes, which range from the lower limits of the colloidal state to the coarsest fractions of sand and gravel. The different particles, especially those of colloidal dimensions, may be found in states ranging from almost complete dispersion to nearly perfect aggregation. In most soils, however, there is only a partial aggregation of the individual particles.

The completely dispersed or primary particles are usually referred to as textural separates. The aggregates, formed from a consolidation of individual particles, are generally regarded as the structural units.

1.1 CHARACTERISTICS OF THE PRIMARY PARTICLES

The solid phase is composed of the products of weathering of the parent rocks or material and the minerals they contain. Foremost among the important characteristics of the dispersed phase are the sizes and shapes of the individual particles

1

resulting from this weathering, their chemical and mineralogical composition, and the physical and chemical properties of their surfaces.

1.1.1 Characterization of Particle Size

Experimental characterization of the size of individual soil particles is achieved by a combination of two procedures: sieving and sedimentation. A soil sample whose mineral phase is to be characterized is first pretreated to remove organic material and to disperse aggregates and then is passed through a series of coarse screens of specified opening. The smallest screen for which this operation is feasible has an aperture of about 0.05 mm (Gee and Bauder, 1986). Care must be taken to ensure that none of the small particles will be excluded from passage through the screen at this stage.

The sizes of the remaining dispersed soil separates are characterized indirectly by sedimentation procedures using either the hydrometer method or the pipette method. Both methods depend fundamentally on Stokes's law, which is used to establish a relationship between the settling velocity V (the speed with which a particle falls through a solution under the influence of gravity) and the particle size.

When a spherical particle of density ρ_s and radius R falls through a liquid of density ρ_l and viscosity η, it is subject to three forces: gravitation, buoyancy, and viscous drag. The force of gravity F_g is calculated from Newton's law:

$$F_g = m_s g = \rho_s \left(\frac{4\pi R^3}{3} \right) g \qquad (1.1)$$

where m_s is the particle mass. This force acts downward. The buoyancy force F_b is calculated by Archimede's principle, which states that a solid object fully immersed in a fluid is buoyed upward by a force equal to the weight of the fluid displaced by the solid volume. Thus

$$F_b = m_l g = \rho_l \left(\frac{4\pi R^3}{3} \right) g \qquad (1.2)$$

where m_l is the liquid mass displaced by the solid volume. This force is upward.

The viscous drag force F_d that the particle feels from the surrounding liquid as it falls at velocity V is calculated with Stokes's law (Bird et al., 1960), which states that

$$F_d = 6\pi R \eta V \qquad (1.3)$$

This force is always opposite to the direction of the velocity V of the spherical particle with respect to the fluid. Thus, it is upward for a particle falling downward.

These three forces rapidly equilibrate, and the particle reaches a final velocity that may be calculated by setting the net force equal to zero:

$$\sum F_i = 0 = F_g - F_b - F_d \tag{1.4}$$

When (1.1)–(1.3) are inserted into (1.4) and the resulting equation is solved for V, we obtain

$$V = \frac{(\rho_s - \rho_l)D^2 g}{18\eta} \tag{1.5}$$

where $D = 2R$ is the particle diameter.

Equation (1.5) gives the velocity of a spherical particle of diameter D and density ρ_s falling through a fluid of density ρ_l and viscosity η. Thus, in a solution containing a perfectly mixed suspension of particles of different diameters, the individual particles will settle to the bottom at different rates. For example, in a time t after stirring of suspension stops, all particles of diameter D greater than D' will have fallen a distance greater than or equal to $X = V't$, where V' is the value of (1.5) when $D = D'$. Therefore, there will be no particles of diameter larger than D' between the water surface and depth X in the suspension beaker at time t. In the hydrometer method of measuring particle size, a precalibrated floating object rests in the solution that is undergoing sedimentation. As the particles fall out of the upper part of the solution, the float sinks lower into the solution because the density of the displaced solution is decreasing. The position of the float at any given time can be used with the calibration curve for the device to calculate the density of the solution and hence the mass of the particles of a given size range that are still in the upper part of the suspension beaker.

The pipette method involves direct sampling of the solution in suspension at various times. Further details of this procedure are given in Gee and Bauder (1986).

1.1.2 Classification of Textural Size Fractions

The different particles in soils are classified into groups of various sizes on the basis of their equivalent diameter, which is an experimental measurement of the effective size of the particles. Thus, the equivalent diameter of the larger particles that can be removed by mechanical sieving (i.e., > 0.5 mm) refers to the diameter of a sphere that will pass through a hole of a given size in a sieve. For the smaller particles, which must be separated by sedimentation techniques, the equivalent diameter of a particle refers to the diameter of a sphere that has the same density and settling velocity in the dispersing medium.

Several systems of soil texture classification have evolved over a period of years. Two of the five classification systems shown in Fig. 1.1 [the U.S. Department of Agriculture (USDA) system and the International Society of Soil Science (ISSS) system] are used by agricultural scientists. Note that only the upper size limit of clay, 0.002 mm, is common to both classification systems. The ISSS classification was suggested by Atterberg (1912), who established the upper limit of the clay

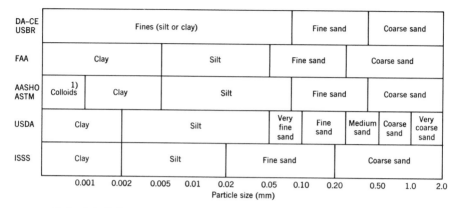

1) Reported with clay
DA--CE = Department of Army, Corps of Engineers
USBR = U.S. Bureau of Reclamation
FAA = Federal Aviation Authority
AASHO = American Association of State Highway Officials
ASTM = American Society for Testing and Materials
USDA = U.S. Department of Agriculture
ISSS = International Society of Soil Science

Figure 1.1 Classification of soil separates less than 2.0 mm on the basis of particle size (DA-CE, Department of Army, Corps of Engineers; USBR, U.S. Bureau of Reclamation; FAA, Federal Aviation Authority; AASHO, American Association of State Highway Officials; ASTM, American Society of Testing and Materials; USDA, U.S. Department of Agriculture; ISSS, International Society of Soil Science). (Adapted from USDA, 1951, and Portland Cement Association, 1962.)

fraction of soil on the basis that particles smaller than 0.002 mm (2 μm) exhibited Brownian movement in aqueous suspension and thus would not fall freely in solution under the influence of gravity. This definition for the clay particle size was given further justification by mineralogical studies that showed (see Section 1.1.3) that relatively few unweathered primary minerals existed in fractions smaller than 0.002 mm. Thus, the clay fraction may be regarded as a mixture of secondary minerals or weathering products, in contrast to the coarser fractions that are typically primary minerals. Prior to 1938, the USDA classification system used 0.005 mm (5 μm) as the upper limit of clay particle size; this limit had been established arbitrarily in 1896, because it represented the smallest size that could be detected by the microscopes that were used then to observe the size of the particles. In 1938, the 2-μm limit of clay particle size from the Atterberg classification was adopted by the USDA system without changing the size limits of the other separates. The two systems are not too dissimilar, except that the classification used by the USDA makes more separations in the sand fractions than the one used by the ISSS.

The clay fraction has a significant influence on many physical and chemical processes that occur in soil, primarily because the small particles have such a large and reactive surface area. In contrast, the sand and silt fractions typically do not have much influence on chemical processes, and their relatively smaller surface areas do not adsorb or retain water nearly to the degree that clays do. Consequently, the sand and silt portion of the soil matrix may be regarded as a passive entity whose influence on soil water is manifested primarily by the geometric arrangement of the particles.

The distinct properties of the clay fraction as compared to the sand and silt fractions will allow some generalizations to be made about various characteristics of field soils that contain dominant amounts of one or the other of these particle size fractions. For example, sandy soils, because they do not retain water significantly, require more frequent water additions to avoid plant water stress than do soils in the same climate that contain significant amounts of clay. In the course of this book we will make many such generalizations about "sandy" and "clayey" soils based on the dominant properties of the respective particle size fractions.

1.1.3 Chemical and Mineralogical Properties

For the most part, this book will only discuss chemical and mineralogical characteristics of soil properties that are relevant to the transport and retention of heat and mass in soil. There are adequate references in the literature that provide comprehensive information about soil mineralogy (Brown, 1961; Dixon and Weed, 1988; Grim, 1953, 1962; Jackson, 1964; Marshall, 1964; van Olphen, 1963.) However, in order to understand the physical and chemical properties of the soil clay fraction, it is necessary to describe its mineralogical composition in some detail. For the more inert silt and sand fractions, it will suffice merely to indicate which minerals comprise the bulk of each particle size group.

Sand and Silt Fraction   The sand and silt fractions of soil contain many primary minerals that are important in soil weathering and development processes and may exert an influence on certain chemical processes in the soil as well. Some of these primary minerals directly determine the mineralogical nature of the clay particles formed in the weathering process. For example, soils derived from schists that have a high content of mica in the sand fraction may contain a considerable amount of vermiculite along with mica in the clay fraction. This is in agreement with the recognized weathering sequence of mica → vermiculite → montmorillonite (Douglas, 1988). Because the larger particles that make up the sand and silt fractions have relatively small specific surface area (surface area per unit mass), their influence on many chemical and physical properties of soil is limited.

Most of the minerals in the coarser soil fractions consist of quartz and aluminosilicates, primarily feldspars. Mineralogical analyses have shown that the feldspars are not found in abundance in particle size fractions smaller than 2 μm, although there are quartz particles as well as weathering products smaller than this size. Mineralogical analyses of New Zealand soils (Fieldes, 1962) indicated the presence of 15–21 mineral species in the coarser fractions and 8–12 different types in the clay fraction. Although quartz was the most abundant mineral found in the sand fraction of nonvolcanic soils in Fieldes's (1962) study, there were many other minerals present as well, most of which were aluminosilicates and oxides. Feldspars were a primary constituent of the sand fraction.

Clay Fraction   The clay fraction is composed primarily of Si, Al, Fe, H, and O along with varying amounts of Ti, Ca, Mg, Mn, K, Na, and P. Although the elemental chemical composition of clay has been known for many years, mineralogical and structural units formed from these primary compounds have been identified only recently.

Prior to the development of quantitative means for the identification of crystalline structure, clay was thought to be made up of a mixture of hydrated oxides of silicon, aluminum, and iron, with the bases present in the adsorbed state. Only when X-ray diffraction techniques were applied to mineralogical studies of clays was this concept replaced with a quantitative model of clay structure. (Hendricks, 1942; Hofmann et al., 1933; Kelley et al., 1931; Marshall, 1935a; Pauling, 1930; Ross and Shannon, 1926). X-ray analyses proved that clays are primarily made up of distinctly crystalline minerals, although there may be certain amounts of noncrystalline material present. The major constituents of the minerals are silicon, aluminum, ferrous and ferric iron, magnesium, and oxygen atoms plus hydroxyl groups.

Two basic structural units are responsible for the patterns of construction of the different clay minerals. The first unit is the silicon tetrahedron, in which oxygen atoms are arranged in such a way that each forms a corner of a tetrahedron held together by a silicon atom in the center (see Fig. 1.2*a*). It is possible for the aluminum atom to substitute for silicon in the tetrahedron.

In an idealized structure, the tetrahedra thus formed are linked together in a planar formation known as the silica or tetrahedral sheet. In this sheet, the three oxygen atoms that form the base of the tetrahedron are jointly shared by three adjacent tetrahedra. Figure 1.2*b* shows a basal view of the silica sheet with the oxygen atom attached to the apex of the tetrahedron pointed downward below the oxygens in the plane of the paper. Thus, the holes in this sheet are ringed by six oxygen atoms in a symmetrical hexagonal arrangement. When this structure repeats in a plane, the unit cell within the long sheet consists of four silicon atoms and six oxygen atoms, as indicated by the dashed rectangle in Fig. 1.2*b*. Two of the silicon atoms are located entirely within the cell (in tetrahedra 3 and 4) whereas four more (in tetrahedra 5, 6, 7, and 8) are on the cell boundary and hence are shared with the adjacent cell. The six oxygen atoms of the unit cell consist of three from tetrahedron 4 and one-half of each of the two atoms from tetrahedra 3, 7, and 8 that are on the edge of the cell. Since the silicon atoms lie within the tetrahedra, the base of the unit cell is an oxygen surface.

The second basic structural unit of clay is the aluminum octahedron, in which six hydroxyl groups (OH) or oxygen atoms are arranged so that each forms a corner of an octahedron held together by an aluminum atom in the center (see Fig. 1.2*c*). It is possible for magnesium and ferrous and/or ferric iron atoms to substitute for aluminum in the center of the octahedron. The octahedra thus formed are linked together in a planar formation known as the alumina or octahedral sheet, in which the six OH groups that form the octahedron are jointly shared by three adjacent octahedra. This arrangement is illustrated in Fig. 1.2*d*, which shows a top view of the octahedral sheet where six OH groups form a symmetrical hexagonal pattern

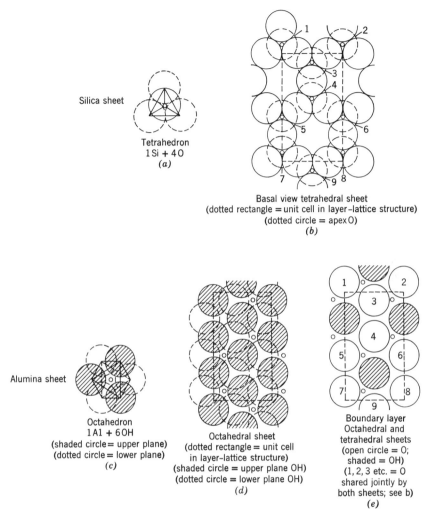

Figure 1.2 Structural arrangements in clay minerals. (Adapted from van Olphen, 1963. Used with permission of John Wiley & Sons.)

that produces a closely packed arrangement with another OH in the center. In a long plane formed by repeated groups having this structure, the unit cell consists of four aluminum atoms and six OH groups, as shown in the dashed rectangle in Fig. 1.2d. Both the top and bottom of the alumina sheet are hydroxyl surfaces.

The silica and alumina sheets join together to form a layer silicate clay mineral. To visualize how this occurs, the two sheets in Fig. 1.2 are altered as follows. First, the OH groups in the upper plane of the alumina sheet shown in Fig. 1.2d, corresponding to positions 1–9 in Fig. 1.2e, are removed. Second, the silica sheet shown in Fig. 1.2b is placed on top of the alumina sheet, so that the oxygen atoms at the apex positions 1–9 occupy the corresponding positions in Fig. 1.2*e*. Thus, the two sheets are held together by jointly shared oxygen atoms that are on the apexes of the silica tetrahedra. Figure 1.2*e* shows only the oxygen–hydroxyl boundary layer between the alumina and silica sheets along with the aluminum atoms of the accompanying octahedra. When so joined, the aluminum in each octahedron is now surrounded by four OH groups and two oxygen atoms. The OH groups at the boundary are found directly opposite the holes in the silica sheet. When the top silica–oxygen and the bottom alumina–hydroxyl sheets are added, the resultant structure is a single layer or 1:1 clay mineral.

A schematic drawing of the basic structural unit of a 1:1 clay mineral lattice is shown in Fig. 1.3. The oxygen surface of the tetrahedral sheet has six atoms, the hydroxyl surface of the octahedral sheet has six OH groups, and the boundary between the sheets has four oxygen atoms and two OH groups. This figure shows that the unit cell of the clay lattice has 28 positive and 28 negative charges to make it electrostatically neutral.

A second silica sheet can be added to the bottom of the alumina sheet in the manner just described to create 2:1 clay minerals. A schematic drawing of such a mineral is shown in Fig. 1.3. There are now two boundaries and two oxygen surfaces. The unit cell has 44 positive and 44 negative charges to make it electrostatically neutral.

The structures of the clay minerals shown in Fig. 1.3 are idealized. In reality, many clays have important irregularities that impart different properties to the mineral. First, isomorphic substitution of one atom for another of similar size in the crystal can change the properties of the clay mineral (Marshall, 1935b). The trivalent aluminum atom, which has a radius of 0.57 Å (1 Å = 10^{-10} m) can substitute in the silica sheet for the tetravalent silicon atom, which has a radius of

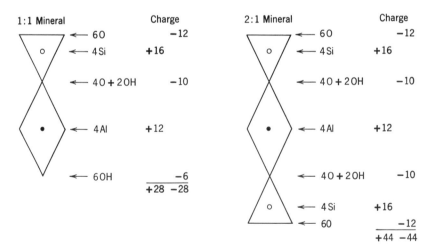

Figure 1.3 Diagrammatic sketches of sheets in clay minerals. (Courtesy of L. P. Wilding.)

0.39 Å. When this occurs, the larger aluminum atom forces the four oxygen atoms of the tetrahedron farther apart than those that surround the silicon atoms, thus introducing a strain in the crystal. This substitution also decreases the positive charge on that particular tetrahedron by 1, thus creating a net negative charge that must be balanced either by a nearby positive charge or by adsorbing a cation from the surrounding solution.

Similarly, the divalent magnesium and ferrous iron atoms and trivalent ferric iron, which have radii of 0.78, 0.83, and 0.67 Å, respectively, can substitute for the aluminum atom in the alumina sheet. These larger atoms will push the hydroxyl groups farther apart than those that surround the aluminum atoms, thereby causing strains in the octahedral units. Also, the divalent atoms will increase the negative charge of the octahedron where the isomorphic substitution takes place, which is usually balanced by an extra cation adsorbed on the surface. Cations adsorbed on the surface of charged clay minerals can be replaced by other cations in the surrounding soil solution. Therefore, they are often called "exchangeable cations." The total number of exchangeable cations that a surface will adsorb (generally expressed as me g^{-1} or cmol kg^{-1} of clay) is called the *cation exchange capacity* (CEC) and reflects the degree of isomorphic substitution within the crystals of the clay.

In addition to strains resulting from isomorphic substitution of larger, lower valence atoms within the crystal, clay minerals can have other physical properties that differ from those of a mineral possessing the ideal structure in Fig. 1.3. It has been suggested that a few of the silica tetrahedra may be inverted instead of having their apex pointing into the octahedral sheet (Edelman and Favejee, 1940). Such a variation in arrangement would change the nature of the crystal surface, creating crystal strains and making the mineral surface rough rather than smooth. Moreover, electron microscope observations have indicated that the edges of minerals may have both beveled and frayed structure (Jackson, 1964).

Radoslovich (1960; Radoslovich and Norrish, 1962) has postulated that the silica tetrahedra can rotate freely within the crystal, thus distorting the hexagonal symmetry of the ideal structure and imparting more of a ditrigonal symmetry to the surface. These distortions and the strain they create in the crystal are partially relieved by a contraction of the tetrahedral layer caused by rotation of up to 30° of the basal tetrahedra. The silica tetrahedra can rotate until the six nearest oxygen atoms are in contact with the interlayer cation.

Because there are many different ways in which silica and alumina sheets can combine to form a lattice structure and because there are several different substituting atoms in the crystal, there are a number of different types of clay minerals, many of which have distinctly different properties. Layer silicate clay minerals may be divided into five major groups (Dixon and Weed, 1988): (i) kaolin group with a 1:1 crystal lattice; (ii) mica ("illite") group with 2:1 crystal lattice; (iii) vermiculite group with a 2:1 expanding crystal lattice; (iv) smectite group with a 2:1 expanding crystal lattice; and (v) chlorite group with a 2:2 crystal lattice. Regular and random interstratified combinations of these groups of clay minerals have also been identified (Sawhney, 1988). A fibrous clay group, consisting of palygorskite and sepiolite with alternating 2:1 ribbons comprising the crystal lattice, is less abundant in soils.

The mineral structure that is characteristic of the kaolin group consists of one silica and one alumina sheet. The chief minerals which belong to this group are kaolinite, dickite, nacrite, and halloysite (Brindley, 1961). There is little, if any, isomorphic substitution in the crystal lattice of these clay minerals. Unit layers are held together by hydrogen bonding between the hydrogen atoms in the hydroxyl surface of one unit and the oxygen atoms in the oxygen surface of the adjacent unit. The unit layers are bonded together so tightly that ions or water molecules cannot permeate the interlayer positions between adjacent kaolinite layers. Consequently, the colloidal properties of kaolin minerals are determined by the external mineral surfaces only. There is little swelling or shrinkage and low plasticity. The charge imbalance in this mineral group is principally caused by broken bonds on the edges of the clay mineral, but the CEC is generally less than 0.1 me g^{-1} of clay. A diagrammatic edge view of two unit layers of kaolinite is shown in Fig. 1.4.

The mineral structure characteristic of the mica group consists of one alumina sheet between two silica sheets. There is isomorphic substitution of aluminum for silicon in the silica sheets and magnesium or iron for aluminum in the alumina sheet. The two major types of minerals belonging to this group are biotite and muscovite (Fanning et al., 1988). The major isomorphic substitutions in muscovite occur in the silica sheets, whereas three Fe^{2+} and three magnesium ions replace aluminum in the octahedral sheet of biotite. The resulting negative charges are balanced by potassium ions, whose radius (1.33 Å) just fits into the holes within the hexagonal oxygen rings of the silica sheets.

If potassium ions fit perfectly in the space, they would hold the structure together while being surrounded by 12 oxygen atoms. However, the potassium ion apparently does not fit completely in the basal network because of some rotated

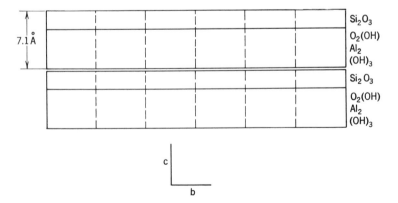

Figure 1.4 Diagrammatic edge view of two layers of kaolinite six unit cells in width. (Courtesy of Marshall, 1964. Used with permission of John Wiley & Sons.)

tetrahedra (Radoslovich and Norrish, 1962) and consequently is in a 6-fold rather than a 12-fold coordination with the oxygen atoms. Nevertheless, the unit layers of mica minerals are held together by the potassium ions, even though their size does not permit as close a spacing as would be expected if the ions fitted completely in the holes. The CEC of mica minerals varies between 0.2 and 0.4 me g^{-1}. Since the unit layers are held so tightly together by the potassium ions, the exchange capacity resides mainly on the external surfaces or on frayed or scrolled crystal edges, except where there may be an interstratification with an expanded, hydrated layer that would create some internal adsorbing surface area. Mica is one mineral in which the CEC does not reflect the degree of isomorphic substitution in the crystal.

Vermiculite minerals differ from the mica minerals primarily because magnesium, rather than potassium, is the interlayer exchangeable cation that helps to balance the negative charges on the lattice arising from a sizable substitution of aluminum for silicon in the silica sheet. The magnesium ions are highly hydrated. Consequently, the interlayer space contains both exchangeable cations and a double layer of water molecules that hold the unit layers together. The lattice has only limited expansion, depending primarily upon the size of the exchangeable ions present in the interlayer. The CEC of vermiculites varies between 1.0 and 1.5 me g^{-1}, which includes both external and internal surfaces.

The mineral structure characteristic of the smectite group (Borchardt, 1988) consists of one alumina and two silica sheets. The spacing between the 2:1 unit layers may expand or contract depending on the amount of water and interlayer cations present. The most important members of this group are montmorillonite, beidellite, and nontronite. In the ideal montmorillonite, substitution consists of iron or magnesium for aluminum in the alumina sheet. Thus, negative charges originate primarily in the octahedral sheets and are balanced by exchangeable cations that reside between the unit layers. The hydration of these cations, along with the adsorption of water molecules on the oxygen surfaces of the silica sheets through hydrogen bonding, causes interlayer swelling and an expansion of the crystal lattice. The extent of expansion or contraction of the lattice varies with the nature of the exchangeable cation and the degree of hydration of the internal surfaces. A diagrammatic edge view of montmorillonite with the interlayer of water and cations between the basic units of the crystal is shown in Fig. 1.5. The CEC of montmorillonite varies between 0.8 and 1.5 me g^{-1}, including both internal and external surfaces.

Beidellite is an end member of the smectite group in which, ideally, most of the negative charge arises from substitution of aluminum for silicon in the silica sheet. Thus, the site of the negative charge is closer to the interlayer cation, making beidellite less expansive than montmorillonite. In nontronite, iron replaces aluminum in the alumina sheet.

The chlorite mineral has a structure that is very similar to that of vermiculite, except that the exchangeable magnesium and water interlayer between the basic unit layers is replaced by an interlayer crystalline sheet in which magnesium instead of aluminum is in octahedral coordination with OH groups. In the chlorite

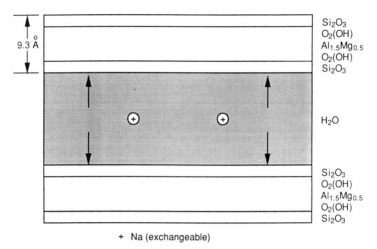

$9.3 \overset{\circ}{A}$

Si₂O₃ → Si_2O_3
O₂(OH) → $O_2(OH)$
Al₁.₅Mg₀.₅ → $Al_{1.5}Mg_{0.5}$
$O_2(OH)$
Si_2O_3

H_2O

Si_2O_3
$O_2(OH)$
$Al_{1.5}Mg_{0.5}$
$O_2(OH)$
Si_2O_3

+ Na (exchangeable)

Figure 1.5 Diagrammatic edge view of montmorillonite. (Courtesy of Marshall, 1964. Used with permission of John Wiley & Sons.)

mineral, substitution of aluminum for silicon in the silica sheets gives it a negative charge, while substitution of aluminum and Fe^{3+} for magnesium in the interlayer crystalline sheet gives it a positive charge. Thus, the three layers of this mineral are bound together electrostatically. Consequently, there are two silica, one alumina, and one magnesia sheet in the chlorite structure. Because there is an overall electrostatic balance within the crystal, the CEC only ranges between 0.1 and 0.4 me g^{-1}. Its physical properties are similar to those of mica.

The palygorskite or fibrous clay group has a 2:1 lattice structure in which the silica tetrahedra are inverted in alternate strips on both sides of the basal oxygen plane. This produces a box or chainlike structure (Bradley, 1940). There are two members of this group, palygorskite and sepiolite.

The clay minerals discussed in the preceding, with the exception of members of the fibrous clay group, all have layered structures consisting of either tetrahedral or octahedral arrangements of oxygen atoms or OH groups with atoms of Al, Si, Mg, or Fe. There are also clay minerals made up of a mixed stacking of unit layers in the crystal instead of a uniform stack of the same units. When this occurs, the minerals are called interstratified or mixed-layer clays. Such interstratification can create surface properties that are different from those of the original clay mineral, particularly with respect to ion exchange and hydration.

The reader is referred to the references on clay mineralogy cited previously for further information about the structure and chemistry of the crystalline materials that constitute the clay fraction of soils. It suffices for the purposes of this text to realize that differences in the crystal lattice structure of various clays will have an impact upon their physical and chemical properties. Some of these will be discussed further in this chapter.

There are also some noncrystalline materials within the clay size fraction that might be part of the composition of a soil depending upon the nature of the soil and the conditions under which it was formed. For example, the volcanic ash soils often contain the noncrystalline material allophane. Allophane is a general name applied to hydrous aluminosilicates with a predominance of Si–O–Al bonds, variable composition, and short-range order (Wada, 1988). Since it lacks rigid structural characteristics, it possesses a very large specific surface, has high cation and anion exchange capacities, and contains tremendous quantities of water. Imogolite, a paracrystalline assemblage of a one-dimensional structural unit, is closely associated with allophane in terms of occurrence, composition, and properties.

The free oxides (including hydroxides) of aluminum, iron, and manganese are present in varying amounts in most soils. They are most abundant in Oxisols, where advanced weathering has depleted the soil of more soluble minerals. Aluminum hydroxide is typically present in soils as the mineral gibbsite ($AL(OH)^3$), whereas goethite ($FeOOH$) and hematite (Fe_2O_3) are the most common iron oxides in soils. Crystalline and, especially, noncrystalline aluminum and iron oxides play an important role in the stabilization of soil aggregates (Hsu, 1988; Schwertmann and Taylor, 1988). Their small particle size and positive charge at low pH make them effective as binding agents of negatively charged clay minerals. Goethite and hematite are also responsible for the yellow-brown and red colors, respectively, commonly observed in soils.

Opaline (amorphous) silica is a common constituent in soils with an abundant source of readily soluble silica, such as volcanic ash or basic igneous rocks, in subhumid mediterranean and arid climates that do not completely leach silica from the soil (Drees et al., 1988). Opaline silica can act as a cement to form extensive duripans and may be important in the weaker cementation of quartz-rich tillage pans.

1.1.4 Shape of Soil Particles

Although many conceptual models of soil assume that the individual soil particles are spherical in shape, most of the experimental evidence indicates that the particles making up the smaller soil fractions are distinctly nonspherical. This evidence has been compiled principally from ultramicroscopic observations of clay soils, the double refraction of clay particles, the layering of particles during deposition, the nature of the clay crystals, and electron microscopy.

When a beam of light is passed through a colloidal system, it is partly transmitted and partly diffracted. The diffracted light is visible in the suspension. (This is the so-called Tyndall effect). Under observation with an ultramicroscope, the diffracted light from a clay soil is irregular in intensity, and light from individual particles seems to flicker on and off. The flickering effects within the diffracted light are produced by the sudden orientation of irregularly shaped particles in a visible position and their equally sudden disappearance from view as they assume a different position.

The advent of the electron microscope has made it possible to determine the shape of clay particles quite accurately by direct observation (Humbert and Shaw,

1941; Marshall et al., 1942; Shaw and Humbert, 1941). Photomicrographs of different clay particles such as those shown in Fig. 1.6 reveal the platelike character of the minerals. In Fig. 1.6*a*, showing beidellite in the 100–500-nm fraction of Putnam clay, the particles are very well defined and are obviously platelike, with a disk thickness of about 16 nm. Montmorillonite (Fig. 1.6*b*) shows a variety of structures ranging from extremely thin plates to material that appears to be amorphous.

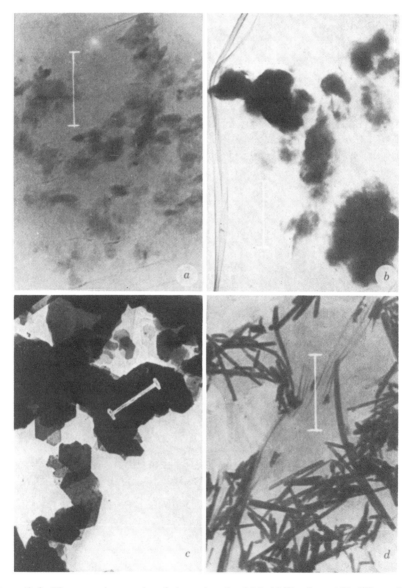

Figure 1.6 Electron micrographs of clay minerals: (*a*) beidellite from 100–500-nm Putnam clay; (*b*) montmorillonite; (*c*) kaolinite; (*d*) halloysite. (Unit of measurement is 1 μm.)

The hexagonal character of the plates of kaolinite, with sharp, well-defined crystal edges, is shown in Fig. 1.6c. In contrast, halloysite, a $1:1$ lattice-type mineral like kaolinite, has definite rod-shaped particles that appear to be composed of twin sections (Fig. 1.6d). From the large body of information provided by electron micrographs, it has been established that the majority of the clay minerals found in soils are platelike or disk shaped.

1.1.5 Surface Area of Soil Particles

Relationship to Particle Size   Systems composed of dispersed particles of small size have a large amount of surface area per unit mass of material. Colloids, which possess very large surface area per unit mass, have enormously reactive properties. The amount of surface area of a dispersed system is usually expressed in terms of specific surface, which is defined as the surface area per gram or per unit volume of the dispersed phase.

The relationship of specific surface to particle size is explored in Example 1.1.

EXAMPLE 1.1:   Calculate the specific area per unit mass s of a spherical particle of density ρ and radius R.

The surface area a of a sphere may be expressed in terms of its radius by the formula $a = 4\pi R^2$. Similarly, the mass m of the sphere may be expressed as $m = \rho V = \rho 4\pi R^3/3$, where V is the volume of the sphere. Thus, the specific surface area s, or the area per unit mass, is given by

$$s = a/m = 3/\rho R \tag{1.6}$$

Note that this is also equal to the specific surface area S of a total mass $M = Nm$ of identical spheres, since this mass has a total area $A = Na$.

As shown in Example 1.1, the specific surface area S of an arbitrary quantity of identical spheres (1.6) is inversely proportional to the radius of the sphere. This inverse relationship between particle size and specific surface area will also be true for other shapes of particles, as shown in Example 1.2.

EXAMPLE 1.2:   Calculate the specific surface area of a circular, disk-shaped plate-let of density ρ, radius R, and thickness T. Assume that $T \ll R$.

The total surface area of the disk is the sum of the two circular surface areas $2\pi R^2$ and the edge surface area $2\pi RT$. Thus, $a = 2\pi R^2 + 2\pi RT$.

The mass m of the platelet is the product of its density ρ and volume V, or $m = \rho V = \rho \pi R^2 T$.

Consequently, the specific surface of the disk-shaped platelet is

$$s = a/m = 2(R + T)/\rho RT \approx 2/\rho T \tag{1.7}$$

since $R \gg T$.

Thus, the specific surface area of a thin disk-shaped platelet (1.7) is inversely proportional to the thickness of the platelet. The formulas in Examples 1.1 and 1.2 may be used to estimate specific surface areas for the soil particles associated with various textural classes. This procedure is illustrated in Example 1.3.

EXAMPLE 1.3: Calculate the specific surface area in centimeters squared per gram of the following idealized soil particles: a 2-mm-diameter gravel sphere, a 0.05-mm-diameter sand sphere, a 0.002-mm-diameter silt sphere, and a circular clay platelet of diameter 0.002 mm and thickness 10^{-6} mm. For simplicity, assume that all particles have a mineral density $\rho = 2.7$ g cm^{-3}.

Table 1.1 summarizes the calculated specific surface area of these prototype particles using formulas (1.6) and (1.7) from the previous two examples.

Thus, the specific surface area of the clay platelet is almost 17,000 times larger than the specific surface area of the sand particle.

Disk-shaped particles also combine to create different bulk properties than spherical particles. For example, disk-shaped particles can be arranged in more intimate contact than spherical ones, creating more contact surface area when closely packed together and hence more cohesion. In addition, flat particles that are stacked together can slide over each other or shear under an applied force, whereas spheres that are in a closely packed hexagonal orientation will not shear until sufficient force has been applied to roll them farther apart.

The soil organic fraction often has a highly reactive surface that may have properties characteristic of material with an extremely high specific surface area (i.e., as high as 1000 m^2 g^{-1}). Consequently, the apparent surface area of soil may be influenced significantly by the organic as well as the clay mineral fraction.

Relationship to Clay Mineralogy The evidence obtained on the structure of clay minerals indicates that all groups have both external planar and external edge surfaces. Because their lattices expand, smectites and vermiculites also have internal planar surfaces. Data in Table 1.2 and Table 1.3 show the variations that occur in total surface area both between and within the clay mineral groups. Because they do not have interlayer surfaces and tend to form stacks containing many layers, kaolin minerals have a very low specific surface area compared to the other groups. The nonexpanding micas have a specific surface area about ten times that of the kaolin group, whereas the specific surface area of the vermiculite minerals is about midway between the micas and the expanding-lattice smectites. Although there may be as much as 100% or more variation within the same group, the difference in surface area between groups is large enough to be significant.

TABLE 1.1 Specific Surface Area of Prototype Particles Found in Soil

Particle	Effective Diameter (cm)	Mass (g)	Area (cm^2)	Specific Surface Area (cm^2 g^{-1})
Gravel	2×10^{-1}	1.13×10^{-2}	1.26×10^{-1}	11.1
Sand	5×10^{-3}	1.77×10^{-7}	7.85×10^{-5}	444.4
Silt	2×10^{-4}	1.13×10^{-11}	1.26×10^{-7}	1.11×10^4
Clay[a]	2×10^{-4}	8.48×10^{-15}	6.28×10^{-8}	7.4×10^6

[a]Thickness = 10^{-7} cm.

TABLE 1.2 Specific Surface Area, Cation Exchange Capacity, and Density of Charge of Various Clay Minerals[a]

Clay Mineral	Specific Surface Area ($m^2\ g^{-1}$)	CEC ($me\ g^{-1}$)	Density of Charge, $\times 10^3$ ($me\ m^{-2}$)
Kaolinites	5–20[b]	0.03–0.15[c]	6–7.5[d]
Micas (illites)	100–200[b]	0.10–0.40[c]	1.0–2.0[d]
Vermiculites	300–500[b]	1.00–1.50[c]	3.0–3.3[d]
Smectites	700–800[b]	0.80–1.50[c]	1.1–1.9[d]

	Specific Surface Area ($m^2\ g^{-1}$)			CEC ($me\ g^{-1}$)	Density of Charge, $\times 10^3$ ($me\ m^{-2}$)	
	N_2	CBP	E.G.		N_2	CPB
Illite-1[e]	93	96	91	0.26	2.8	2.7
Illite-2[e]	132	138	144	0.41	3.1	2.9
Kaolinite-1[e]	17	21	21	0.043	2.5	2.0
Kaolinite-2[e]	36	36	—	0.050	1.4	1.4
Kaolinite-3[e]	40	9	13	0.030	0.75	3.3
Montmorillonite-1[e]	47	800	—	0.98	21.0	1.22
Montmorillonite-2[e]	49	600	—	0.98	21.0	1.6
Montmorillonite-3[e]	101	800	—	0.99	9.8	1.22

[a] Abbreviations: CPB, cetyl pyridinium bromide; EG, ethylene glycol.
[b] From Fripiat (1964).
[c] From Grim (1962). Used with permission of McGraw-Hill Book Co.
[d] Calculated from the data in second and third columns.
[e] From Greenland and Quirk (1964).

TABLE 1.3 Heat of Wetting, Specific Surface Area, Cation Exchange Capacity, and Density of Charge of Various Clay Minerals

Clay Mineral	Heat of Wetting ($cal\ g^{-1}$)	Specific Surface Area ($m^2\ g^{-1}$)	CEC ($me\ g^{-1}$)	Density of Charge, $\times 10^3$ ($me\ m^{-2}$)
Kaolinite				
1	1.6	14.8	0.035	2.4
2	1.4	12.0	0.023	1.9
3	1.4	11.0	0.019	1.9
4	2.1	25.0	0.043	1.7
Hydrous mica				
1	7.6	150	0.25	1.7
2	4.8	110	0.17	1.5
3	7.9	160	0.30	1.9
4	16.5	250	0.43	1.7
Montmorillonite				
1	16.5	690	0.92	1.3
2	17.4	640	0.83	1.3
3	22.2	700	1.13	1.6

Source: Greene-Kelly (1962).

The methods for the measurement of specific surface area usually make use of the adsorption properties of the mineral surfaces. A brief description of some of these methods follows the discussion of the surface properties of clay particles.

1.1.6 Surface Properties of Clay Particles

Density of Charge The total charge on a mineral surface is called the intrinsic surface charge density, which is the sum of the structural surface charge density resulting from isomorphic substitutions in the soil mineral and the proton surface charge density resulting from proton-selective surface functional groups (Sposito, 1984). The structural charge density is a permanent charge determined by the degree of substitution in the crystal lattice and is independent of the conditions surrounding the mineral. In contrast, the proton surface charge density is due to the imbalance of complexed proton and hydroxyl charges on the surface, primarily on the exposed periphery of the mineral, and varies with the pH of the surrounding solution.

Most soils have a negative total surface charge density because of the negative charges on layer silicates and organic matter, but some highly weathered soils that contain substantial amounts of allophane and hydrous oxides can actually develop a net positive charge at sufficiently low pH. The pH dependence of colloids can be characterized experimentally by measuring the zero point of charge (ZPC), which is defined as the pH value at which the total surface charge vanishes.

The pH-dependent surface charge density will contribute significantly to mineral properties only for those clays that have a high surface area of positive edges of hydroxy aluminum. The vermiculite and smectite groups have a large permanent negative structural charge resulting from isomorphic substitution, which accounts for about 80% of the total charge density (Grim, 1962). In contrast, kaolinite has little isomorphic substitution and therefore a significant variation in its total charge density as pH is varied. Highly weathered soils, which are abundant in iron and aluminum oxides, also display significant pH dependence in their surface charge.

Because both the structural and proton-dependent surface charge depend strongly on mineral composition, the pH dependence of the charge density of different soils varies greatly. Figure 1.7 shows a graph of net surface charge versus pH for several clay mineral types. The negative charges of the 2:1 layer silicate clays, montmorillonite, Rothamsted subsoil (illite), and Taita soil clay (illite) vary little with pH, although there are significant increases at the higher pH values. In these clays, particularly montmorillonite, the permanent negative charges overshadow the pH-dependent ones. There is a fairly sizable increase in net negative charge above pH 5.5 for Whatatiri soil clay (metahalloysite, gibbsite). Egmont soil clay, which contains allophane, shows the largest increase in net negative charge of any of the minerals shown, occurring principally between pH 6.0 and 7.6. Moreover, the Whatatiri and Egmont clays have a ZPC around pH 5.7–6.0, so that anion adsorption becomes the dominant exchange reaction at low pH.

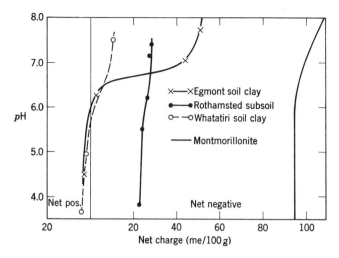

Figure 1.7   Net electric charges of clays in relation to pH. (After Fieldes and Schofield, 1960.)

Ionic Adsorption   A detailed discussion of the physical chemistry of ion exchange is beyond the scope of this book and is reviewed elsewhere (Bolt, 1955a, 1982; Jenny, 1932; Kelley, 1948; Marshall, 1964; Mattson, 1929; Sposito, 1981; Wiklander, 1964). However, since many of the physical properties of soils are affected by the amount, nature, and activity of adsorbed ions, it is important to characterize the amount of exchangeable ions per unit mass and the energy with which different ions are held on the surface of soil minerals.

Cation Adsorption   The negative charges present on soil minerals produce a positive electrostatic attraction for cations in soil solution. As a result, the fluid in the immediate vicinity of a negatively charged mineral surface has an excess of cations and a deficit of anions compared to the bulk solution farther away from the surface. Cations present in this surface fluid layer are retained by electrostatic attraction to the charged surface unless they exchange with other cations in solution. The total quantity of cations that may be retained in this manner is called the cation exchange capacity (CEC) of the soil. The CECs of different clay mineral groups are influenced principally by the lattice construction and the degree of isomorphic substitution. The order of CEC for the various mineral groups is vermiculite > smectite > mica > kaolinite (Table 1.2).

The CEC is an increasing function of the amount of internal and external surface and therefore should increase with decreasing particle size. The dependence on particle size is most pronounced for those clay minerals in which broken bonds are responsible for many of the exchange sites. One would not normally expect the expanding-lattice clays to exhibit increasing CEC with decreasing particle size, since 80% of the adsorption sites are on the planar surfaces (Grim, 1962). The

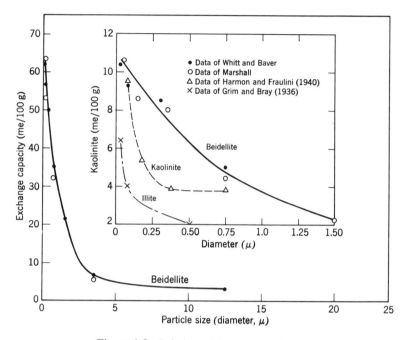

Figure 1.8   Relation of CEC to particle size.

relationship between CEC and particle size for various clay minerals is illustrated in Fig. 1.8.

Diffuse Double Layer   Several models have been proposed to describe the distribution of cations and anions in the water layer adjacent to a clay mineral. The simplest representation, called the Helmholtz model, assumes that all balancing cations are held in a single layer between the mineral surface and the bulk solution. However, such a picture is oversimplified, because cations possess thermal energy that will cause a dynamic concentration gradient to form away from the surface, creating a diffuse double layer (Bolt, 1982; van Olphen, 1963).

Standard but cumbersome calculations using potential theory may be performed to calculate cation and anion concentrations within the double layer between the mineral surface and the bulk solution. If the entire double layer is regarded as diffuse, the model is called a Gouy–Chapman double layer (Bohn et al., 1979) and the calculated concentration gradients change smoothly from the mineral surface to the bulk solution, as in Fig. 1.9a. If the double layer is treated as having a rigid region next to the mineral surface and a diffuse layer joining with the bulk solution, the model is called a Stern double layer (Bohn et al., 1979). Then, the calculated concentration gradients are less steep in the diffuse layer since the rigid layer (commonly called the Stern layer) acts to lower the surface charge, which must be neutralized by the solution (Fig. 1.9b). Since anions are also attracted into the

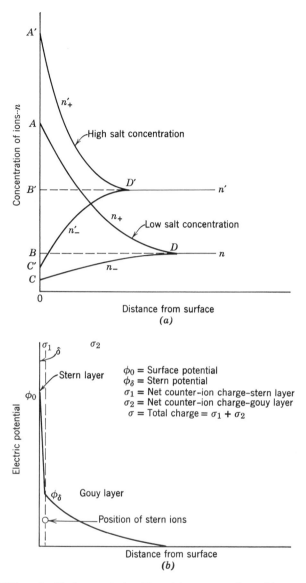

Figure 1.9 Diffuse double-layer relationships: (*a*) concentration of ions versus distance from surface; (*b*) electric potential of Stern and Gouy layers. (Adapted from van Olphen, 1963. Used with permission of John Wiley & Sons, Inc.)

diffuse region by the cation excess, the total cation concentration must neutralize both the anion charge and the surface charge. The cations that neutralize the surface charge are exchangeable with other cations in solution.

Addition of more electrolytes to the system will produce a compression of the double layer and a decrease in the electric potential. This is especially true at high concentrations, as seen by comparing areas ABD and $A'B'D'$ and diffuse layer thicknesses BD and $B'D'$ in Fig. 1.9a.

The thickness of the liquid layer around the soil particle is assumed to be infinite in the double-layer model calculations. If the calculated double-layer thickness is greater than the actual thickness of the liquid layer, then the theory is not applicable. The double layer can never extend beyond the surface of the liquid layer since the cations necessary to neutralize the surface charge must remain within the liquid layer. Therefore, when the liquid layer is small, the concentration never reaches a constant value; i.e., there is no equilibrium solution. If water is added to such a system, the double layer will expand and produce swelling in the medium, unless the particle binding forces are strong enough to oppose the swelling pressure. A sodium–montmorillonite has few binding forces and swells freely as water is added. Conversely, a calcium–montmorillonite is strongly bound and tends to swell only to a very limited extent.

Corrections to the Gouy–Chapman and Stern theories, which take into account the influence of ionic interactions, polarization of ions, and dielectric satuation are small for colloidal clay suspensions if the density of charge does not exceed 2×10^{-7}–3×10^{-7} me cm^{-2} (Bolt, 1955b). However, the size of the ion and its extent of hydration have a considerable influence on the distribution of cations between the rigid Stern layer and the diffuse portion of the double layer (Shainberg and Kemper, 1966a, 1967). This effect may be visualized by making the following assumptions: (1) The exchange sites in the surface are fixed charges; (2) part of the surface charges are neutralized by ions in the Stern layer; (3) the remaining surface charge is balanced by cations in the diffusion portion of the double layer; and (4) the balancing cations may be either fully hydrated or unhydrated with no water molecules between them and the adsorbing surface. The total density of charge Q_t on the mineral surface is equal to the sum of the charge densities of the hydrated, Q_h, and unhydrated, Q_u, charge balancing cations:

$$Q_t = Q_h + Q_u \tag{1.8}$$

The unhydrated ion is assumed to be in the Stern Layer directly on the surface and the hydrated ions, because of their larger sizes, in the diffuse phase. The hydration shell around a sodium ion, for example, would increase the distance from the center of the ion to the plane of negative charge in the octahedral sheet of the crystal lattice of montmorillonite by over 50% above that of the dehydrated cation. Computations of the distribution of the charge densities of monovalent cations in the double layer in relation to their hydration and polarization energies are shown in Table 1.4. These results confirm the well-known lyotropic series describing the ease of replacement of adsorbed cations: Li > Na > K > Cs. Only 17% of the highly hydrated lithium ion is in the Stern layer as compared with 76% of the unhydrated cesium ion. The difference between sodium and potassium ions is of the same order of magnitude as the variations in their maximum bonding energies.

TABLE 1.4 Distribution of Monovalent Cations between Stern and Diffuse Layers

Ion	Radius (Å)	Charge Density (C m^{-2})		Amount in Stern Layer (%)
		Q_u	Q_h	
Li	0.60	0.017	0.080	17.3
Na	0.95	0.037	0.090	41.2
K	1.33	0.057	0.039	59.1
Cs	1.69	0.070	0.026	76.0

Source: Shainberg and Kemper (1967).

Cation Exchange Equations Many different equations have been proposed to model cation exchange reactions. Common features of these models are that they assume that the mineral surface has a constant total CEC at a given pH and that the exchange reaction is considered to be stoichiometric and reversible. The most general type of reaction model is the mass action equation, which describes Ca^{2+}–Na^{2+} exchange as follows (Bohn et al., 1979):

$$CaX + 2Na^+ \rightleftarrows 2NaX + Ca^{2+} \qquad (1.9)$$

where X represents the exchange phase of the cation. The equilibrium reaction coefficient for the reaction in (1.9) is therefore

$$K_K = \frac{(NaX)^2 (Ca^{2+})}{(CaX)(Na^+)^2} \qquad (1.10)$$

where ( ) denotes activity.

All cation exchange models are limited by the fact that there are no models or measurements that characterize the activity of the adsorbed phases exactly. As an approximation, the activities in (1.9) are frequently replaced by concentrations. This approximate representation is called a Kerr model and may be reasonably valid over narrow concentration ranges. However, the reaction coefficient varies substantially over the entire range of cation concentration in the exchange phase (Marshall and Garcia, 1959).

A modified mass action equation was developed by Gapon in 1933, who expressed the exchange reaction in terms of chemically equivalent quantities in solution and adsorbed phases and used concentrations in place of activities. These assumptions produce the following reaction coefficient.

$$K_G = \frac{[NaX][Ca^{2+}]^{1/2}}{[Ca_{1/2}X][Na^+]} \qquad (1.11)$$

where [ ] denotes concentration. The Gapon equation (1.11) is inaccurate over large ranges of exchange concentrations, but it is reasonably accurate within the range of values of sodium exchange concentration found in most agricultural settings (U.S. Salinity Laboratory Staff, 1954).

Other exchange models have been proposed that differ primarily in the way they treat the activity of the exchange phases. For example, Vaneslow (1932) assumed that the activities of the exchangeable cations were proportional to their mole fractions.

Anion Adsorption Anions react with mineral surfaces in two different ways. The first reaction is the nonspecific electrostatic attraction to or repulsion from the positively or negatively charged mineral surface. Since most minerals are negatively charged at normal pH, the net anion adsorption is actually a repulsion from the surface, sometimes called negative adsorption or anion exclusion. This process has an influence on chemical transport through soil, as it embodies anions with a somewhat higher mobility than neutral species moving with water by excluding the anions from the slowest moving portion of the water volume closest to the stationary solid surfaces.

The second anion reaction with soils is ion specific and involves positive adsorption to oxide surfaces. This reaction is known as ligand exchange and is postulated to occur when the oxygen ions on a hydrous oxide surface are replaced by anions that can coordinate with Al^{3+} or Fe^{3+} ions. The reaction can occur on surfaces with positive or negative charge and is selective for different anions in the order $SiO_4 > PO_4 \gg SO_4 > NO_3 \approx Cl$ (Bolt and Bruggenwert, 1976). At normal pH, the adsorption of SO_4, NO_3, or Cl is usually negligible, and these ions experience a negative net adsorption from electrostatic interactions.

Adsorption of Nonelectrolytes Nonelectrolytes in solution may also be adsorbed by soil solids, either as a result of partial ionization in solution or by other mechanisms to be discussed in what follows. Some nonelectrolytes will accept protons in acid solutions, forming a cationic complex that will react with the negatively charged mineral surfaces. In contrast, organic acids such as 2,4-D, dinoseb, and picloram will dissociate into anionic complexes in solution (Green, 1974).

Nonionic species that do not ionize in solution can still react with mineral surfaces by hydrogen bonding and van der Waals reactions. The hydrogen bond reaction between a nonelectrolyte and a mineral surface occurs between organic functional groups and either siloxane oxygen atoms or surface hydroxyl group but does not appear to be a significant factor in the adsorption of dissolved organics to soil (Sposito, 1984). The reaction may be much more important for adsorption to organic matter surfaces (Burchill et al., 1978).

For nonpolar organic compounds such as DDT, van der Waals forces induced by instantaneous dipole moments in nearby nonpolar molecules are the principal mechanisms causing adsorption, which predominantly occurs on organic surfaces. The structure of soil organic matter is so complex, however, that the relative importance of specific reaction mechanisms is poorly understood.

Ion Adsorption and Flocculation   The extent of hydration of exchangeable cations has long been known to be a dominant factor in the stability of clay suspensions (Jenny and Reitemeier, 1935; Tuorila, 1928; Wiegner, 1925). Divalent exchangeable cations result in flocculated clay systems while monovalent exchangeable cations produce dispersed systems.

Polyvalent cations have a high flocculating power (Beavers and Marshall, 1951; Jenny and Reitemeier, 1935). This is in accordance with the Schulze–Hardy rule, which states that the flocculating power of active ions of charge opposite to that of the mineral surface increases with their valence. The rule is not universally applicable to clay flocculation, however, because that processs is affected also by the nature of the adsorbed cation, the hydrogen ion concentration, and the concentration of the solution (Bradfield, 1928). When a salt is added to a clay suspension that is saturated with the same cation, the double layer contracts further, effectively lowering the net negative charge on the mineral, which reduces the repulsion between adjacent minerals and promotes flocculation. If a salt containing a different cation is added, there will be both ionic exchange and compression of the double layer (Jenny, 1938; Kahn, 1958; van Olphen, 1963).

The positively charged edges contribute to flocculation, because of the electrostatic attraction between positive edges and negative planar faces of adjacent minerals (Schofield and Samson, 1954). In the presence of electrolytes, the thickness of both diffuse double layers is diminished and flocculation can be enhanced by face-to-face or edge-to-edge association of different minerals because of van der Waals forces.

Adsorption Isotherms   The adsorption of a substance in solution on soil particle surfaces is usually pictured as occurring over one molecular layer on the surface. Beyond that distance, the substance is assumed to be in the bulk solution. The relation between the amount of substance adsorbed and the concentration of the substance in solution at any given temperature is known as the adsorption isotherm, one of the fundamental experimental characteristics of a dissolved chemical in soil.

Isotherms may have a variety of shapes depending on the characteristics of the adsorbant and adsorbing surface and sometimes on other constituents in solution. Although many isotherm shapes have no direct physical interpretation, the so-called Langmuir isotherm (Langmuir, 1918) relating the adsorbed C_a and solution C_l concentrations,

$$C_a = \frac{aQC_l}{1 + aC_l} \qquad (1.12)$$

may be derived from a set of reasonable assumptions. This derivation is illustrated in Example 1.4.

EXAMPLE 1.4:   Derive the relationship between C_a and C_l for a surface that has a finite number Q of identical adsorption sites per unit mass, assuming that the adsorbing molecules do not interact with each other.

At equilibrium the rate of adsorption r_A must equal the rate of desorption r_D. If the molecules do not interact with each other, then the rate of desorption r_D should be proportional to the number of adsorbed molecules per unit mass of surface C_a, which we may write as

$$r_D = k_1 C_a$$

where k_1 is a rate constant. Similarly, the rate of adsorption r_A should be proportional to the concentration C_l in solution but also to the number $Q - C_a$ of unfilled sites on the surface. Thus,

$$r_A = k_2 (Q - C_a) C_l$$

At equilibrium, $r_A = r_D$, so that we may equate

$$k_1 C_a = k_2 (Q - C_a) C_l$$

This expression may be solved for C_a, with the result

$$C_a = \frac{aQC_l}{1 + aC_l}$$

where $a = k_2/k_1$.

The Langmuir isotherm is shown in Fig. 1.10a for various values of a. This type of isotherm, in which C_a increases linearly with C_l at low concentrations and approaches a constant (Q) at high concentrations, is most appropriate in soil for processes such as cation exchange, which have a finite adsorption capacity.

Many compounds do not adsorb to soil according to the Langmuir isotherm but instead act as though the surface contained different types of adsorption sites. For these compounds, the shape of the isotherm may often be described by the Freundlich isotherm,

$$C_a = K_f C_l^{1/N} \tag{1.13}$$

where K_f and N are constants with $1/N \leq 1$. Plots of the Freundlich isotherm are shown in Fig. 1.10b. Sposito (1981) has shown that (1.13) may be derived by assuming that the soil surface is composed of a distribution of adsorption sites, each of which obeys a Langmuir isotherm. A special case of the Freundlich isotherm for $N = 1$,

$$C_a = K_d C_l \tag{1.14}$$

is called the linear isotherm, where K_d is the distribution coefficient.

The adsorption isotherm is extremely important in assessing the mobility of dissolved chemicals in soil. This subject will be developed fully in Chapter 7.

Adsorption of Gases   The adsorption of gases on solid surfaces has been studied in a number of different research disciplines. Although many surfaces have only a finite number of sorption sites for gases to adsorb onto, the Langmuir adsorption isotherm does not describe the process well, because more than one monolayer of

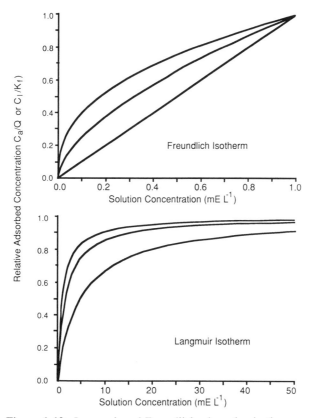

Figure 1.10 Langmuir and Freundlich adsorption isotherms.

gas can be condensed onto a surface. Brunauer et al. (1938) extended the Langmuir approach to multilayer adsorption by assuming that the heat of adsorption of the first layer of gas has a higher value than the heat of adsorption of all succeeding layers, the latter of which is set equal to the heat of vaporization of the liquid adsorbate. This leads to an equation of the form

$$\frac{x}{V(1 - x)} = \frac{1}{V_m C^*} + \frac{(C^* - 1)x}{V_m C^*} \tag{1.15}$$

where $x = P/P_0$ is the ratio of the vapor pressure of the gas to the saturated vapor pressure of the liquid at the specified temperature, V is the volume of gas adsorbed, V_m is the volume adsorbed when the surface is covered with one monolayer, and C^* is a constant related to the heat of adsorption. Equation (1.15) is known as the Brunauer–Emmett–Teller (BET) equation, whose parameters can be evaluated from adsorption data if the isotherm has a shape compatible with the model. Commonly, the left side of (1.15) is plotted versus x to determine if the BET equation is valid.

If the plot is linear, the model is regarded as correct and the BET parameters V_m, $C*$ are calculated from the slope and intercept. The BET equation appears to work reasonably well for the adsorption of nitrogen on soil minerals at the temperature of liquid N_2.

Measurement of Specific Surface Area Several methods have been developed for measuring the external and internal surfaces of clays. The BET equation has been used to interpret the adsorption of nitrogen at liquid nitrogen temperatures on soils, from which one may estimate surface area (Nelson and Hendricks, 1943). External surface area may be measured by first heating the minerals to high temperature (e.g., 600°C) to collapse the internal surfaces. On both preheated and unheated samples the adsorption of N_2 is measured and fitted to the BET equation. The surface area is then calculated from the BET parameter V_m describing the volume of a monolayer of gas and the known surface area per unit volume of N_2. A limitation of the method is that N_2 molecules do not penetrate the interlayer freely.

A second method for calculating surface area uses adsorption of polar molecules such as ethylene glycol, glycerol, cetyl pyridinium bromide (CPB), or ethylene glycol monoethyl ether (EGME) on the surfaces of the clay minerals.

The principle of this adsorption technique is to saturate a dry surface with ethylene glycol in a vacuum desiccator. The excess of this polar liquid is removed under vacuum over either anhydrous $CaCl_2$ or a $CaCl_2$–ethylene glycol mixture. When the weight of the clay mineral–glycol mixture reaches a constant value, the surface area is calculated from the equation

$$A = \frac{W_g}{W_a \times 0.00031} \tag{1.16}$$

where A is the specific surface in square meters per gram, W_g is the weight of the ethylene glycol in the sample, W_a is the weight of the air-dried sample, and 0.00031 represents the grams of ethylene glycol that are required to form a monolayer per square meter of surface [see Mortland and Kemper (1965) for details].

Comparisons of specific surfaces measured by various methods are shown in Table 1.2. There is general agreement between the BET, CPB, and ethylene glycol methods for illites and kaolinites. It is obvious that the BET technique does not determine the interlayer surfaces of montmorillonite.

When expansive clay minerals are heated to 600°C, the interlamellar layers collapse irreversibly and ethylene glycol is not adsorbed except on external surfaces. A comparison of the BET and ethylene glycol techniques for measuring specific surface under these conditions shows rather close agreement (Bower and Gschwend, 1952).

The internal surface of expansive clays can be determined either by comparing ethylene glycol adsorption before and after heating the sample or by measuring external surface by the BET technique and subtracting this value from the total surface obtained by the ethylene glycol method. For example, the external surfaces of the three montmorillonites shown in Table 1.2 vary between 47 and 101 m^2 g^{-1}; the internal surfaces range between 551 and 753 m^2 g^{-1}.

Measurement of Cation Exchange Capacity   There are a number of different methods used for the determination of the CEC of soil, but most of them use the same principle, which consists of adding a sufficient amount of new cation to completely saturate the exchange sites and measuring the total quantity of cations displaced from the surfaces by the replacing cation. Two common reagents used for this purpose are ammonium acetate (pH 7.0) and sodium acetate (pH 8.2) (Chapman, 1965). Various sources of error in this determination are dicussed in Bohn et al. (1979).

1.2 CHARACTERISTICS OF BULK SOIL

Many of the important transport and retention processes in the soil are influenced strongly by the composite properties of the soil matrix, which are sometimes called bulk soil properties. These properties are commonly characterized with samples that contain many individual soil particles, void spaces, and water films. Hence, the bulk soil properties are said to be volume averaged.

1.2.1 Volume Fractions

A soil volume V containing a total mass M of solid, liquid, and gaseous material may be divided into the sum of the contributions from various phases of interest. For example, the volume may be subdivided into

$$V = V_g + V_l + V_s \tag{1.17}$$

where g, l, s refer to gaseous, liquid, and solid phases, respectively. If the total volume V is divided into each side of (1.17), the result may be written as

$$\phi = 1 - \frac{V_s}{V} = a + \theta \tag{1.18}$$

where ϕ is the volume of void space per total volume and is called the porosity, a the volume of gas space per total volume and is called the volumetric air content, and θ the volume of liquid per total volume and is called the volumetric water content. Note that the porosity is equal to 1 minus the solid volume fraction.

1.2.2 Bulk and Mineral Densities

It is often useful to characterize the density of the solid soil matrix. The bulk density ρ_b, or the density of the solid matrix in place, is defined as the mass of dry soil per volume of soil. The mineral density ρ_m, or the density of the solid material comprising the soil matrix, is defined as the mass of dry soil per volume of soil solids. The relationship between ρ_b and ρ_m is derived in Example 1.5.

EXAMPLE 1.5: Derive a relationship between ρ_b and ρ_m using the preceding definitions.

By definition,

$$\rho_b = \frac{m_s}{V} = \frac{m_s}{V_s} \frac{V_s}{V}$$

where m_s is the mass of dry soil in V. Thus, since $\rho_m = m_s/V_s$ and $1 - \phi = V_s/V$ [(1.18)], then

$$\rho_b = \rho_m(1 - \phi) \tag{1.19}$$

Measurement of Bulk Density   Measurement of dry bulk density requires separate estimation of both the mass of dry soil and the volume the soil occupied prior to measurement. Dry soil is often defined operationally as the state of a soil sample after 24 h of drying in an oven at 105°C, even though this does not remove interlayer water adsorbed onto clay surfaces. The volume of a soil sample obtained by coring is assumed to be the volume of the inside of the coring tube. However, care must be taken to ensure that the sampling tube does not compress the sample excessively. For this reason, bulk densities calculated from samples obtained by pounding coring tubes into the soil are likely to be inaccurate, particularly below the surface layer of soil.

Bulk densities may also be determined on intact clods of soil by coating their surfaces with a thin layer of paraffin and measuring the volume of displaced water when the clod is lowered into a full beaker.

1.2.3   Soil Structure

Soil structure and its stability play an important role in a variety of processes in the soil, such as erosion, infiltration, root penetration, aeration, or mechanical strength. Since these processes all have observable characteristics, soil structure is often evaluated by methods that correlate it to the properties of the process of interest. An agricultural scientist may be interested in soil *tilth*, which is a general term that signifies the ability of the granules or aggregates to withstand destruction by the impact of implements, raindrops, or running water. Tilth also involves water retention and the workability of soils, both of which depend to a large extent upon the basic textural nature of the soil.

In contrast, a plant scientist may primarily be interested in the effect of soil structure on root penetration and may therefore seek to correlate penetrometer resistance or root growth with aggregate size or some other index of structure.

Soil structure can be evaluated by determining the extent of aggregation, the stability of the aggregates, and the nature of the pore space. These characteristics, which significantly influence plant response to water management practices, will change with tillage practices and cropping systems.

Aggregation   In evaluating the aggregation of soils, one is interested in the size distribution, quantity, and stability of the aggregates. These parameters of aggregation are important in determining both the amount and distribution of the pore spaces associated with the aggregates and the susceptibility of the soil to water and wind erosion.

Aggregate Analysis   An aggregate analysis aims to measure the percentage of water-stable secondary particles in the soil and the extent to which the finer mechanical separates are aggregated into coarser fractions. In general, three techniques are employed to accomplish such an analysis. They are wet and dry sieving, elutriation, and sedimentation.

Direct dry sieving of soils in their natural state has been used to evaluate the distribution of clods and aggregates (Cole, 1939; Keen, 1933). Air-dry sifting is considered to give a better picture of aggregation of some arid-zone soils than wet sieving, since the aggregates are so weakly held together when moist that the mechanical action of sieving is sufficient to destroy them. Clogging of the flat sieves and breaking up of weak aggregates by the mechanical action required in the sieving operation are major problems with this technique. These difficulties have been overcome by the use of a rotary sieve (Chepil, 1962). Dry sieving of aggregates provides an important index for characterizing the susceptibility of soils to wind erosion. No special preparation of the sample is required for consistent results.

The wet-sieving technique of Tiulin (1928) is the best known of the early techniques designed to find a measure of soil aggregation. In this method, the soil is slowly wetted from below the sample for 30 min and is then transferred onto a nest of sieves immersed in water with the finest sieve (0.25 mm) on the bottom. The sieves are slowly raised and lowered in the water 30 times by standard mechanical procedures (Kemper, 1965; Yoder, 1937). The weight of soil on each sieve is then determined. Wet sieving is well adapted to the separation of large aggregates. Although it can be used to screen out aggregates as small as 0.1 mm, 0.25 mm is a more practical lower limit.

The greatest problem in wet sieving is achieving a consistent method of wetting the sample for analysis. Air drying decreases the percentage of large aggregates in the sample (Russell, 1938; Yoder, 1937), especially if done rapidly. In addition, rapid wetting of dry soil samples tends to destroy large aggregates. Immersing the soil in water causes more destruction of the larger aggregates than wetting by capillarity. Spraying water on the aggregates with an atomizer produces the least destruction of any of these three methods. However, the destruction can be minimized by placing the sample in a vacuum during wetting. Pretreatment of aggregates with ethyl alcohol to displace the air before wet sieving will also stabilize the system (Hénin et al., 1955). A complete discussion of wet-sieving procedures is presented in Kemper and Chepil (1965).

Elutriation may be used for separating aggregates with diameters between 1 and 0.02 mm. It is particularly useful for making separations below the limit where wet sieving cannot be employed (Baver and Rhoades, 1932; Demolon and Hénin, 1932).

Sedimentation methods such as the hydrometer and pipette procedures (see Section 1.1.1) have been used to determine the aggregate distribution in the finer fractions that cannot be separated by sieving. They are limited to aggregate sizes < 1 mm. However, the interpretation of the data is clouded by two factors: there is a varying aggregate density, especially with the larger secondary particles, and there is the possibility of flocculation during sedimentation because of the downward motion of the larger aggregates.

Stability of Structure   The stability of structure refers to the resistance that the soil aggregates offer to the disintegrating influences of water and mechanical manipulation. The stability of the aggregate is of utmost importance in forming and preserving good structural relationships in soils. It is recognized that "aggregate stability" is not necessarily synonymous with "structural stability" because soil aggregates may be altered by a variety of destructive forces. Moreover, it is often not obvious which of these forces are involved in the use of "aggregate stability." It is known that water content is often a crucial factor in structural stability. It is almost always a factor in determining the degree to which particular mechanical forces will cause structural breakdown. For example, rather compact and coherent aggregates may be found in the dry state, but if these secondary particles disintegrate in water, the aggregation is not very stable. Water may cause the deterioration of aggregration in two ways. First, there is the hydration effect of water, which causes a disruption of the aggregate through the processes of swelling and the exploding of entrapped air. The second manner in which water destroys aggregation and deteriorates soil structure is by falling rain. The impact of falling drops of water on exposed soil exerts a significant dispersive action on the aggregates. The dispersed particles are then carried into the soil pores, causing increased compaction and decreased porosity. Intense rains destroy the granulation and open structure of the top inch or more of soil to form a dense, impervious surface known as a crust. This type of structure degradation is least common with those aggregates that are stabilized with humus or iron compounds. Insufficient emphasis has been placed upon the deteriorating action of falling raindrops; in many instances raindrops are the major cause of the dispersion of soil aggregates. Their immediate influence is confined to a shallow layer in the surface, but the structure of this layer may be broken down to limit the air and moisture relations of the entire profile.

Cultivation and other tillage operations generally cause a continued decrease in the stability of aggregates unless the organic matter level of the soil is kept relatively high and mechanical manipulation of the soil is performed at optimum moisture contents.

PROBLEMS

1.1   Use Equation (1.5) for the settling velocity to calculate the amount of time required for particles of diameter D equal to 2.0, 0.05, 0.002, and 0.001 mm to fall 10 cm in water solution. Assume $\rho_s = 2.7$ g cm^{-3}.

1.2 Fifty grams of soil containing 10% clay by weight are added to a beaker containing 1000 ml of water and are mixed thoroughly. Calculate the density of the suspension before and after the sand and silt settle out. Ignore any volume change in the solution as the soil is added and removed.

1.3 Since most of the surface area of a disk lies on the circular faces, why does the formula for specific surface [Equation (1.7)] not depend on the circular radius R?

1.4 If the distribution coefficient K_d defined in (1.14) for a linear equilibrium adsorption process is regarded as the slope of the C_a–C_l curve, calculate the "effective K_d" of the Freundlich isotherm (1.13) and show that it is a function of the dissolved concentration. For a compound with a Freundlich coefficient of $1/N = 0.8$, calculate the ratio of the effective K_d at $C_l = 1$ mg l^{-1} and $C_l = 1000$ mg l^{-1}.

$\widehat{2}$ Water Retention in Soil

The two most important characteristics of the soil water phase are the amount of water in a specified quantity of soil and the force holding the water in the soil matrix. The amount of water in the soil influences many processes, including gas exchange with the atmosphere, diffusion of nutrients to plant roots, soil temperature, and the speed with which solutes move through the root zone during irrigation or rainfall. The force with which water is retained by the soil matrix also affects many processes occurring in soil, including the efficiency of plant water extraction, the amount of drainage occurring under gravity, and the extent of upward movement of water and solutes against gravity. This chapter will introduce a formalism for quantifying water content and water retention in soil and will describe how the parameters and functions that describe the water properties are measured.

2.1 PROPERTIES OF WATER

Many of the bulk properties of soil water are a consequence of the molecular characteristics of water. Therefore, prior to discussing the fluid properties of water, it is worthwhile to describe the properties of the water molecule that are most relevant to understanding the behavior of the bulk fluid in soil.

2.1.1 Molecular Properties of Water

The water molecule consists of two hydrogen atoms bonded to an oxygen atom by sharing electrons. The oxygen and hydrogen protons in the water molecule are about 0.97 Å apart, whereas the hydrogen protons are 1.54 Å apart, and the angle formed by the H—O—H bond is about 105° (Pauling, 1948). Since the H_2O molecule contains the same number of positively charged protons as negatively charged electrons, it is electrically neutral. However, because the center of positive charge is displaced from the center of negative charge, the molecule possesses a dipole moment, which produces an electric field in the vicinity of each molecule. As a result, water molecules interact with each other, with dissolved ions in solution, and with the electric field of solid minerals and organic material in soil.

The electric fields of the dipole moments of adjacent water molecules create an attractive force, forming a relatively weak intermolecular hydrogen bond between the proton of the hydrogen atom of one molecule and the oxygen atom of the other. The effect of the weak attraction is to form an effective particle made up of many bonded molecules.

The hydrogen bonds linking water molecules together are significantly weaker than the covalent bonds joining a given molecule. Hence, liquid water does not form a stable crystalline structure, because the energy of agitation of individual molecules is occasionally high enough to break the intermolecular bonds. The crystalline structure is more complete in ice, where each water molecule bonds to four neighbors in a hexagonal structure like that shown in Fig. 2.1. The straight lines represent hydrogen bonds, the small open circles represent hydrogen protons, and the filled circles represent oxygen nuclei. Hydrogen proton pairs oriented perpendicular to the plane of the paper are shown as concentric circles. Three adjoining molecules associated with one hexagonal ring beneath the plane of the paper are shown with dashed lines. These lines show the position of the fourth molecule associated with each molecule in the structure.

When ice crystals melt, the hydrogen bond length increases to about 2.90 Å. In contrast to most liquids, the density of liquid water actually increases with increasing temperature up to a temperature of nearly 4°C, above which the density decreases monotonically. The reason for this density increase above the melting

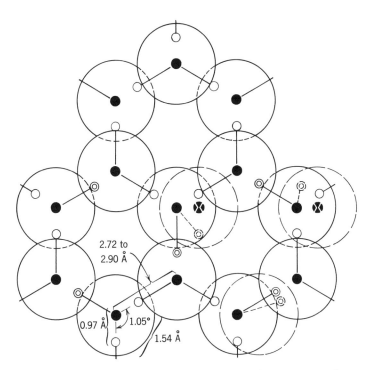

Figure 2.1 Postulated structure of ice showing a hexagonal arrangement of water molecules and their orientation. The straight lines indicate hydrogen bonds, open circles, and hydrogen protons and the filled circles oxygen nuclei. When ice melts, the bonds are stretched so that the distance between oxygen nuclei increases. (After Bernal and Fowler, 1933.)

point is that some hydrogen bonds break as the bonds are stretched upon melting and warming, so that a crystalline structure develops that is less ordered than the ice structure and is of higher density. As the thermal agitation of the liquid water increases at higher temperature, the lattice of this structure expands. However, some crystalline properties are maintained in the liquid state at temperatures well above freezing. In the liquid state, each water molecule can be surrounded at any one time by five or more adjacent molecules joined by hydrogen bonding. Despite fluctuations in the positions of the molecules because of thermal motion, the structure is sufficiently stable that aggregates of water molecules behave in some respects like molecules, causing water to exhibit polymerlike characteristics. Only in the vapor state do water molecules completely lose their intermolecular hydrogen bonds.

Because the covalent electron bond between the hydrogen and oxygen atoms of an individual molecule is so strong, the hydrogen nucleus has a finite probability of acquiring sufficient thermal energy to break the attraction to its own electron and to ionize, forming a hydrated hydronium (H_3O^+) ion with the water molecule to which it is hydrogen bonded. The remnant hydroxyl (OH^-) ion from the nucleus, having lost the positively charged proton from one of the hydrogen atoms forming the neutral water molecule, is negatively charged. This reversible reaction,

$$2H_2O \rightleftarrows H_3O^+ + OH^- \tag{2.1}$$

ionizes a small fraction of the water molecules; these charged species have electric fields that differ considerably from the dipole field of the undissociated molecules. Consequently, the dissociated fraction of the water significantly influences the behavior of charged particles (e.g., dissolved ions, mineral surfaces) in contact with the solution and has a critical effect on many chemical reactions occurring within the solution.

The dissociation constant K_w for the ionization reaction given in (2.1) is equal to

$$K_w = [H^+][OH^-] = 10^{-14} \tag{2.2}$$

where $[H^+]$ and $[OH^-]$ are concentrations in moles per liter (M). Equation (2.2) has been derived from the more general formula appropriate for the reaction in (2.1) by assuming that the concentration of neutral water molecules is constant and that the dissociated species are dilute (Robinson and Stokes, 1959).

Because the dissociation is so slight, there is a negligible difference between the activity and concentration of the hydrogen and hydroxyl ions. The measure of the degree of ionization of water is called pH and is equal to the negative of the base 10 logarithm of the molar concentration of the hydrogen ion $[H^+]$:

$$pH = -\log_{10}[H^+] \tag{2.3}$$

The pH of pure water at 25°C is 7, implying that

$$[H^+] = [OH^-] = 10^{-7} \, M \tag{2.4}$$

This ionization state is called neutral. Whenever there are more $[H^+]$ ions than $[OH^-]$ ions, the solution has a pH less than 7 and is called acidic. Conversely, a solution with less $[H^+]$ ions than $[OH^-]$ ions has a pH greater than 7 and is called basic or alkaline.

2.1.2 Fluid Properties of Water

Thermal and Mechanical Properties Water is an unusual fluid. It has a very high boiling point, a high melting point, a low fluid density, and a liquid phase that is denser than the solid phase. It requires a substantial quantity of heat, called the heat of fusion H_f, to melt a unit mass of ice and an even larger quantity of heat, called the heat of vaporization H_v, to evaporate a unit quantity of water. It has a high dielectric constant, which makes it a good electrical insulator, and has a high specific heat, causing it to experience a smaller temperature rise as it absorbs a unit quantity of heat than other fluids. It is an excellent solvent.

These properties, whose characteristics are summarized in Table 2.1, are all a consequence of the molecular properties discussed in the preceding. Several of the fluid properties of water that are most important in understanding its behavior in soil will be discussed further in the sections that follow.

Surface Tension and Interfacial Curvature When a fluid such as water forms an interface with a different fluid or with a solid, molecules near the interface are exposed to different forces than molecules within the fluid. For example, at an air–water interface, water molecules within the bulk fluid away from the interface are hydrogen bonded to adjacent molecules and experience no net attraction from the

TABLE 2.1 Some Physical Properties of Pure Water

Property	Value	Units	Temperature (°C)
Density			
(liquid)	0.998	$g \, cm^{-3}$	20
(solid)	0.910	$g \, cm^{-3}$	0
(vapor)	1.73×10^{-5}	$g \, cm^{-3}$	20
Heat of fusion	3.34×10^8	$erg \, g^{-1}$	0
Heat of vaporization	2.45×10^9	$erg \, g^{-1}$	20
Specific heat	0.999	$erg \, g^{-1} \, °C^{-1}$	20
Dielectric constant	80	—	20
Thermal conductivity	6.03×10^3	$erg \, cm^{-1} \, s^{-1} \, °C^{-1}$	20
Viscosity	1.00×10^{-2}	$g \, cm^{-1} \, s^{-1}$	20
Surface tension	72.7	$erg \, cm^{-2}$	20

water in any direction. However, molecules at the air–water interface feel a net attraction into the liquid because the density of molecules on the air side of the interface is lower than on the liquid side. This unequal attraction deforms the hydrogen bonds of the molecules at the interface and imparts some "membrane-like" properties to the surface, which stretches over the water volume like a skin. As a consequence, water molecules require extra energy to remain at the interface. The extra energy per unit surface area possessed by molecules at the interface is called the surface tension σ. It may alternately be defined as the energy per unit area required to increase the surface area of the interface or as the force per unit length holding the surface together (Shortley and Williams, 1965).

The curvature of an air–water interface at equilibrium is related to the pressure difference across the interface. If the water is pure and the interface flat, the pressure is the same above and below the interface. However, if the interface is curved, the pressure is greater on the concave side of the interface by an amount that depends on the radius of curvature and the surface tension of the fluid. For a hemispherical interface of curvature R between pure water and air, the pressure difference ΔP between the air and liquid sides of the interface is given by

$$\Delta P = \frac{2\sigma}{R} \qquad (2.5)$$

where $\Delta P = P_a - P_l$ when the interface curves into the liquid (i.e., an air bubble in water) and $\Delta P = P_l - P_a$ when the interface curves into the gas (i.e., a water droplet in air). Figure 2.2 illustrates the relationship between curvature and pressure difference for various interfaces.

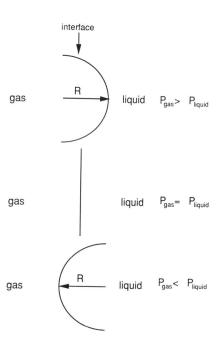

Figure 2.2 Relationship between interfacial curvature and pressure difference between the liquid and gas phases.

Equation (2.5) may be derived from a force balance for certain types of ideal interface geometry. The derivation is illustrated for a spherical water droplet in Example 2.1.

EXAMPLE 2.1: Derive Equation (2.5) for a spherical water droplet of radius R in air.

To derive the relationship between the air pressure P_a pushing on the droplet from outside, the water pressure P_l pushing outward, and the surface tension holding the surface together, imagine that the droplet in Fig. 2.3a is cut in half and the right hemisphere is replaced by a network of forces that exactly equal those exerted by the right hemisphere when it was in place (Fig. 2.3b). Since the forces balance exactly, the left hemisphere remains unchanged and in static equilibrium with zero net force on it. The force pushing to the left,

$$F_1 = \pi R^2 P_l \quad \text{(to left)}$$

is that which was exerted by the water from the right hemisphere on the cross-sectional area $A = \pi R^2$ of the circle where the cut was made.

There are two forces acting to the right. The first is the net force exerted from the air pressure P_a on the outside area of the hemisphere. The components pushing up and down cancel by symmetry, leaving a net force

$$F_2 = \pi R^2 P_a \quad \text{(to right)}$$

(This may be derived exactly using calculus.)

The third force, acting to the right, is the force required to keep the surface in place. Since the surface tension σ is a force per unit length of surface and the perimeter of the circle along the cut is $2\pi R$, then

$$F_3 = 2\pi R\sigma \quad \text{(to right)}$$

Thus, at equilibrium $F_1 = F_2 + F_3$ or

$$P_l - P_a = 2\sigma/R$$

Note that $P_l > P_a$ in accordance with Fig. 2.2.

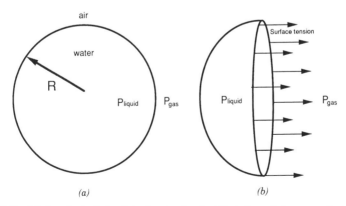

Figure 2.3 (*a*) A spherical droplet of water in air. (*b*) A force balance on the left hemisphere of the droplet.

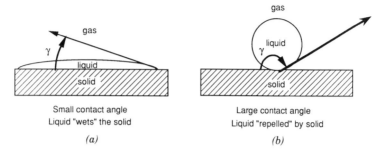

Small contact angle
Liquid "wets" the solid

(a)

Large contact angle
Liquid "repelled" by solid

(b)

Figure 2.4 Illustration of the contact angle γ for the case where (a) the liquid wets the solid (small γ) and (b) it repels the solid (large γ).

Contact Angle   When liquid is present in a three-phase system containing air and solids, the angle measured from the liquid–solid interface to the liquid–air interface is called the contact angle γ. When the liquid is preferentially attracted to the solid phase compared to its cohesive attraction to other liquid molecules, this angle is small and the liquid is said to "wet" the solid (Fig. 2.4a). Conversely, when the cohesive force of the liquid is much stronger than the attractive force to the solid, the liquid is said to "repel" the solid (Fig. 2.4b).

Capillary Rise   When a small cylindrical glass capillary of radius R is placed in contact with a water reservoir open to the atmosphere, water (which has a small contact angle with glass) will spread over the inside walls of the capillary, curving the air–water interface within the tube. This curvature lowers the water pressure at the interface below atmospheric pressure [see (2.5)], causing water to rush into the capillary and to rise upward into the tube. A new equilibrium is reached when the force exerted on the column of water across the air–water interface is balanced by the weight of water in the tube.

> *EXAMPLE 2.2:* Calculate the height of rise of water in a clean glass capillary of radius R ($\gamma = 0$) at equilibrium.
>
> Since the water completely wets the glass surface, the radius of curvature of the air–water interface will be equal to the radius R of the capillary, and the pressure difference will be, using (2.5),
>
> $$\Delta P = P_l - P_a = 2\sigma/R$$
>
> The column of water will be approximately cylindrical, with a volume $V = AH$, where $A = \pi R^2$ is the cross-sectional area and H is the height of rise in the capillary tube. Thus the upward force will be
>
> $$F_{up} = \Delta P\, A = \pi R^2\, \Delta P = 2\pi R\sigma$$
>
> The downward force is the weight of the column, Mg, where $M = \rho_w V$ is the mass of water,
>
> $$F_{down} = \rho_w V g = \pi R^2 \rho_w H g$$

At equilibrium, these two forces must be equal, from which we obtain

$$H = \frac{2\sigma}{\rho_w g R} \tag{2.6}$$

Equation (2.6) gives the height that water will rise in a capillary tube of radius R with which it forms a zero contact angle. When the contact angle is nonzero, the height reached will be less than this value, as shown in Example 2.3.

EXAMPLE 2.3: Repeat Example 2.2 for the case of water rising in a capillary of radius R made of a substance with which water forms a contact angle γ.

As shown in Fig. 2.5, the radius of curvature of the interface is $r = R/\cos \gamma$. Thus, the pressure difference is, from (2.5),

$$\Delta P = \frac{2\sigma}{r} = \frac{2\sigma \cos \gamma}{R}$$

and the force balance this time produces

$$H = \frac{2\sigma \cos \gamma}{\rho \omega g R} \tag{2.7}$$

In this case, the water rises less than in Example 2.2 since $\cos \gamma < 1$. In fact, for capillaries made of substances that repel water, γ is greater than $90°$, $\cos \gamma < 0$, and water in the capillary will be pushed below the height of water in the reservoir.

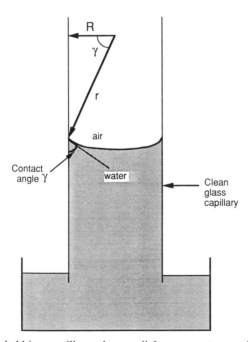

Figure 2.5 Water held in a capillary whose wall forms a contact angle γ with the water.

Viscosity   Since adjacent water molecules are attracted to each other, they will resist any attempt to accelerate a quantity of water within the fluid that has a force exerted on it. This resistive force is called a drag or shear force. A hypothetical example will illustrate the effect of this interaction. Assume that a large volume of water is in a basin filled to a height L (Fig. 2.6). A large massless plate is laid over the top of the water and is pushed to the right with a force F, which because of the resistive forces will attain a terminal velocity V_{max} when the applied force is balanced. The water attached to the plate will move at a velocity V_{max}, and the water attached to the bottom of the basin will not move, setting up a linear change of water velocity in the vertical direction. The ratio between the force per unit area of plate (the tangential force per unit area, or shear force τ) and the velocity gradient perpendicular to the motion (equal to V_{max}/L in this example) is called the coefficient of viscosity ν (units of mass per length per time):

$$\nu = \frac{F/A}{V_{max}/L} = \frac{FL}{AV_{max}} \tag{2.8}$$

The general expression of this relation between drag force and velocity is called Newton's law of viscosity:

$$\tau = F/A = -\nu \, dV/dy \tag{2.9}$$

where y is the direction perpendicular to the fluid flow.

Osmotic Pressure   Since water molecules have a dipole moment, other ions in solution are attracted by the electric field around individual water molecules and tend to cluster near them. The effect of this clustering, as will be seen later in this chapter, is to lower the energy state of the water. If a membrane permeable to water but impermeable to solutes is used to separate pure water from a solution containing ions, water from the pure side of the membrane will cross over into the solution side. This mass transfer will continue indefinitely unless stopped by an opposing force. If the solution is sealed inside a flexible volume, such as a rubber

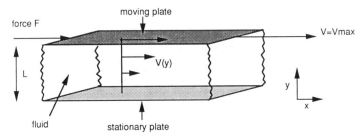

Figure 2.6   Massless plate on water surface pushed by force F to a final velocity V_{max}. The velocity increases linearly in the y direction.

diaphragm, then the pure water entering the volume will expand it and cause the hydrostatic pressure to rise, eventually stopping the water flow. The final hydrostatic pressure of the solution, which balances the ionic attraction of the water at equilibrium, is called the osmotic pressure π of the solution. For dilute solutions, this pressure is given by the approximate formula (Slatyer, 1967).

$$\pi \approx C_s RT \qquad (2.10)$$

where π is the osmotic pressure in ergs per square centimeter, C_s is the concentration in moles per square centimeter, T is the Kelvin temperature, and R is the universal gas constant (8.32×10^7 erg/mol deg). Equation (2.10) refers to the contribution made by a single ion species.

EXAMPLE 2.4: Calculate the osmotic pressure of a 0.01 *M* solution of HCl at $T =$ 300 K. Using (2.10), one obtains

$$\pi = (10^{-5} \text{ mol/cm}^3)(300)(8.32 \times 10^7)(\text{erg/mol})$$

$$= 2.5 \times 10^4 \text{ erg/cm}^3 \cong 0.25 \text{ atm}$$

for each ion species. Thus the total osmotic potential is approximately 0.5 atm.

2.1.3 Water Near Particle Surfaces

In the same way that water molecules are attracted to each other by hydrogen bonding, they are able to associate with charged ions in the electric double layer and with charged solid surfaces. Water molecules that surround charged ions impede their movement somewhat, and the attractive force between water molecules and ions increases the force with which water is held in the solution. Water is attracted to charged ions in the electrical double layer, and the layer acts in some ways like a semipermeable membrane, giving unusual properties to water near the layer. Detailed discussions of this phenomenon may be found in Bolt and Peech (1953), Bolt (1955b), and Gast (1977).

As discussed in Chapter 1, colloidal clays are characterized by the presence of oxygen atoms and hydroxyl units on their surfaces, particularly at broken edges. Hydrogen bonds readily form an association with the outer shell electrons of oxygen and with the hydrogen protons. The strength of this hydrogen bond to oxygen in the lattice approaches the strength of covalent bonding, because lone-pair electrons of the oxygen atom are somewhat repelled by excess electrons in the negatively charged crystal lattice; this makes possible a closer tie with a hydrogen proton of the water molecule.

Bonding of water molecules to surfaces leads to "structured" water close to such surfaces. Evidence exists that water close to particle surfaces is less dense than normal water because of the increased structural organization (Low, 1961). It has been demonstrated also that the viscosity of water in the first few molecular layers may be several times greater than that in water unaffected by such surfaces (Kemper, 1961a,b; Low, 1959).

Evidence of strong attractive forces at particle surfaces has been obtained through experimental measurements such as those performed by Goates and Bennett (1957), who measured the heat necessary to absorb the first molecular layer of water vapor to the particle surface. They found the energy of adsorption of water on a kaolinite clay to be about -14.5 kcal mol^{-1} with 50% of the surface covered and about -13 kcal mol^{-1} with the entire surface covered by a monolayer. (The negative sign indicates that work must be done to remove the adsorbed water.) This is about 3 or 4 kcal mol^{-1} more energy than was required to condense vapor on a free water surface. The energy of adsorption or desorption of successive layers of water after one monolayer has covered the surface rapidly approaches values characteristic of the bulk solution. The energy of adsorption is also influenced to some extent by the nature of the particle surface and by the kind of exchangeable cations present on the surface.

Montmorillonite, which has a high cation exchange capacity and a large specific surface area, has a high capacity to adsorb water. In contrast, kaolinite has the lowest adsorption capacity of the common soil minerals. Illite has adsorption properties intermediate between the two. At 50% relative humidity (at which the soil is very dry), the mass of water adsorbed per mass of soil is of the order of 21, 4.5, and 0.5% for montmorillonite illite, and kaolinite, respectively. These differences arise from the makeup of the crystal lattices and the sites of the negative charges. As shown by Thomas (1928) and Baver and Horner (1933) and by numerous investigators more recently, the nature of the exchangeable cation has a sizable impact upon water adsorption. Generally, clays saturated with divalent cations exhibit greater water adsorption than do those saturated with monovalent cations. An exception is the Li$^+$ ion, which behaves more like a divalent ion. Adsorption of water vapor decreases with increasing size of the ion within both the monovalent and divalent groups.

Inasmuch as the freezing temperature of water is affected by its structure, the presence of ions and the proximity of solid surfaces that affect water structure will also depress its freezing-point temperature. Water at a considerable distance from soil particles freezes first as the temperature is reduced, and the highly structured water adjacent to particle surfaces freezes last. Freezing-point depression has been used as a measure of the energy state of soil water (Babcock and Overstreet, 1957; Day, 1942). However, water can be appreciably supercooled before freezing. The degree of supercooling depends heavily upon the presence or absence of freezing nuclei. Consequently, the complex nature of the soil solid and solution phases make the degree of supercooling difficult to predict, and therefore measurements are sometimes difficult to interpret (Low, 1961).

As a consequence of water structuring near particle surfaces and near hydrated ions, water or ion movement close to particle surfaces is considerably more difficult to describe than flow farther from such surfaces. Water film thicknesses, ranging from a few molecular layers (e.g., about 8 Å in kaolinite) to many layers (e.g., up to 68 Å in montmorillonite), have been reported to be affected by clay surfaces. The extent of this effect depends upon the nature of exchangeable cations on the surfaces. As porous materials dry out, increasingly greater proportions of the remaining water are found in thin films strongly attached to particle surfaces. Thus water in dry soils has a very low energy state.

2.2 SOIL WATER CONTENT

2.2.1 Definitions

The quantity of water in soil is expressed in two different units, as the volumetric water content θ_v and the gravimetric water content θ_g. The volumetric water content is the volume of liquid water per volume of soil, and the gravimetric water content is the mass of water per mass of dry soil.

> *EXAMPLE 2.5:* Calculate a relation between θ_v and θ_g.
>   The mass of water per volume of soil is obtained by multiplying θ_v by ρ_w, the density of water. This is also equal to the product of θ_g and ρ_b, the dry soil bulk density. Thus $\rho_w \theta_v = \rho_b \theta_g$, or
>
> $$\theta_v = \rho_b \theta_g / \rho_w \qquad (2.11)$$

2.2.2 Measurement of Soil Water Content

Measurement of θ_g The gravimetric water content θ_g of a sample of moist soil is measured by weighing the moist soil sample, drying it to remove the water, and reweighing it. The customary method of drying is to place the sample in an oven at 105°C for 24 h. This removes the interparticle water but not the water molecules trapped between clay layers (Gardner, 1986).

> *EXAMPLE 2.6:* Calculate θ_g in the following experiment.
>   A can of moist soil is brought to the lab and weighed, dried, and reweighed. The following data are taken:
>
> | | |
> |---|---|
> | Mass of can with moist soil | 140 g |
> | Mass of can with dry soil | 120 g |
> | Mass of empty can | 20 g |
>
> $$\theta_g = \frac{\text{mass of water}}{\text{mass of dry soil}} = \frac{140 - 120}{120 - 20} = 0.2$$

Measurement of θ_v by Mass and Volume Estimation The volumetric water content may be estimated from θ_g with (2.11) if the bulk density ρ_b is known. In order to measure ρ_b, an undisturbed sample of soil must be taken and its volume estimated. For intact soil peds, the soil volume may be estimated by covering the block with a thin layer of paraffin, lowering it into a full beaker of water, and measuring the volume of displaced water (Blake and Hartge, 1986). Alternatively, a soil core could be taken by pushing a cylinder into the soil, which encloses a sample in the known volume of the coring tube. However, it is extremely difficult to take such samples without compacting the soil and thus changing its density.

Measurement of θ_v by Gamma Ray Attenuation   The volumetric water content may be measured nondestructively for enclosed soil samples by gamma ray attenuation (Gardner, 1986). In this method a narrow beam of gamma radiation is sent through a soil sample of known thickness and is collected beyond its exit from the sample by a detector. Because the detector is beyond a narrow slot that is aligned with the incident beam, it records only those gamma rays that pass through the soil without scattering off of an atom along the way. The gamma radiation has a narrow range of wavelengths and has a characteristic probability of interacting with any obstacle in its path, which depends on the type of substance and its density (see Section 2.7). In a soil that does not swell, all of the solid phase material (soil, column walls, etc.) influence the absorption identically during each measurement, and any change in the reading of transmitted gamma radiation from one time to the next is attributed to a change in water content. As shown in Section 2.7, the gamma ray equation may be written as

$$n(L) = n_0 \exp\left(-\nu_m \rho_b L - \nu_w \rho_w \theta_v L\right) \tag{2.12}$$

where $n(L)$ is the number of counts of gamma radiation per unit time recorded at the detector, n_0 is the background count rate when soil is removed, ν_m is the soil mineral gamma ray absorption coefficient, ν_w is the water gamma ray absorption coefficient, and L is the thickness of the soil sample. The absorption coefficients may be measured for a given system as shown in the next example.

> *EXAMPLE 2.7:*   Devise an experiment to measure ν_w.
> Place an empty tray of inner width L in the path of the beam and measure the number n_0 of counts in 30 s. Fill the tray with water and measure the number of counts n_1 in 30 s. Equation (2.12) then may be written as
>
> $$n_1 = n_0 \exp\left(-\nu_w \rho_w L\right)$$
>
> since $\rho_b = 0$ and $\theta_v = 1$ when no soil is present. Thus
>
> $$\nu_w = -\ln(n_1/n_0)/L\rho_w$$
>
> The mineral absorption coefficient may be measured in a similar way, as illustrated in Problem 2.8.

It is also possible to use gamma ray scanning with two beams of different energies to measure both ρ_b and θ_v in swelling systems in which bulk density changes over time (Reginato, 1974).

Measurement of θ_v by Neutron Attenuation   The neutron attenuation method (Gardner, 1986; Gardner and Kirkham, 1952) for measuring volumetric water content is used exclusively in the field. The device consists of a compact radiation source and detector that is small enough to move inside of a hollow access tube in the ground. The radiation source, usually radium–beryllium or americium–beryllium, emits high-energy neutrons in the range of 5 MeV, which collide with the nuclei of atoms in the surrounding soil. Since the nuclei of most atoms are substantially heavier than the neutrons, most collisions will not slow the neutrons down from their initial energy (see Problem 2.1). However, when neutrons collide with hydrogen nuclei, they are slowed substantially and reach velocities characteristic of the thermal motion of the hydrogen atoms in the soil after a few collisions. The detector, which is located alongside the radiation source, is sensitive only to neutrons moving at "thermal velocities," which are the velocities characteristic of the hydrogen atoms in the soil. Since it requires an enormous number of collisions to slow down neutrons when any atom other than hydrogen is struck, the thermalized neutron counts essentially are proportional to the density of hydrogen atoms in the vicinity of the source. Thus, a calibration curve may be obtained, giving the number of counts per unit time received by the detector versus the amount of hydrogen present, which is predominantly in the form of liquid water. Background hydrogen present in, for example, organic matter or kaolinite will just register as a constant factor in the intercept of the calibration curve. When simultaneous measurements are made of water content (by soil coring) and neutron count rate, the calibration curve may be converted to a relationship between volumetric water content and thermal neutron count rate. The sphere of influence surrounding the radiation source varies between about 15 cm (wet soil) and about 70 cm (very dry soil) (Van Bavel et al., 1956). To work correctly, this method obviously requires that the background counts do not change with time. Therefore, it would be unsuitable in swelling soil unless there was a unique relationship between the soil bulk density and water content. In this case, the count rate could still be calibrated against θ_v, although the curve would probably be nonlinear.

Measurement of θ_v by Time Domain Reflectometry   Time domain reflectometry (TDR) is a recent method by which volumetric water content is estimated indirectly. The method consists of the measurement of the permittivity or dielectric number ϵ of the soil and the subsequent calibration of this property with the volumetric water content (Dalton and van Genuchten, 1986; Topp et al., 1980). Measurement of the dielectric number consists of placing a prong with two arms (usually about 30 cm long or less) forming two parallel waveguides into the soil and sending a step pulse of electromagnetic radiation along with guides. This pulse is reflected at the end of the prong and returned to the source, where its travel time and velocity can be estimated with an oscilloscope. The permittivity of the material (the soil) between the waveguides causes the velocity of the pulse to deviate in a known manner from the velocity of light in vacuum. Hence, the permittivity can be estimated from the travel time, and the water content can be calculated by a regression equation from the permittivity.

Various regression methods have been proposed for this purpose. Topp et al. (1980) used an empirical third-order polynomial to calculate θ_v based on measurements from several soils. Dobsen et al. (1986) used a theoretically based expression for the dielectric of a composite medium, from which θ_v could be calculated. This latter model was found to estimate the relation between ϵ and θ_v accurately for a variety of different soils (Roth et al., 1990).

2.3   ENERGY STATE OF WATER IN SOIL

Next to water content, the energy state of the water is probably the most important single soil physical characteristic. Our understanding of the relationship between soil water content, its energy state, and processes that involve energy gradients in soil–water systems is still far from perfect. Enough is known, however, to make soil water phenomena constitute the subject area in soil physics for which the largest body of mathematical theory is now available.

Both kinetic and potential energy concepts have importance in dealing with soil water. Kinetic energy is that energy which matter has by virtue of its motion; quantitatively it is expressed as $\frac{1}{2} mv^2$, where m is the mass of a body that has a velocity v. On atomic and molecular scales, all matter above absolute zero of temperature is in motion and has kinetic energy. Therefore, water flowing through soil has kinetic energy by virtue of its motion. Additionally, changes of state and variations in temperature involve changes in kinetic energy. However, a great many processes that involve water in soil and plant systems may be dealt with merely by characterizing potential energy changes; kinetic energy enters into equations only implicitly. Use of potential energy concepts is restricted to those situations where temperature may be regarded to have a negligible effect upon the process under consideration. Thus all of the subsequent theory applies only to isothermal soil water.

2.3.1   Potential Energy of Water in Soil

Potential energy is the energy a body has by virtue of its position in a force field. For example, a mass possesses greater potential energy in a gravitational field than an identical mass lying below it, because the latter would have to have work performed on it to move it up to the former's position. Numerous forces act upon water in a porous material like soil. The gravitational field of the earth pulls vertically downward upon the water. Force fields that are caused by the attraction of solid surfaces for water pull water in various directions. The weight of water and sometimes the additional weight of soil particles that are not constrained in the soil matrix also exert downward forces upon water lying underneath. Ions dissolved in water have an attractive force for water and resist attempts to move it. An especially important force existing at water surfaces is associated with the attraction of water molecules for each other and the imbalance of these forces that exists at an air–water interface.

The variety of forces and the directions in which they act make description of force networks in soil water difficult. However, it is possible to calculate the potential energy of a unit quantity of water as a consequence of the forces acting upon it. Potential energy differences from point to point in isothermal systems determine the direction of flow, the amount of work available for causing flow, or the amount of work that must be done from the outside to cause flow. This is the merit of describing soil water in terms of the potential energy associated with it.

2.3.2 Reference or Standard State

The potential energy of water in soil must be defined relative to a reference or standard state, since there is no absolute scale of energy. The standard state is customarily defined to be the state of pure (no solutes), free (no external forces other than gravity) water at a reference pressure P_0, reference temperature T_0, and reference elevation Z_0 and is arbitrarily given the value zero (Bolt, 1976). The soil water potential energy is defined as the difference in energy per unit quantity of water compared to the reference state. This statement may be made more rigorous through the definition given in the next section.

2.3.3 Total Soil Water Potential

The total soil water potential of the constituent water in soil at temperature T_0 is the amount of useful work per unit quantity of pure water that must be done by means of externally applied forces to transfer reversibly and isothermally an infinitesimal amount of water from the standard state to the soil liquid phase at the point under consideration (Bolt, 1976).

There are several systems of units in which the total potential and its components may be described, depending on whether the quantity of pure water mentioned in the preceding definition is expressed as a mass, a volume, or a weight. Table 2.2 summarizes these systems and their units.

The first system of units is used extensively in chemical thermodynamics. We will use the latter two systems in this course.

EXAMPLE 2.8: Derive Relations between μ_T, ψ_T, and h_T.

Since pure water is the standard quantity in all cases, the relationship between the mass m_w and the volume V_w of pure water is given by $m_w = \rho_w V_w$, where ρ_w is the density of water. Thus $\mu_T = \rho_w^{-1} \psi_T$. The relationship between the mass m_w and the weight W_w of water is $W_w = g m_w$ where g is the acceleration of gravity. Thus $\mu_T = g h_T$ and $\psi_T = \rho_w g h_T$.

TABLE 2.2 Systems of Units of Total Soil Water Potential

Units	Symbol	Name	Dimensions	CGS Unit	SI Unit
1. Energy/mass[a]	μ_T	Chemical potential	L^2/T^2	erg/g	J/kg
2. Energy/volume	ψ_T	Soil water potential	M/LT^2	erg/cm^3	N/m^2
3. Energy/weight	h_T	Soil water potential head	L	cm	m

Source: Adapted from Sposito (1981).

[a] Or energy/mole.

2.3.4   Components of Water Potential

The transformation of water from the reference state to the soil water state may be broken up into a series of steps, each of which partially converts water to the final state. As long as each step is performed reversibly and isothermally, the total change in potential energy of the water may be set equal to the sum of the potential energy changes corresponding to each of the steps. These sequential potential energy changes will be called components of the total water potential.

In accordance with the recommendations of the 1976 soil physics terminology committee of the ISSS (Bolt, 1976), we will define the transition from the reference pool to the soil water state in three steps.

Gravitational Potential ψ_z (z in Head Units)   The gravitational potential ψ_z is the energy per unit volume of water required to move an infinitesimal amount of pure, free water from the reference elevation z_0 to the soil water elevatioin z_{soil}. It has the value $\psi_z = \rho_w g (z_{\mathrm{soil}} - z_0)$.

Solute Potential ψ_s (s in Head Units)   The solute potential ψ_s is the change in energy per unit volume of water when solutes identical in composition to the soil solution at the point of interest in the soil are added to pure, free water at the elevation of the soil.

The remaining components of the water potential have been defined somewhat differently by various authors, in part because the remaining effects on water caused by the presence of the soil matrix are difficult to isolate from each other.

In 1976 commission of the ISSS completed the transition from the reference state to the soil water with the following definition.

Tensiometer Pressure Potential ψ_{tp}   The tensiometer pressure potential ψ_{tp} is the energy per unit volume required to transfer reversibly and isothermally an infinitesimal amount of solution containing solutes (which are identical in composition to the soil water) from a reservoir of solution (which is at reference pressure and is located at the elevation of the soil) to the point of interest in the soil.

This component encompasses all effects on soil water other than gravity and solute interactions. Its influence may include the effects of binding to soil solids, interfacial curvature, air pressure, weight of overlying soil solid material, and hydrostatic water pressure in saturated soil. This definition is not in common use in the literature, because soil scientists have traditionally divided the energy change associated with the transfer of a unit quantity of water from a pool of solution to soil water into several other components, which account separately for the effects of air pressure, weight of soil solids, hydrostatic water pressure, and binding by the soil matrix (matric effects). As pointed out by the 1976 commission, however, these divisions must be made carefully because the various forces on a soil water element can interact with each other. This interaction will become apparent from the discussion that follows.

A potential component that has been used since the formulation of soil water potential relations began is the matric potential, which describes the potential energy change associated with the addition of the soil matrix.

Matric Potential ψ_m (h in Head Units)   The matric potential is the energy per unit volume of water required to transfer an infinitesimal quantity of water from a reference pool of soil water at the elevation of the soil to the point of interest in the soil at reference air pressure.

Thus, the matric potential differs from the tensiometer pressure potential in that the soil air pressure is maintained at reference pressure. The transition to the final state is achieved with the following definitions, depending on whether the soil water is unsaturated and has an air phase that can exert pressure on the water or the soil water is saturated and experiences hydrostatic pressure from an overlying water phase.

Air Pressure Potential ψ_a (a in Head Units)   The air pressure potential ψ_a is defined as the change in potential energy per unit volume of water when the soil air pressure is changed from the pressure P_0 of the reference state to the pressure P_{soil} of the soil. If this change does not alter the geometry of the soil liquid phase, then ψ_a is approximately equal to $P_{soil} - P_0 = \Delta P_a$, the gauge pressure of the soil air relative to the standard state air pressure.

Hydrostatic Pressure Potential ψ_p (p in Head Units)   The hydrostatic pressure potential ψ_p is defined as the water pressure exerted by overlying unsupported (saturated) water on the point of interest in the soil. By definition, it is equal to the water pressure exerted by the height of water between the point of interest z_{soil} and the water table (saturated–unsaturated soil interface) z_{wt}; $\psi_p = \rho_w g (z_{wt} - z_{soil})$.

The potential components discussed thus far are sufficient to describe soil water energy states whenever the soil solid matrix is rigid and self-supporting. For these so-called nonswelling soils, the weight of soil solid material above the point of interest is entirely borne by the rigid matrix. No external forces exerted on the solid boundary may influence the energy state of the soil water. In this case, the matric potential described in the preceding may be characterized by the amount of water in the soil, since the soild framework remains fixed during changes in wetting or drying.

The situation is far more complex in swelling soils. Since the solid particles are not in complete contact with each other, part or all of the weight of overlying solid material may be exerted on an element of soil water. Furthermore, changes in water content cause reorientation of the solid particles, which create large changes in interfacial curvature. In swelling soils, air pressure variations may cause the soil geometry to change, thereby altering the interfacial curvature of water in soil.

For these reasons, the matric potential is a far less useful concept in swelling media than in rigid soils, and it may not be possible to characterize it adequately from a knowledge of water content alone. For such applications, it is customary

TABLE 2.3   Components of Total Soil Water Potential for Various Applications

Application	Components
1. Unsaturated swelling soil	$\psi_T = \psi_z + \psi_s + \psi_{tp}$
	$\psi_T = \psi_z + \psi_s + \psi_a + \psi_m$
	$\psi_T = \psi_z + \psi_s + \psi_a + \psi_b + \psi_w$
2. Unsaturated rigid soil	$\psi_T = \psi_z + \psi_s + \psi_{tp}$
	$\psi_T = \psi_z + \psi_s + \psi_a + \psi_m$
3. Saturated swelling soil	$\psi_T = \psi_z + \psi_s + \psi_{tp}$
	$\psi_T = \psi_z + \psi_s + \psi_p + \psi_b$
4. Saturated rigid soil	$\psi_T = \psi_z + \psi_s + \psi_{tp}$
	$\psi_T = \psi_z + \psi_s + \psi_p$

to break the matric potential up into two additional components, one that describes the effect of the soil geometry (i.e., adsorptive forces, interfacial curvature) and one that describes the effect of the mechanical forces exerted by the pressure of the solid material.

Overburden Pressure Potential ψ_b (b in Head Units)   The overburden potential ψ_b is defined as the change in energy per unit volume of soil water when the envelope pressure P_e (mechanical pressure exerted by the unsupported solid material on the soil water) is changed from zero to P_e.

Wetness Potential ψ_w (w in Head Units)   The wetness potential ψ_w is the value of the matric potential at zero external air pressure and zero envelope pressure.

Table 2.3 summarizes the various ways in which the total potential is divided for different applications.

2.4   ANALYSIS OF SYSTEMS AT EQUILIBRIUM

When two systems at the same temperature containing water at different energy states are brought into contact with each other, water will flow from the region of higher total potential to the region at lower total potential. This flow will occur until the total potential energy of the two systems are equal. At this time, the flow will cease and the systems are said to be at equilibrium (technically, at thermal and mechanical equilibrium). Equivalently, all points of a water system at equilibrium have the same value of total water potential energy.

The equilibrium principle may be used to estimate components of the total water potential using various measurement devices. The analysis of a system at equilibrium by this procedure should proceed in the orderly sequence of steps outlined in Table 2.4.

TABLE 2.4   Analysis of Components of Water Potential at Point of Interest Using Principle of Equilibrium

Step 1	Verify that the system is at equilibrium. Water flow must be zero or negligible everywhere and the temperature must be constant.
Step 2	Define a reference elevation and reference air pressure.
Step 3	Find a location in the system where the total water potential ψ_T may be estimated. Call this value ψ_{T_0}.
Step 4	Set $\psi_T = \psi_{T_0}$ at the point of interest.
Step 5	Divide ψ_T into appropriate components using Table 2.3.
Step 6	Directly evaluate those components for which appropriate information at the point of interest is available.
Step 7	Determine the remaining component or sum of components using the equation in step 4.

This procedure may be illustrated by reanalyzing the capillary rise problem discussed earlier.

EXAMPLE 2.9: Calculate the height of rise of water in a clean glass cylindrical capillary tube of radius R placed in an open vessel of pure water.

Assume that the contact angle of the water on glass is zero and that the air above the interface is at the same pressure as the air above the open vessel. Assume that evaporation of water into the air has been suppressed or is negligible.

Proceeding according to Table 2.4, we execute the following steps:

Step 1 The system shown in Fig. 2.7 will reach equilibrium (if water evaporation is suppressed) as soon as the water in the capillary stops rising.

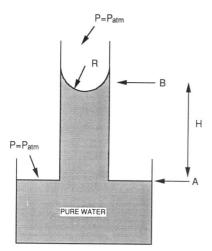

Figure 2.7 Capillary tube containing water at equilibrium everywhere.

Step 2 Define $z = 0$ at point A and $P_0 = P_{atm}$.

Step 3 At point A all components of ψ_T may be evaluated.

$$\psi_z = 0 \quad \text{since } z = 0$$

$$\psi_s = 0 \quad \text{pure water}$$

$$\psi_a = 0 \quad P = P_0 = P_{atm}$$

$$\psi_m = 0 \quad \text{no soil}$$

$$\psi_p = 0 \quad \text{no overlying hydrostatic pressure}$$

Thus relative to the reference state, $\psi_T = 0$ at A.

Step 4 At point B, $\psi_T = 0$ since point B is at equilibrium with point A where $\psi_T = 0$.

Step 5 Regarding point B as a location in a rigid unsaturated "soil," we may use the relation $\psi_T = 0 = \psi_z + \psi_s + \psi_a + \psi_m$. The matric potential ψ_m accounts for the decrease in liquid water pressure caused by interfacial curvature resulting from adsorption of water to the interior capillary walls.

Step 6 By definition, at point B the components are

$$\psi_s = 0 \quad \text{pure water}$$

$$\psi_a = 0 \quad P = P_0 = P_{atm}$$

$$\psi_z = \rho_w g (z - z_0) = \rho_w g H$$

$$\psi_m = P_l - P_a = -2\sigma/R \quad \text{by (2.5)}$$

Step 7 $\psi_T = 0 = \rho_w g H - 2\sigma/R$ or $H = 2\sigma/\rho_w g R$.

Notice that solute potential differences could not be present at equilibrium in this system, even if there were solutes in the water, since diffusion from one point of the system to another would create a uniform distribution at equilibrium (except for a small density stratification because of gravity), provided that no solute barriers were present. Therefore, the solute potential would be the same everywhere and could not affect the value of any of the other components. Thus, the only time that solute potential needs to be included in the analysis is when such barriers are present within the soil–water system. This point is illustrated in the next example.

EXAMPLE 2.10: Repeat Example 2.9 for the case where the open vessel contains a solution of osmotic pressure π (energy per volume) and the capillary has a perfect semipermeable membrane at the bottom that restricts the passage of solutes but allows water to flow into the capillary tube (Fig. 2.8). Assume that the surface tension and contact angle are unaffected by the addition of solutes.

Repeating the steps of Table 2.4, we may write:

Step 1 Equilibrium is reached when the capillary rise ceases at height H.

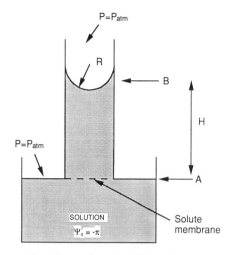

Figure 2.8 Capillary tube with an internal solute membrane, containing water at equilibrium with a solution of osmotic pressure π.

Step 2 $z = 0$ at A and $P_0 = P_{atm}$

Step 3 At point A

$$\psi_z = 0 \qquad z = z_0$$

$$\psi_a = 0 \qquad P = P_0$$

$$\psi_m = 0 \qquad \text{no soil}$$

$$\psi_p = 0 \qquad \text{no hydrostatic pressure}$$

$$\psi_s = -\pi \qquad \text{by definition}$$

Thus, $\psi_T = \psi_{T_0} = -\pi$ at A.

Step 4 $\psi_T = -\pi$ at point B.

Step 5 $\psi_T = -\pi = \psi_z + \psi_s + \psi_a + \psi_m$ at point B.

Step 6 By definition, at point B

$\psi_s = 0$ pure water in the capillary

$\psi_a = 0$ $P = P_0 = P_{atm}$

$\psi_z = \rho_w g H'$

$\psi_m = P_l - P_a = -2\sigma/R$ since, by assumption, the contact angle remains zero

Step 7 $\psi_T = -\pi = \rho_w g H' - 2\sigma/R$.

Thus,

$$H' = 2\sigma/\rho_w g R - \pi/\rho_w g \qquad (2.13)$$

In this case, the water will not rise as high as it did in the previous example, since the solutes in the vessel attract the water molecules on the capillary side of the membrane. In fact, if $\pi > 2\sigma/R$, then no water will enter the capillary through the membrane.

The previous example illustrates the important effects that solute membranes may have on a water system. The most important solute membranes in soil are found in plant roots.

2.5 MEASUREMENT OF COMPONENTS OF WATER POTENTIAL

Many of the devices that measure components of the water potential operate on the equilibrium principle in that they are placed in the soil water system and exchange water until they come to equilibrium with the point in the soil where the measurement is made. These devices are used to measure those components of the system that are not completely specified in terms of parameters directly susceptible to measurement.

2.5.1 Direct Measurement of Potential Components

Several of the components discussed in the preceding may be individually measured by standard devices:

1. *Gravitational Potential.* Measure the vertical distance from the reference elevation to the point of interest.
2. *Solute Potential.* Extract soil solution and measure its osmotic pressure or analyze for the solute concentration and use (2.10). The osmotic pressure of a solution may be measured by standard methods such as equilibrium dialysis or vapor pressure depression (Rawlins and Campbell, 1986).
3. *Air Pressure Potential.* Measure the soil air pressure with a barometer and subtract the reference air pressure.
4. *Hydrostatic Pressure Potential.* Measure the vertical height of saturated water above the point of interest. This requires finding the water table location, which may not be easy to detect in swelling soil systems (see Fig. 2.9 in the next section).

2.5.2 Measurement Devices

The remaining components (matric potential, overburden potential, wetness potential) may not be evaluated directly from their definitions but must be inferred from the readings of devices that equilibrate with soil water. The principal devices used for such purposes are described in what follows.

Piezometer Tube   The piezometer tube (Fig. 2.9) is a hollow tube that is placed into the soil with a water entry point at the place where a measurement is desired. The other end of the tube extends vertically from the measurement point and is open to the atmosphere. If water in the soil is under positive pressure, it will enter the tube and rise to a height whose gravitational potential relative to the soil at the

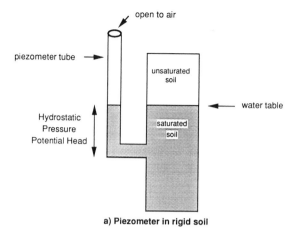

a) Piezometer in rigid soil

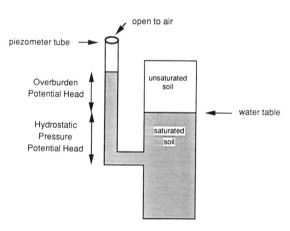

b)Piezometer in swelling soil

Figure 2.9 Illustration of a piezometer tube used to measure components of the water potential in (*a*) a rigid and (*b*) a swelling soil.

point of interest is equal to the tensiometer pressure potential of the soil water. (For practice, the reader may verify this using the equilibrium analysis.) Excluding unsaturated soil, in which water would not enter the piezometer, the tensiometer pressure potential in a saturated system is equal to the sum of the overburden and the hydrostatic pressure potentials. Thus, in a rigid soil, the water in the piezometer will rise to a height equal to the water table height. In a swelling soil, the height reached will exceed the height of the water table by an amount equal to the overburden potential head (see Fig. 2.9).

Tensiometer   A tensiometer consists of a water-saturated porous ceramic cup connected to a manometer through a water-filled tube (see Fig. 2.10). The ceramic,

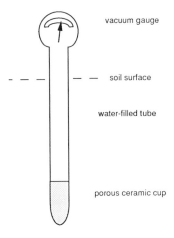

a)Vacuum gauge tensiometer

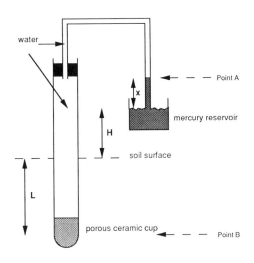

b) Mercury manometer tensiometer

Figure 2.10 Diagram of a tensiometer, measuring energy state of soil water with (*a*) a vacuum gauge and (*b*) a mercury manometer.

consisting of very fine pores, remains water saturated even when placed in contact with soil at relatively low water potentials. Upon contact, water moves from the tube to the soil, creating a suction at the manometer end until equilibrium is reached and the total potential of the water system is equal everywhere.

The equilibrium principle may be applied to the mercury manometer tensiometer shown in Fig. 2.10 to determine which components of the soil are measured and the relationship between these components and the height of rise X of the mercury in the manometer.

Assuming that equilibrium is reached rapidly, we shall regard points A and B as being at the same total water potential. The reference height is taken as $z = 0$ at the soil surface, and the reference pressure is taken as the air pressure around the mercury reservoir. Since solutes will pass freely through the ceramic and through the tube, we will ignore ψ_s in the subsequent analysis. At point A,

$$\psi_z = \rho_w g (H + X) \quad \text{relative to } z = 0 \text{ at the soil surface}$$

$$\psi_a = 0 \quad \text{no air inside the tube}$$

$$\psi_m = 0 \quad \text{no soil inside the tube}$$

$$\psi_p = -\rho_m g X \quad \text{where } \rho_m \text{ is the density of mercury}$$

The hydrostatic pressure of the water must be less than the reference pressure or mercury will not rise in the tube. The weight of suspended mercury per unit area $\rho_m g X$ must be equal to the pressure difference across the ends of the column. Thus the mercury pressure at the top, which is equal to the water pressure at the interface, is $-\rho_m g X$.

Therefore, at point A, $\psi_T = \rho_w g (H + X) - \rho_m g X$, which must be equal to the total potential at point B in the soil. Assuming that the soil is unsaturated, we may write

$$\psi_T = \rho_w g (H + X) - \rho_m g X = \psi_{zB} + \psi_{mB} + \psi_{aB}$$

where $\psi_{zB} = -\rho_w g L$.
Thus,

$$\psi_{mB} + \psi_{aB} = \rho_w g (H + L) - (\rho_m - \rho_w) g X \tag{2.14}$$

This relationship shows that the manometer reading is a function of the distance $H + L$ between the ceramic and the mercury reservoir and that the tensiometer measures the sum of the matric potential (including overburden influences if any) and air pressure potential. However, the soil air pressure will usually not be different from atmospheric ($\psi_a \sim 0$) and

$$\psi_m \approx \rho_w g (H + L) - (\rho_m - \rho_w) g X \tag{2.15}$$

or in head units

$$h \approx H + L - 12.6X \tag{2.16}$$

Often the vacuum gauge and mercury manometer tensiometers will be offset to read matric potential directly. For the mercury manometer, this is accomplished by placing the $h = 0$ point of the scale reading at $X = (H + L)/12.6$.

EXAMPLE 2.11: Calculate the matric potential measured by the tensiometer in the following application.

A mercury manometer tensiometer is placed so that the ceramic is 100 cm below the mercury reservoir. The mercury rises to a height of 20 cm above the reservoir. Calculate the matric potential of the soil in the vicinity of the ceramic. Assume ψ_a = 0.

In this example $H + L = 100$ and $X = 20$. Therefore, by (2.16) we obtain

$$h = 100 - (12.6)(20) = -1.52 \text{ cm}$$

or in units of energy per volume

$$\psi_m = -1.49 \times 10^5 \text{ erg/cm}^3 \quad (-14.9 \text{ kPa})$$

The tensiometer has a practical range of about -800 cm, below which gases dissolved in the water will begin to form bubbles and the liquid column will break up (Cassel and Klute, 1986).

Soil Psychrometer The soil psychrometer measures the relative humidity (RH) of the water vapor in the soil, from which the water potential of the vapor phase may be calculated using (Edlefsen and Anderson, 1943)

$$\text{RH} = P_v/P_v^* = \exp\left(M_w\psi_w/\rho_w RT\right) \tag{2.17}$$

where P_v is vapor pressure, P_v^* is saturated vapor pressure, and M_w is the molecular weight of water. At equilibrium, the vapor water potential is equal to the liquid water potential. Since the vapor and liquid phases are at essentially the same elevation, the components of the soil water potential measured by the psychrometer are the sum of the matric and osmotic potentials, assuming that the air pressure is atmospheric.

The thermocouple psychrometer is the most widely used psychometric device in soil. A small (0.025-mm-diameter) chromel–constantan thermocouple inside a thin-walled ceramic cup is buried in the soil, so that the vapor in the atmosphere that surrounds the thermocouple is at equilibrium with the soil solution. By using a small direct current, the temperature of the thermocouple is reduced to the dew point by Peltier cooling so that water is condensed on the junction. Cooling is then stopped and the temperature of the junction is measured with a microvoltmeter while the junction is cooled by evaporation. The temperature of the junction depends upon the evaporation rate, which in turn depends upon the relative humidity of the atmosphere. The device, when calibrated over osmotic solutions of known humidities, permits measurement of water potentials in a range from about -0.1 bar down to the order of -70 bars, although it is least accurate in wet soil (Bruce and Luxmoore, 1986). Even lower values of water potential may be measured with special techniques (Rawlins and Campbell, 1986). The psychrometer also may be used in the laboratory on plant tissues and on soil and plant samples.

2.6 WATER CHARACTERISTIC FUNCTION

2.6.1 Measurement

In rigid porous media, the matric potential as defined in the preceding represents the effect of adsorptive soil solid forces and interfacial curvature on water potential energy. The functional relationship between the matric potential and the gravimetric or volumetric water content is called the water characteristic function or matric potential–water content function $\psi_m(\theta)$. This function may be evaluated by measuring matric potential and water content simultaneously with the methods already discussed during a succession of water content changes. In the laboratory, $\psi_w(\theta)$ may be measured on replicated prepared samples over a large range of water contents. Virtually the entire range from water-saturated soil to very dry soil may be covered by using a hanging water column, a pressure membrane, and equilibration over salt solutions. These devices will be illustrated using the equilibrium principle.

Hanging Water Column (**Range** -100 cm $< h < 0$)   A hanging water column consists of a water-saturated, highly permeable porous ceramic plate connected on its underside to a water column terminating in a reservoir open to the atmosphere. Water-saturated samples of soil held in rings are placed in contact with the flat plate when the water reservoir height is even with the top of the plate. Then the reservoir is lowered to a new height a distance H below the top of the plate (Fig. 2.11).

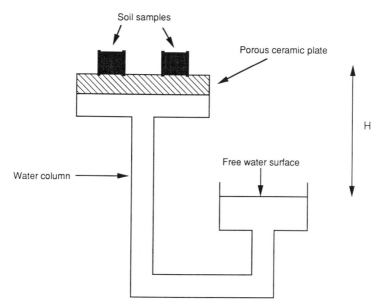

Figure 2.11 Desaturation of soil water samples to a desired energy state with a hanging water column.

By the equilibrium principle, water will flow from the soil samples through the ceramic to the reservoir until the total water potential of the system is constant. At this time the potential of the free reservoir may be set equal to zero, and at the soil sample height H we may write ($z = 0$; $P = P_{atm}$; neglect solutes) $\psi_m + \psi_z = 0 = \psi_m + \rho_w gH$, or $\psi_m = -\rho_w gH$.

When equilibrium has been restored, some of the samples may be removed and their gravimetric or volumetric water content measured. The tube may then be lowered further and a new set of samples measured.

If there is good contact between the soil and the ceramic, equilibrium will be reached rapidly (i.e., several hours) since the samples are quite moist. The range of the device is limited chiefly by the space available for lowering the water column.

Pressure Plate (**Range** $-1500\,\text{cm} \le h \le -300\,\text{cm}$)   The pressure plate consists of an air-tight chamber enclosing a water-saturated, porous ceramic plate connected on its underside to a tube that extends through the chamber to the open air. Saturated soil samples are enclosed in rings and placed in contact with the ceramic on the top side. The chamber is then pressurized, which squeezes water out of the soil pores, through the ceramic, and out the tube (Fig. 2.12).

At equilibrium, flow through the tube will cease. We may set the total potential equal to zero at the point where the water exits the tube. Inside the plate we may write ($P = P_{atm}$; $z = 0$; neglect solutes) $\psi_a + \psi_m = 0 = \psi_m + \Delta P$, or $\psi_m = -\Delta P$.

When equilibrium is reached, the chamber may be depressurized and the water content of the samples measured. An assumption is made in this method that the matric potential of the sample does not change as the air pressure is lowered to atmospheric.

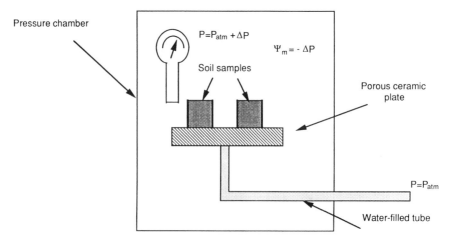

Figure 2.12 Desaturation of soil water samples to a desired energy state with a pressure plate.

This method may be used up to air gauge pressures of about 15 bars if special fine-pore ceramic plates are used. Since these devices have a very high flow resistance, it may require a substantial amount of time to remove the last small amount of water from the soil. Thus, the time of equilibrium is difficult to estimate.

Equilibration over Salt Solutions ($h < -15{,}000$ cm)   By adding precalibrated amounts of certain salts, the energy level of a reservoir of pure water may be lowered to any specified level. If this reservoir is brought into contact with a moist soil sample, water will flow from the sample to the reservoir. If the sample and the reservoir are placed adjacent to each other in a closed chamber at constant temperature, water will be exchanged through the vapor phase by evaporation from the soil sample and condensation in the reservoir until equilibrium is reached.

Since the reservoir is a pool of salt solution, at equilibrium the total potential will be $\psi_T = \psi_{so}$ of the solution. In the soil $\psi_T = \psi_m = \psi_s$ since the air–water interface acts as a solute membrane. Thus $\psi_m = \psi_{so} - \psi_s$, the difference between the solute potentials of the reservoir and the soil. In practice, the soil will usually not be saline enough for its solute potential to be significant compared to ψ_{so} in the range where these measurements are made.

The equilibration time for this method can be shortened by creating a partial vacuum in the chamber (Campbell and Gee, 1986). Care should be taken that the sample and the reservoir are at the same temperature, because even small temperature differences will cause the soil and salt solution to equilibrate at very different potentials (Campbell and Gee, 1986).

Figure 2.13 shows typical matric potential–volumetric water content curves for a sandy soil and a finer textured soil high in clay measured from soil initially at water saturation. The water characteristic function for a soil desorbed from saturation may be roughly divided into three regions, as shown in the figure. The air entry region corresponds to the region at saturation where the matric potential changes but the water content does not. The minimum suction that must be applied to a saturated soil to remove water from the largest pores is called the air entry suction, which varies from about 5 to 10 cm for sands to much higher values in unaggregated, fine-textured soils.

After air begins to enter the system, incremental increases in suction on the soil sample will drain progressively smaller pores, and the water content will drop. This intermediate part of the curve is called the capillary region.

When essentially all of the water held in pores has been drained, only the tightly bound water adsorbed to particle surfaces remains. Large changes in matric potentials in this region, called the adsorption region, are associated with small changes in water content.

The differences in the shapes of the water characteristic function for the prototype sandy and clay soils in Fig. 2.13 may be explained by considering the properties of the bulk solid phases. The clay soil generally has a lower bulk density and hence a higher water content at saturation. The clay soil has very few large pores and a broad distribution of particle sizes. Hence, it decreases gradually in water content with decreases in matric potential. The sandy soil, on the other hand,

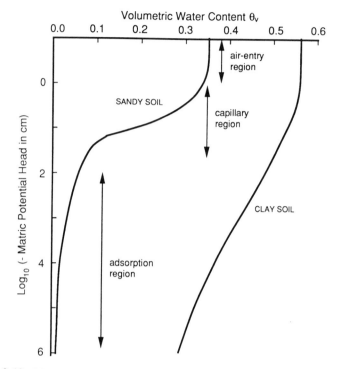

Figure 2.13 Matric potential–water content function (water characteristic function).

has much of its water held in large pores that drain at modest suctions. Hence it will have a very rapid decrease in water content in the capillary region. Finally, the clay soil has a very large surface area compared to the sand and will have a large amount of water adsorbed to the surfaces.

EXAMPLE 2.12: Calculate the gravimetric water content of a sand and a clay containing only one monolayer of adsorbed water using the following information:

Montmorillonite, surface area/mass	$S = 800 \text{ m}^2 \text{ g}^{-1}$
Quartz sand, surface area/mass	$S = 5 \times 10^{-2} \text{ m}^2 \text{ g}^{-1}$
Area occupied by one adsorbed water molecule	$A_m = 1.05 \times 10^{-15} \text{ cm}^2$
Mass of one water molecule	$m_w = 3 \times 10^{-23} \text{ g}$

The gravimetric water content is the mass of water per mass of soil. In a total mass M_s of soil there is:

Total area of soil	$A_s = M_s S$
Number of water molecules adsorbed	$N = A_s/A_m = M_s S/A_m$
Mass of water adsorbed	$M_w = N m_w = m_w M_s S/A_m$
$\theta_g = M_w/M_s = m_w S/A_m$	

Thus for montmorillonite $\theta_g = 0.23$, and for the quartz sand $\theta_g = 1.4 \times 10^{-5}$.

2.6.2   Hysteresis in Water Content–Energy Relationships

Water content and the potential energy of soil water are not uniquely related because the potential energy state is determined by conditions at the air–water interfaces and the nature of surface films rather than by the quantity of water present in pores. Soil pores are highly variable in size and shape and interconnect with each other in a variety of ways. Common to porous media are so-called bottleneck pores, which have large cavities but narrow points of connection to adjacent pores. Water is held most tenaciously in small pores, which fill first when water is admitted to a system. But they do not always empty again during drying in the same order as they were filled.

The factors involved in hysteresis may be discussed most clearly by assuming that the soil is initially completely devoid of water and subsequently has no air phase (just liquid water and its vapor). If water is added at this point to the system, small pores fill first, followed by successively larger and larger pores until all pores are filled and the matric potential is zero. At intermediate values of saturation, with enough water in the system so that vapor–water interfaces can exist between particles and in small pores, the curvature of such interfaces is given by (2.5), where the pressure difference ΔP refers to the difference in pressure between the vapor and the liquid water. Water content and water potential in such a system will follow the wetting curve in Fig. 2.14. Some small pores could be isolated during wetting, so that they might remain dry while larger pores are filled. However, this would not be the case at equilibrium in the absence of air, inasmuch as

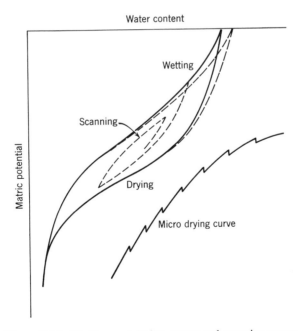

Figure 2.14   Wetting and drying curves and scanning curves.

vapor transfer would assure the wetting of all pores small enough to retain water at a particular matric potential.

If the system is dried either by evaporating water or by bringing the soil into contact with a dry, porous material that pulls water away from the system, pores will begin to empty, generally from large to small. However, liquid water may now be trapped in large pores in such a way that they will not empty in the order that they filled. Water will be held in large pores until conditions are reached where at least one interconnecting smaller pore can empty; at this time the larger pore quickly empties. The sudden release of a relatively large amount of water from a large pore floods surrounding pores and increases the matric potential in them temporarily. If matric potential were monitored in a small porous system having discrete differences in pore size, the matric potential–water content relationship for drying might be saw-toothed, as indicated by the drying curve in Fig. 2.14.

In real soil systems, the pore size distribution contains many pores in all size ranges, and the water content and potential distributions tend to average out so that a smooth curve is obtained. However, the water content for a given matric potential is higher than for the wetting system, as is shown by the drying curve in Fig. 2.14. This principle is illustrated by the pore–water system in Fig. 2.15a. Here it may be observed that the curvature of the vapor–water interface in the small pores of two identical systems can be in equilibrium with each other even though their water contents are grossly different. The matric potential is determined by the curvature of the liquid interface (2.5), which at equilibrium would be precisely the same in the small pores connecting with the large pore in each case. This ideal representation commonly is called the "ink bottle principle," which refers to the fact that an ink bottle has a small opening into a large cavity.

Large pores that are interconnected by smaller pores are not required for hysteresis to occur. It is possible for a single pore to contain the same amount of water at two different water potentials, as is shown in Fig. 2.15b. Water that condenses initially into such a capillary from a humid environment is shown by the diagonal cross-hatched area. However, as condensation proceeds, water at the center finally coalesces and a concave meniscus is formed as a consequence of surface tension forces (vertical cross-hatching). Whereas positive pressure existed in the system before coalescence, the system suddenly goes under negative pressure as a consequence of its new configuration. In the example in Fig. 2.15b, water in a cylindrical pore about 1 cm in length and 0.1 cm in radius would have a slight positive potential of about 0.15 mbar immediately before coalescence and a potential of -1.5 mbars immediately afterward.

Surface wetting can also induce hysteresis. Unless particle surfaces are meticulously clean, they will form a nonzero contact angle with water when wetted (Fig. 2.15c). This results in thicker films than would be present in the drying phase where water films are drawn tightly over the surface by adsorptive forces.

Thus far the discussion of hysteresis has not involved the presence of air in the system, which can introduce additional differences between water content at a given matric potential during wetting and drying. As small pores and interstices between particles fill with water, air may become entrapped in large pores. Continued water

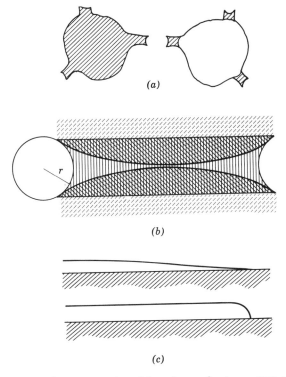

Figure 2.15 Diagrammatic representation of three forms of water content–matric potential hysteresis.

entry into such pores will cause a buildup of air pressure. Since air is slightly soluble in water, pressures in such pores may gradually be relieved, which will sometimes allow more complete pore filling. However, the order of filling and access to pores will be influenced by entrapped air, so that water content is still permanently affected despite the fact that some air may go into solution and disappear.

In the absence of air, the water potential–water content relationship for complete wetting and complete drying will follow approximately the dashed line–solid line loop shown in Fig. 2.14. However, this loop is not exactly reproducible because of the inherent difficulty associated with repetition of the exact order of pore filling over each cycle. When air is present in the system, the curves are offset somewhat toward the dry side (solid curve, Fig. 2.14). If a soil is completely wetted so that no air is present and then is dried, it will follow the dashed–solid curve down; upon rewetting in the presence of air, it will follow the solid wetting curve and will not return to the starting point because of the presence of entrapped air. If the process is reversed at any time during wetting or drying, curves like those in the interior (dotted curves) of the hysteretic envelope are produced. These interior curves have been called scanning curves; the curves that form the hysteretic envelope have been called characteristic curves or the soil moisture characteristic (Childs, 1969). The wetting curve of the hysteretic envelope is commonly known as a sorption curve and the drying curve a desorption curve. Since air is almost always present when such curves are produced experimentally, the solid curves shown in Fig. 2.12 are obtained. However, some ambiguity exists, because the starting point for many measurements is a wet and sometimes even puddled sample in which the degree of air removal is unknown.

Hysteretic phenomena also exist in soil materials as a consequence of shrinking and swelling, which can affect microscopic pore size geometry as well as overall bulk density. Both factors would lead to a volumetric water content for a given energy state that differs from that which would exist if the soil matrix remained fixed. Shrinking and swelling often take place slowly and usually irreversibly, particularly when organic matter is involved; this complicates the evaluation of their contribution to hysteresis. Experimental observations do not always reveal that measurements involve true hysteresis and permanent or semipermanent changes in the porous system.

2.7 APPENDIX: GAMMA RAY ATTENUATION

2.7.1 Transmission through a Pure Substance i

A gamma ray photon moving in the x direction through a homogeneous material of density ρ_i has a characteristic probability P_i of interacting with the substance in each infinitesimal length increment Δx. This probability P_i is proportional to the density of the material. Thus,

$$P_i = \nu_i \rho_i \qquad (2.18)$$

where ν_i is the mass absorption coefficient of substance i, which is a function of the energy of the gamma radiation. If a flux $n(x)$ of gamma rays per unit area per unit time is entering a thickness Δx of the material, the flux $n(x + \Delta x)$ that leaves the thickness Δx will be less than $n(x)$ because a certain fraction of the gamma rays in the beam will interact with the absorbing medium. The number of these interactions is equal to the probability $P_i \, \Delta x = \rho_i \nu_i \, \Delta x$ that a single gamma ray will be absorbed multiplied by $n(x + \Delta x/2)$, the average number of gamma rays in thickness Δx. Therefore,

$$n(x + \Delta x) - n(x) = -\nu \rho_i n(x + \Delta x/2) \, \Delta x \qquad (2.19)$$

In the limit as $\Delta x \to 0$ this may be written as

$$\lim_{\Delta x \to 0} \frac{n(x + \Delta x) - n(x)}{\Delta x} = \frac{dn}{dx} = -\nu_i \rho_i n \qquad (2.20)$$

Assuming that $n(x) = n_0$ at $x = 0$, the number of gamma rays per unit time leaving the adsorber may be calculated by integrating (2.20):

$$\int_{n_0}^{n(L)} \frac{dn}{n} = -\int_0^L \nu_i \rho_i \, dx \tag{2.21}$$

which integrates to $\ln [n(L)/n(0)] = \nu_i \rho_i \, dx$, or

$$n(L) = n_0 \exp (-\nu_i \rho_i L) \tag{2.22}$$

2.7.2 Transmission through a Heterogeneous Material

By superposition, the total probability per unit length of a gamma ray having a collision in Δx is the sum of its probability of interacting with each substance i in the material,

$$P = \sum_i P_i = \sum_i \nu_i \rho_i \tag{2.23}$$

As before, the decrease in $n(x)$ passing through Δx is

$$n(x) - n(x + \Delta x) = -P \, \Delta x = i \sum_i \nu_i \rho_i \, \Delta x \tag{2.24}$$

which becomes

$$\frac{dn}{dx} = -\sum_i \nu_i \rho_i n \tag{2.25}$$

and after integration

$$n(L) = n_0 \exp \left(-\sum_i \nu_i \rho_i L \right) \tag{2.26}$$

2.7.3 Transmission through Soil

Equation (2.26) may be adapted to soil by regarding the medium as consisting of water, air, and solid material. The absorption of photons by air is negligible, so that (2.26) becomes

$$n(L) = n_0 \exp [-\nu_m \rho_b L - \nu_w \rho_w \theta_v L] \tag{2.27}$$

where ν_m is the effective mass absorption coefficient for the solid material and ρ_b is the bulk density of solid material (mass of solid per volume of space). The density of water in space is $\rho_w \theta_v$.

PROBLEMS[1]

2.1   A sphere of mass m_1 moving at velocity v_1 strikes a stationary sphere of mass m_2 directly, so that all subsequent motion will occur along the line of flight. Using conservation of energy,

$$\sum_{j=1}^{2} \frac{1}{2} m_i v_i^2 = \text{const}$$

and conservation of momentum,

$$\sum_{j=1}^{2} m_i v_i = \text{const}$$

calculate the final velocities of m_1 and m_2. Evaluate the special cases $m_2 \gg m_1$ and $m_2 = m_1$.

2.2   Calculate the height of rise of water in a clean glass capillary tube of radius $R = 0.001$ cm and height $L = 200$ cm that is sealed at the top before placing it in contact with the free-water reservoir. Assume that the gas pressure inside the capillary is initially atmospheric and obeys Boyle's law, $PV = \text{const}$ thereafter.

2.3   Find the total potential head components at points A, B, and C for the system in Fig. 2.16.

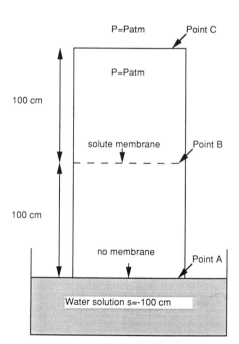

Figure 2.16   Equilibrium system.

[1]Problems marked with a dagger are more difficult.

2.4   A tensiometer is buried in an unsaturated soil sample at a depth of 15 cm below the soil surface. The surface of the mercury reservoir is 10 cm above the soil surface, and the mercury–water interface in the tube connecting the tensiometer to the mercury is 20 cm above the soil surface. What is the matric potential of the soil sample in head units (centimeters of water)?

2.5   A curious student has constructed a manometer tensiometer (Fig. 2.17) that uses alloy X instead of mercury. Alloy X has the following properties: it is insoluble in water, is colored liquid at room temperature, and has a density of 8.0 g cm^{-3}. A tensiometer is buried 50 cm deep in the soil where the matric potential head of the soil water is $h = -150$ cm. The surface of the reservoir of alloy X is 10 cm above the soil surface. How high above the soil surface will alloy X rise in the manometer tube when equilibrium is reached in the system?

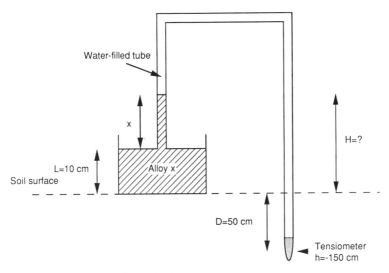

Water-filled tube

x

H=?

L=10 cm
Soil surface

Alloy x

D=50 cm

Tensiometer
h=-150 cm

Figure 2.17   Equilibrium system.

2.6   A 10-cm^3-volume of soil sample weighs 15 g before drying and 13 g after oven drying. Calculate (a) bulk density, (b) volumetric water content, (c) gravimetric water content, (d) total porosity, and (e) air-filled porosity of the sample. (Assume a particle density of 2.6 g cm^{-3}.)

2.7   Give the components of the soil water potential at A, B, and C for the systems in Fig. 2.18 at equilibrium.

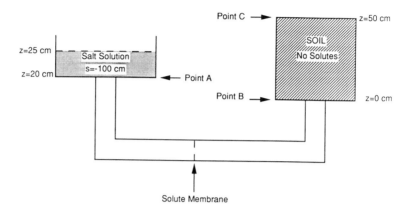

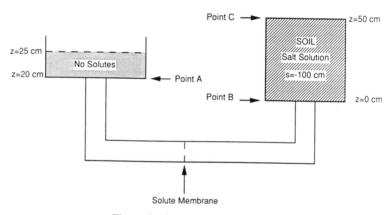

Figure 2.18 Equilibrium systems.

2.8 In a gamma ray scanning experiment, an empty tray of width L is placed in the path of the radiation beam, and a number N_0 of counts are recorded in 15 s. The tray is then repacked with oven-dried soil to a specified bulk density ρ_b, and a number N_1 of counts are recorded in 15 s. Calculate the mineral mass absorption coefficient ν_m of the soil as a function of L, N_0, N_1, and ρ_b.

†2.9 A dual gamma radiation beam uses radiation sources of different energy that have different mass absorption coefficients for both water and soil. Assuming that each beam is attenuated according to (2.12) (but with different ν_w and ν_m), show that changes in both bulk density and water content may be detected as a function of time.

3 Water Movement in Soil

The soil water systems analyzed in the previous chapter were in a state of thermodynamic equilibrium. In a system at equilibrium, the total water potential is constant everywhere and no water flow occurs. In this chapter, nonequilibrium problems will be examined, in which the total potential of water varies from point to point in the system. In such cases, water will flow from regions of higher potential to regions of lower potential at a rate that depends on the hydraulic resistance of the medium.

When two points at different potentials are brought into contact with each other, water flowing from high to low potential will tend to restore equilibrium, unless external control is maintained over the water or its potential. For example, when rain is falling on a soil surface and infiltrating into the profile, the system will not return to equilibrium because the water that flows from the region of highest potential (the soil surface) is replaced by water entering from outside the system. The control exerted on a system from its boundary is called a boundary condition.

A nonequilibrium flow system in which the flux or water potential depends on time is called transient or time dependent. A special class of nonequilibrium flows in which all variables have fixed values in time is called steady state. A steady-state system is characterized by water flows that do not cause storage changes within the soil.

3.1 WATER FLOW IN CAPILLARY TUBES

Before discussing water flow in soil, it is useful to examine flow through a system with an ideal geometry. One of the simplest systems, which nonetheless has many properties relevant to understanding flow in porous media, is a water-saturated capillary tube (Fig. 3.1). The horizontal cylindrical capillary tube of radius R in Fig. 3.1 has water flowing through it as a result of an imposed hydrostatic pressure difference $\Delta P = P_2 - P_1$ across the length L of the tube. At low flow rates, it is possible to neglect fluid acceleration and assume that the net applied force on the fluid is opposed by a viscous force described by Newton's law of viscosity [Equation (2.9)].

Since the fluid is not accelerating, the net force on any water volume within the tube must be equal to zero. We may therefore derive an expression for the shear stress τ in (2.9) by conducting a force balance on the cylindrical water volume of radius $r < R$ and length L shown in the figure within the capillary tube. Each end of the tube has a force exerted on it from the imposed hydrostatic pressure ($F =$

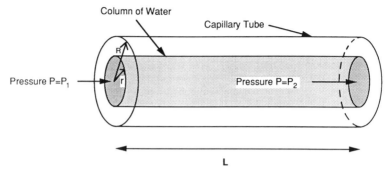

Figure 3.1   Section of capillary tube of radius R and length L filled with water flowing in response to a pressure difference $P_1 - P_2$. A force balance is conducted on a water cylinder of radius $r < R$.

PA), so that the net force F_p caused by the unequal pressure across the two ends is equal to

$$F_p = P_1 \pi r^2 - P_2 \pi r^2 = \Delta P \, \pi r^2 \qquad (3.1)$$

This pressure-induced force is exactly balanced by the fluid resistance or shear force F_s exerted on the water volume by the water molecules in contact with the external surface area $2\pi r L$ of the cylindrical water volume:

$$F_s = \tau(2\pi r L) \qquad (3.2)$$

where τ is the shear stress, or the tangential force per unit area. When fluid acceleration is neglected, $F_s = F_p$, from which we obtain, using (3.1) and (3.2),

$$\tau = \Delta P \, r / 2L \qquad (3.3)$$

Newton's law of viscosity [Equation (2.9)] relates τ to the rate of change of velocity V in the direction perpendicular to the motion of the fluid. Thus, equating (2.9) and (3.3), we obtain

$$\tau = \Delta P \, r / 2L = -\nu \, dV/dr \qquad (3.4)$$

Equation (3.4) may be integrated to produce an expression for the velocity $V(r)$ as a function of r. Since one integration is required, the velocity must be specified at one value of r to eliminate the constant of integration. We will assume that the water molecules at the walls of the capillary tube adhere perfectly to the surface, so that $V(R) = 0$. This is called a no-slip condition in fluid dynamics (Bird et al., 1960). Here it amounts to assuming that the water molecules are attracted more strongly to the solid capillary walls than to other water molecules. To integrate (3.4), we place all factors that depend explicitly on r on the same side of the equation (see Section 3.5):

$$r\,dr = -\frac{2L\nu}{\Delta P}\,dV \tag{3.5}$$

where now the left hand side depends only on r and the right hand side only on V. Integrating (3.5) from r to R, we obtain

$$\int_r^R r\,dr = -\frac{2L\nu}{\Delta P}\int_{V(r)}^0 dV = \frac{R^2}{2} - \frac{r^2}{2} = +\frac{2L\nu V(r)}{\Delta P} \tag{3.6}$$

or

$$V(r) = \frac{\Delta P}{4L\nu}\left(R^2 - r^2\right) \tag{3.7}$$

Equation (3.7) describes a parabolic velocity profile, plotted in Fig. 3.2, where the velocity changes from a value of zero at $r = R$ to a maximum of $\Delta P\,R^2/4L\nu$ at $r = 0$.

3.1.1 Poiseuille's Law

To calculate the volume of water Q flowing per unit time through the capillary, (3.7) must be integrated over the entire cross-sectional area of the tube, since water is flowing at a different rate through each point of the cross section. Because the cross-sectional area has a circular shape, it is most convenient to use cylindrical coordinates to integrate over the area. As shown in Fig. 3.3, the area element $dA = dx\,dy$ is equal to $r\,dr\,d\phi$ in cylindrical coordinates (Arfken, 1985). For a circular

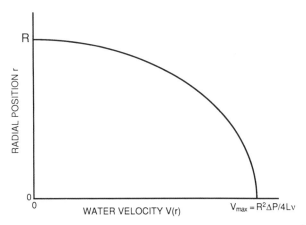

Figure 3.2 Parabolic water velocity distribution in the capillary tube of Fig. 3.1.

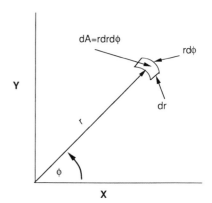

Figure 3.3 Illustration of the area element dA in cylindrical coordinates.

area, r is integrated from zero to R and ϕ from zero to 2π radians. Since the velocity $V(r)$ is the same for all ϕ at any value of r, the integration over $d\phi$ yields 2π. Thus, the volume flow rate is

$$Q = \iint V(r)\, dA = \int_0^R \int_0^{2\pi} V(r)r\, dr\, d\phi = 2\pi \int_0^R V(r)r\, dr$$

$$= \frac{\pi\, \Delta P}{2Lv} \int_0^R (R^2 - r^2)r\, dr = \frac{\pi\, \Delta P}{2Lv}\left(\frac{R^4}{2} - \frac{R^4}{4}\right) \tag{3.8}$$

or finally

$$Q = \frac{\pi R^4\, \Delta P}{8Lv} \tag{3.9}$$

Equation (3.9) is called Poiseuille's law. It says that for a given hydrostatic pressure difference ΔP across a length L of cylindrical capillary, the volume of water flowing per unit time Q will be proportional to the fourth power of the radius. Thus, a tube of radius $2R$ will have 16 times as much water flowing through it per unit time as a tube of radius R if each tube has the same pressure gradient $\Delta P/L$ acting on the water.

As we shall see later in this chapter, water flow equations are commonly expressed in terms of the volume flow rate per unit area, or water flux J_w. The average flux through the preceding capillary tube is $Q/\pi R^2$, or

$$J_w = \frac{R^2\, \Delta P}{8Lv} \tag{3.10}$$

3.2 WATER FLOW IN SATURATED SOIL

Soil contains a large distribution of pore sizes and channels through which water may flow. The exact geometry of these openings is unknown, so that Newton's law of viscosity [Equation (2.9)] may not be used directly to calculate flow rates in response to gradients of water potential. Instead, averages are taken over many pores to define macroscopic flow equations to describe movement of water through porous media. The first person to employ this method was Henry Darcy in 1856 (Darcy, 1856).

3.2.1 Darcy's Law

Darcy, an engineer working for the city of Paris, measured the volume of water Q flowing per unit time through water-saturated packed sand columns of length L and area A when a hydrostatic pressure difference $\Delta P = P_2 - P_1$ was placed across them (Fig. 3.4). After conducting a number of experiments, he developed the following relationship among the variables:

$$Q = K_s A \, \Delta P / L \tag{3.11}$$

where K_s, called the saturated hydraulic conductivity, is a constant for rigid, saturated soil in a given geometric configuration.

Darcy's law may be generalized to apply between any two points of a saturated porous medium provided that the total potential difference of the water between the two points is known. We will assume that the soil is rigid and saturated and that no solute membranes exist within the water flow paths. Under these restrictions, the total water potential in saturated soil consists of the sum of the hydrostatic pressure and gravitational potential components. When head units are used, this combination is called the hydraulic head H,

$$H = p + z \tag{3.12}$$

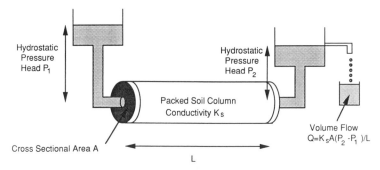

Figure 3.4 Soil column experiment illustrating Darcy's law.

TABLE 3.1   Procedure for Using Darcy's Law in Vertical Flow Problems

Step 1	Define a reference elevation.
Step 2	Determine two points 1 and 2 where the hydraulic head H is known.
Step 3	Calculate the gradient $(H_2 - H_1)/(z_2 - z_1)$
Step 4	Plug into (3.13).

Note that when water flow is in the horizontal direction, H reduces to p, since z is the same at all points (see Fig. 3.4).

As in the case of equilibrium calculations, it is useful to employ a sequence of steps in applying Darcy's law to a flow problem. First, we must develop a sign convention for the flow direction. In this book we will use the z coordinate to indicate the vertical direction and assume that the upward direction is positive. Similarly, we will use the x coordinate for one-dimensional horizontal flow and assume that the direction pointing to the right is positive. Thus, upward water flow (i.e., evaporation) is positive and downward water flow (i.e., drainage) is negative. With this convention, Darcy's law may be written in the flux form $J_w = Q/A$ between two points 1 and 2 as

$$J_w = -K_s(H_2 - H_1)/(z_2 - z_1) \quad \text{(vertical flow)} \qquad (3.13)$$

$$J_w = -K_s(p_2 - p_1)/(x_2 - x_1) \quad \text{(horizontal flow)} \qquad (3.14)$$

where H_1 is the hydraulic head at point z_1 and H_2 that at z_2 (see Figs. 3.5 and 3.6).

Note that these equations have a minus sign, which must be included in all calculations. Also, K_s represents an average over the entire region between points 1 and 2. When the head form [(3.13) and (3.14)] of Darcy's law is used, K_s has units of length per time, as does J_w.

Table 3.1 lists the steps to be followed in using Darcy's law to calculate vertical water flow. A similar procedure is used in horizontal flow. The first example illustrates the application of this procedure.

EXAMPLE 3.1:   A 50-cm-long soil column containing packed sand with a saturated hydraulic conductivity of 100 cm day^{-1} is placed vertically with the bottom open to the atmosphere ($p = 0$) (Fig. 3.5). A constant 10 cm of water is ponded continuously on the top surface. Calculate the steady water flux J_w through the soil.

Step 1 Define $z = 0$ at the bottom of the column.

Step 2 Let point 1 be the bottom of the column where $z_1 = 0$ and $p_1 = 0$. Thus $H_1 = 0$. Let point 2 be the top of the column where $z_2 = +100$ cm and $p_2 = 10$ cm (since 10 cm of unsupported water lies above z_2). Thus $H_2 = 110$.

Step 3 The gradient is $(H_2 - H_1)/(z_2 - z_1) = (110 - 0)/(100 - 0) = 1.1$.

Step 4 The flux is $J_w = -K_s(H_2 - H_1)/(z_2 - z_1) = -100(1.1) = -110$ cm day^{-1}. The negative sign means the water is flowing in the downward direction.

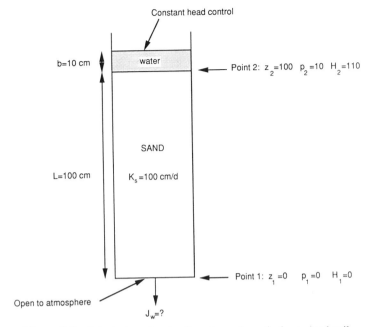

Figure 3.5 Calculation of water flow through vertical saturated soil.

This example illustrates a fundamental difference between equilibrium problems and flow problems. If the bottom of the column was sealed, then at equilibrium the hydrostatic pressure potential head at $z = 0$ would be 110 cm, since the weight of all of the water above $z = 0$ is exerted on that point (see definition of ψ_p in Section 2.3.4). When the bottom is open to the atmosphere, however, water will leave the pores at the bottom of the column as soon as any positive pressure develops. Thus, $p = 0$ there. The difference between this case and the equilibrium one is that the weight of the water in the column is opposed by the viscous resistive forces on the water.

An important difference between vertical and horizontal flow is illustrated in the next example.

EXAMPLE 3.2: Assume that the soil column in Example 3.1 is placed horizontally, with 10 cm of water ponded over the left side while the right side is open to the atmosphere (Fig. 3.6). Calculate the flow through the column. (Ignore any elevation differences within the column.)

Step 1 Define the left side of the column to be $x = 0$.

Step 2 Let point 1 be the left (inlet) end of the column. Here $x_1 = 0$ and $p_1 = 10$ cm. Let point 2 be the right (outlet) end of the column. Here $x_2 = +100$ cm and $p_2 = 0$ (open to the atmosphere).

Step 3 The gradient for horizontal flow is $(p_2 - p_1)/(x_2 - x_1) = (0 - 10)/(100 - 0) = -0.1$.

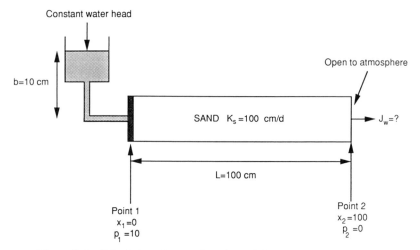

Figure 3.6   Calculation of water flow through horizontal saturated soil.

Step 4   By (3.14), the flux is $J_w = -K_s(p_2 - p_1)/(x_2 - x_1) = -100(-0.1) = +10$ cm day^{-1}. The flux is positive, meaning that water is flowing to the right.

Notice that the same externally applied 10 cm water pressure only produced one-eleventh as much water flow in the horizontal column as in the vertical column. This occurred because gravitational potential was constant everywhere in the horizontal column, so that the hydraulic head difference was only 10 cm, whereas the hydraulic head difference across the vertical column was 110 cm.

3.2.2   Measurement of Saturated Hydraulic Conductivity

In the previous two examples we have assumed that K_s is known and have used Darcy's law to calculate J_w. However, since it is easy to measure J_w in a laboratory column experiment, Darcy's law may be used to measure K_s.

Assume that soil is packed uniformly in a vertical soil column of length L. A constant height b of water is maintained over the upper end by an external manometer, and the bottom end is open to the atmosphere so that $p = 0$. The water volume flow is collected at the bottom and is used to calculate J_w.

For this system, if z is set equal to zero at the bottom, then $H_1 = 0 + 0 = 0$ (bottom), $H_2 = b + L$ (top), the gradient is $(b + L)/L$, and by (3.13)

$$K_s = -J_w L/(b + L) \qquad (3.15)$$

Since J_w is downward, it is negative, so that $K_s > 0$. This procedure is called the constant-head method of measuring K_s (Klute and Dirksen, 1986).

There is an even simpler method of measuring K_s in the laboratory using the

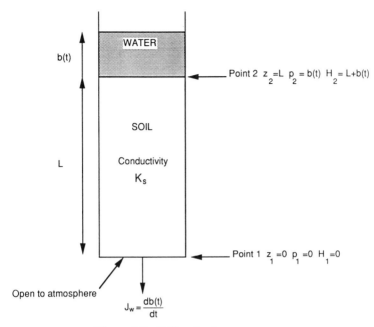

Figure 3.7 Falling-head permeameter.

preceding apparatus. Water is initially ponded to a height b_0 above the saturated column and is allowed to fall with time as water flows through the column and out the bottom where $p = 0$. This device is called a falling-head permeameter (Fig. 3.7).

In contrast to the controlled pressure head in the device discussed before leading to (3.15), here the hydrostatic pressure $p_2 = b(t)$ at the surface is time dependent, as is the gradient $[L + b(t)]/L$. Consequently, the flux J_w is also time dependent and is equal to the time rate of change of the water height $J_w = db/dt$, provided that the water ponded on the surface is contained in a volume of the same cross-sectional area as the soil column. Each small decrease in the height of the water column per unit time represents a flow of the same amount of water per unit area per unit time through the column. Hence, we may write Darcy's law (3.13) as

$$J_w = \frac{db}{dt} = -\frac{K_s}{L}(b + L) \tag{3.16}$$

Equation (3.16) may be rearranged (see Section 3.5) by placing all factors that depend on b on the left side:

$$\frac{db}{b + L} = -\frac{K_s\,dt}{L} \tag{3.17}$$

where $b = b_0$ at $t = 0$.

Equation (3.17) may be integrated from $t = 0$ to some time $t = t_1$ when b has fallen to $b_1 < b_0$. The left side of (3.17) is equal to

$$\int_{b_0}^{b_1} \frac{db}{b + L} = \ln(b + L)\Big|_{b_0}^{b_1} = \ln\left(\frac{b_1 + L}{b_0 + L}\right) \tag{3.18}$$

where ln is the natural (base e) logarithm. The right side of (3.17) may be integrated simply, producing

$$-\int_0^{t_1} \frac{K_s \, dt}{L} = -\frac{K_s t_1}{L} \tag{3.19}$$

Thus, equating (3.18) and (3.19) and recalling that $\ln(A/B) = -\ln(B/A)$, we can solve for K_s:

$$K_s = \frac{L}{t_1} \ln\left(\frac{b_0 + L}{b_1 + L}\right) \tag{3.20}$$

Equation (3.20) demonstrates that K_s may be measured in the falling-head permeameter by measuring b_0, b_1, L, and t_1 only. This procedure is called the falling-head method of measuring K_s (Klute and Dirksen, 1986).

EXAMPLE 3.3: A 100-cm-long soil column is saturated and 10 cm of water is ponded over the top at $t = 0$ in a vessel that has the same cross-sectional area as the column. At $t = 1$ h, the height of overlying water has fallen to 5 cm. Calculate K_s in centimeters per day. In (3.20)

$$L = 100 \text{ cm} \quad t_1 = 1 \text{ h} = 0.042 \text{ day}$$

$$b_0 = 10 \text{ cm} \quad b_1 = 5 \text{ cm}$$

Thus

$$K_s = (100/0.041) \ln (110/105)$$

$$= 111.65 \text{ cm day}^{-1}$$

3.2.3 Calculation of Hydrostatic Pressure in Soil Columns

Thus far, Darcy's law has been used only to calculate K_s or to estimate J_w when K_s is known. It is also possible to use Darcy's law to calculate pressure profiles within the soil if K_s is known everywhere. This procedure is illustrated in the next example.

EXAMPLE 3.4: Assume that the hydraulic conductivity K_s is constant everywhere in the soil column of Fig. 3.5 and Example 3.1. Calculate the hydrostatic pressure at an arbitrary point z in the column.
    Since K_s is constant everywhere, Darcy's law may be written between the new

point z, where p is unknown, and the bottom of the column, where $p = z = 0$. Thus, using (3.13) with $H_2 = z + p$, we obtain

$$J_w = -K_s \frac{p + z}{z}$$

However, by using Darcy's law over the entire column, we already know the flux J_w in terms of the known pressure $p = b$ at $z = L$

$$J_w = -K_s \frac{b + L}{L}$$

Since the flux is constant, these two expressions are equal. Solving for p, we obtain

$$p = bz/L$$

Thus, the pressure decreases linearly from $p = b$ at $z = L$ to $p = 0$ at $z = 0$.

This linear decrease will only occur if the soil column is homogeneous and K_s is constant everywhere. In layered soils that have a nonuniform K_s, the results are more complex.

3.2.4 Water Flow in Saturated Layered Soil

Figure 3.8 illustrates steady water flow through a layered soil column containing N layers of thickness L_J and saturated hydraulic conductivity K_J ($J = 1, \cdots,$

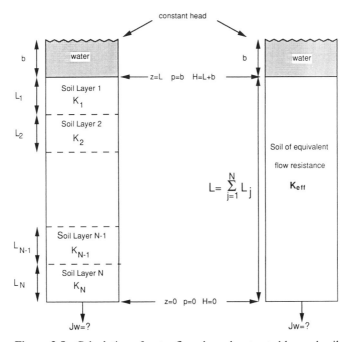

Figure 3.8 Calculation of water flow through saturated layered soil.

N). We would like to calculate the water flux and hydrostatic pressure distribution given the values of K_J, L_J, and b.

An easy way to approach this problem is to make use of the well-known law of resistances in electrical circuits. This law states that when current i is flowing from one point to another because of a voltage difference ΔV through a series of resistors in accordance with Ohm's law ($\Delta V = iR$), the net resistance between the two points is equal to the sum of the individual resistances. We may establish an analog with this law by rewriting Darcy's law in the form

$$\Delta H = H_1 - H_2 = \frac{J_w L}{K_s} \quad \text{(Darcy's law)} \tag{3.21}$$

$$\Delta V = V_1 - V_2 = iR \quad \text{(Ohm's law)} \tag{3.22}$$

where we have set $z_2 = L$ and $z_1 = 0$ in (3.13). Thus, since a current is a flux of electrons and a voltage is an electrical potential difference, Darcy's law and Ohm's law are mathematically identical. The electrical resistance R is by definition a potential difference divided by a flux. By analogy, we may define a hydraulic resistance R_H as

$$R_H = \frac{\Delta H}{J_w} = \frac{L}{K_s} \tag{3.23}$$

Now we may return to the layered soil column in Fig. 3.8. The effective flow resistance $R_{H,\text{eff}}$ of the entire column is, by the law of resistances,

$$R_{H,\text{eff}} = \sum_{J=1}^{N} R_{H_J} \tag{3.24}$$

Using (3.23), the hydraulic resistance R_{H_J} of each layer is L_J/K_J, and the hydraulic resistance of the entire column is $\sum_{J=1}^{N} L_J/K_{\text{eff}}$, where K_{eff} is the effective hydraulic conductivity of the whole column. Thus (3.24) may be written as

$$\sum_{J=1}^{N} \frac{L_J}{K_{\text{eff}}} = \sum_{J=1}^{N} \frac{L_J}{K_J} \tag{3.25}$$

or

$$K_{\text{eff}} = \frac{\sum_{J=1}^{N} L_J}{\sum_{J=1}^{N} \left(\frac{L_J}{K_J}\right)} \tag{3.26}$$

Once we have calculated K_{eff} with (3.26), we may replace the layered column with an equivalent column having a conductivity K_{eff} for the purpose of calculating

TABLE 3.2 Procedure for Calculating Water Flow through Saturated Layered Soil

Step 1	Define a reference elevation.
Step 2	Calculate K_{eff} using (3.26).
Step 3	Calculate J_w using the procedure given in Table 3.1.
Step 4	Write Darcy's law across each layer to calculate p at the interfaces.

J_w. Then, once J_w is known, the pressure drop across any homogeneous layer within the column may be calculated using Darcy's law with the actual saturated hydraulic conductivity of the layer being analyzed. The procedure to be used in calculating flow through layered columns is given in Table 3.2.

This procedure is illustrated in the next example.

EXAMPLE 3.5: A layered vertical soil column (Fig. 3.9) consists of 25 cm of a loam soil of $K_s = 5$ cm day^{-1} overlain by 75 cm of a sandy soil of $K_s = 25$ cm day^{-1}. The top of the column has water ponded to a constant height of 10 cm above the column, and the bottom is open to the atmosphere. Calculate J_w and the hydrostatic pressure distribution.

Step 1 Let $z = 0$ at the bottom of the column.

Step 2 Using (3.26), we calculate the effective conductivity as

$$K_{eff} = \frac{25 + 75}{\dfrac{25}{5} + \dfrac{75}{25}} = \frac{100}{5 + 3} = \frac{100}{8} = 12.5 \text{ cm day}^{-1}$$

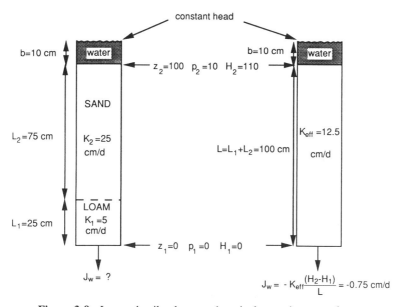

Figure 3.9 Layered soil column and equivalent resistance column.

Step 3 Let point 2 be the top of the column, where $z_2 = 100$, $p_2 = 10$, and $H_2 = 110$ cm. Let point 1 be the bottom of the column, where $z_1 = 0$, $p_1 = 0$, and $H_1 = 0$. Then Darcy's law (3.13) may be written across the entire column to calculate J_w. Thus,

$$J_w = -\frac{K_{\text{eff}}}{100}(110 - 0) = -\frac{110}{8} = -13.75 \text{ cm day}^{-1}$$

This is also equal to the flux at every point in the layered column.

Step 4 Now that J_w is known, we may apply Darcy's law to part of the column to calculate p at point 3, the interface between the layers. We may write Darcy's law between point 3, where $H_3 = p_3 + 25$, and point 1, where $H_1 = 0$, as

$$J_w = -\frac{K_1(H_3 - H_1)}{z_3 - z_1} = -\frac{5}{25}(p_3 + 25) = -13.75$$

or $p_3 = 43.75$ cm. Note that for this calculation we use the saturated hydraulic conductivity $K_1 = 5$ cm day^{-1} between points 1 and 3 and not K_{eff}.

We may check our answer for consistency by writing Darcy's law between points 2 and 3:

$$J_w = -\frac{K_2(H_2 - H_3)}{z_2 - z_3} = -\frac{25}{75}(110 - 25 - p_3) = -13.75$$

or $p_3 = 43.75$.

As shown in the preceding, the pressure will change linearly in each layer where K_s is constant. Hence, the pressure and hydraulic head distributions in the layered soil column are as shown in Fig. 3.10. The hydrostatic pressure linearly increases through the sand layer and then sharply decreases to zero through the loam layer. The hydraulic head continually decreases (as it must for flow to be downward), but more gradually through the sand than through the loam.

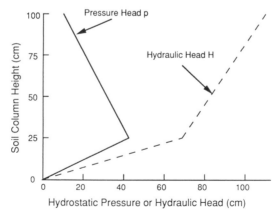

Figure 3.10 Hydrostatic and hydraulic head profiles within layered soil column of Fig. 3.9.

The reason that the hydrostatic pressure builds up in the sand is that the loam offers substantially more resistance to flow than the sand and hence produces most of the force opposing the externally applied hydraulic head difference acting on the water in the column. This pressure must build up in order to "push" the water through the loam at the same rate as it is moving through the sand.

EXAMPLE 3.6:   Repeat Example 3.5 with the column rotated 180° so that the loam lies over the sand.

Step 1   As before, $z = 0$ at the bottom.

Steps 2, 3 Since the resistance law does not depend on the order of the resistances, $K_{eff} = 12.5$ cm day^{-1} as before. [Work it out using (3.26) if you do not agree.] Therefore, $J_w = -13.75$ cm day^{-1} as before, since the boundary conditions on the effective column are identical.

Step 4   Writing Darcy's law across the sand layer, where $H_3 = 75 + p_3$, $H_1 = 0$, we obtain

$$J_w = -13.75 = -\tfrac{25}{75}(75 + p_3)$$

or $p_3 = -33.75$ cm.

In this case a negative hydrostatic pressure will develop at the interface in order to decrease the water flow through the low-resistance sand until it matches the flow through the high-resistance loam.

In practice, this negative pressure will probably exceed the air entry suction of the saturated sand, causing part of the soil in the column to unsaturate. As soon as the column unsaturates, the resistance will change, thereby invalidating the assumptions of the calculation.

3.3 WATER FLOW IN UNSATURATED SOIL

When soil is partially unsaturated, an air phase is present and the water flow channels are drastically modified from those in saturated soil. In unsaturated soil, the water phase is bounded partially by solid surfaces and partially by an interface with the air phase. In contrast to the positive water pressure found in saturated soil, the water pressure within the liquid phase is caused by water elevation, attraction to solid surfaces, and the surface tension of the air–water interface and is lower than the reference liquid pressure at the same elevation. As the water content decreases, the liquid pressure decreases and the water phase is constrained to narrower and more tortuous channels.

3.3.1 Buckingham–Darcy Flux Law

In 1907, Edgar Buckingham proposed a modification of Darcy's law (3.13) to describe flow through unsaturated soil. This modification rested primarily on two assumptions:

1. The driving force for water flow in isothermal, rigid, unsaturated soil containing no solute membranes and zero air pressure potential is the sum of the matric and gravitational potentials.
2. The hydraulic conductivity of unsaturated soil is a function of the water content or matric potential.

In head units, the Buckingham–Darcy flux law may be expressed for vertical flow as

$$J_w = -K(h) \frac{\partial H}{\partial z} = -K(h) \frac{\partial}{\partial z} (h + z) = -K(h) \left(\frac{\partial h}{\partial z} + 1 \right) \quad (3.27)$$

where $H = h + z$ (in centimeters) is the hydraulic head in unsaturated soil and $K(h)$ (in centimeters per day) is the unsaturated hydraulic conductivity. As in saturated flow, the flux J_w (centimeters per day) is the water flow per unit cross-sectional area per unit time.

Several points should be stressed about (3.27). First, it is a differential equation that is written across an infinitesimally thin layer of soil over which h is constant and $K(h)$ is constant. It may not be written across a finite layer of soil unless the water content and matric potential of the layer are uniform, which occurs only under special conditions to be covered later. Second, the derivative in (3.27) is a partial derivative, because in unsaturated soil h may be a function of both z and t. The partial derivative $\partial h / \partial z$ implies that the derivative with respect to z is taken at constant t; it is the instantaneous value of the slope of $h(z)$:

$$\frac{\partial h}{\partial z} \equiv \left(\frac{\partial h}{\partial z} \right)_t = \lim_{\Delta z \to 0} \frac{h(z + \Delta z, t) - h(z, t)}{\Delta z} \quad (3.28)$$

where $(\;)_t$ means that the derivative is evaluated at constant t. Partial derivatives are required for the mathematical description of transient (time-dependent) flow. If the system is at steady state, the partial derivative reduces to an ordinary derivative, since in steady state h depends only on z.

3.3.2   Unsaturated Hydraulic Conductivity

The unsaturated hydraulic conductivity is a nonlinear function of water content or matric potential. Figure 3.11 shows typical $K(h)$ curves for a coarse-textured (sandy) and a fine-textured (clayey) soil. At saturation, the coarse-textured soil has a higher conductivity than the fine-textured soil, because it contains larger pore spaces, which are filled and water conducting at saturation. However, these pores drain at modest suctions, producing a dramatic decrease in hydraulic conductivity in the sandy soil. Eventually, the curves will cross and the sandy soil will actually have a lower hydraulic conductivity than the clayey soil at the same matric potential, because the latter will retain considerably more water and will contain a greater number of filled pores.

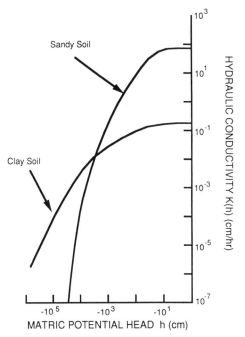

Figure 3.11 Typical hydraulic conductivity–matric potential curves for a sandy and a clayey soil.

Insight into the relation between the pore sizes of a porous medium and its unsaturated hydraulic conductivity may be gained by using a simple capillary tube model of soil. This model is derived in the next section.

3.3.3 Capillary Tube Model of Unsaturated Hydraulic Conductivity

As has been shown in the previous section in this chapter (Fig. 3.11), the value of the unsaturated hydraulic conductivity $K(h)$ is highly dependent upon the amount of water in the soil and also on the thickness of the films or channels that are filled. For most soils, $K(h)$ drops to a small fraction of its value at saturation after the loss of only a small percentage of the total water content. This property of $K(h)$ completely dominates important flow processes under some conditions. An understanding of the reason why $K(h)$ has such a great dependence upon water content or film thickness is essential to an understanding of water flow in unsaturated soil. Although real flow systems in soil are too complex to analyze quantitatively with the Poiseuille equation (3.9), a simple model of soil water flow may be developed using bundles of capillary tubes of different sizes. This so-called capillary bundle model of soil contains many of the same properties as a real soil water system and may be analyzed to produce a relationship between the unsaturated hydraulic conductivity and the geometry of the solid phase.

The number and sizes of the capillary tubes in the bundle are chosen so as to retain water in the same way as the soil they represent. Consequently, the matric potential–water content curve for the bundle of tubes and the real soil will be identical. However, the flow system of this bundle of tubes differs from the water flow network of a real soil in several ways:

1. Each capillary tube is continuous and contains no dead end or stagnant regions, although it may have a twisted or tortuous shape.
2. All capillary tubes in the model have the same length.
3. The water flow boundaries of the model are composed entirely of solid–water interfaces with empty capillaries constituting the air space, whereas in soil the flow boundaries are a combination of solid–water and solid–air interfaces.
4. The radius of the capillary tube entirely governs the thickness of the water films, whereas the water film thickness on surfaces in soil is governed by the matric potential.
5. Flow is steady state, which is contrary to the usual situation in soil.

The hydraulic conductivity of the capillary bundle model is derived as follows. The network of tubes, each of which is assumed to have a length L_c, is twisted internally and ultimately bundled to form a soil column of length $L < L_c$ and cross-sectional area A. A hydraulic head gradient ΔH is placed across the ends of the "column," causing water to flow through each of the capillary tubes in accordance with Poiseuille's law (3.9). The total flux through the column is equal to the sum of the volume flow rates out of each tube divided by the cross-sectional area of the column. Thus, a single capillary of radius R_J has a volume flow rate Q_J given by (3.9) as

$$Q_J = \frac{\pi R_J^4 \rho_w g}{8\nu} \frac{\Delta H}{L_c} \tag{3.29}$$

and the total volume flow rate Q_T through the column when all tubes are filled is

$$Q_T = \sum_{J=1}^{M} N_J Q_J = \frac{\pi \rho_w g}{8\nu} \frac{\Delta H}{L_c} \sum_{J=1}^{M} N_J R_J^4 \tag{3.30}$$

where N_J represents the number of capillaries of radius R_J in the bundle and M is the number of different capillary size classes in the bundle of tubes making up the column. The flux J_w through the column is given by

$$J_w = \frac{Q_T}{A} = \frac{\pi \rho_w g}{8\nu} \frac{\Delta H}{L_c} \sum_{J=1}^{M} n_J R_J^4 \tag{3.31}$$

where $n_J = N_J/A$ is the number of tubes per unit area of radius R_J in the bundle.

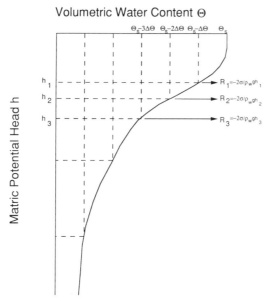

Figure 3.12   Calculation of the distribution of capillary sizes required to produce a matric potential–water content curve that is equivalent to the real one.

A correspondence between a real soil and the capillary bundle model is established by using the matric potential–water content curve of the soil (Fig. 3.12) to calculate the distribution of capillary tubes that would have equivalent water retention characteristics. This is achieved using the following procedure: (1) The matric potential–water content curve $h(\theta)$ of the real soil is divided into a number of equally spaced water content intervals of width $\Delta\theta$. (2) The matric potential associated with each decrease in θ is determined from the $h(\theta)$ relation. Thus $h_1 = h(\theta_s - \Delta\theta)$, $h_2 = h(\theta_s - 2\,\Delta\theta)$, and so on. (3) All tubes of $r > R_J$ given by the capillarity equation

$$R_J = -2\sigma/\rho_w g h_J \qquad (3.32)$$

are assumed to be drained when $h = h_J$. Thus, when the water content changes from θ_s (when all pores are filled) to $\theta_s - \Delta\theta$, we assume that all of the tubes of radius $R_1 = 2\sigma/\rho_w g h_1$ in the capillary bundle model drain. Therefore, the number n_1 of these tubes per unit area must correspond to the decrease $\Delta\theta$ in the water content of the real soil. If we assume that in a unit volume the capillary bundle contains n_1 cylinders per unit area, each one of which has cross-sectional area πR_1^2 and a unit length, then the change $\Delta\theta$ in water volume per total volume when these cylinders drain is

$$\Delta\theta = n_1 \pi R_1^2 \qquad (3.33)$$

Therefore, the number of tubes of radius R_1 per unit area is

$$n_1 = \Delta\theta / \pi R_1^2 \qquad (3.34)$$

Similarly, $n_J = \Delta\theta / \pi R_J^2$ in each water content interval, where R_J is related to the soil by (3.32) and $h_J = h(\theta_s - J \Delta\theta)$.

We may now insert (3.32) and (3.34) into (3.31), so that

$$J_w = \frac{\rho_w g}{8\nu} \frac{\Delta H}{L_c} \Delta\theta \sum_{J=1}^{M} R_J^2 \qquad (3.35)$$

$$= -\left[\frac{\sigma^2 \Delta\theta}{2\nu\rho_w g} \sum_{J=1}^{M} \frac{1}{h_J^2} \right] \frac{L}{L_c} \frac{\Delta H}{\Delta z} \qquad (3.36)$$

where $\Delta z = 0 - L$, in accordance with the sign convention used for Darcy's law (3.13).

Equation (3.36) now is in the form of Darcy's law (3.13), so that we may associate the factors in square brackets in front of the gradient $\Delta H / \Delta z$ with the saturated hydraulic conductivity K_s of the capillary bundle:

$$K_s = \frac{\tau\sigma^2 \Delta\theta}{2\nu\rho_w g} \sum_{J=1}^{M} \frac{1}{h_J^2} \qquad (3.37)$$

where

$$\tau = L/L_c \qquad (3.38)$$

The factor τ, called the tortuosity, is the ratio between the column length and the capillary or water path length.

As the water content decreases from θ_s to $\theta_s - \Delta\theta$, the largest tubes of radius R_1 drain and no longer contribute to the flow. Thus

$$K(\theta_s - \Delta\theta) = \frac{\tau\sigma^2 \Delta\theta}{2\nu\rho_w g} \sum_{J=2}^{M} \frac{1}{h_J^2} \qquad (3.39)$$

and in general,

$$K(\theta_s - i \Delta\theta) = \frac{\tau\sigma^2 \Delta\theta}{2\nu\rho_w g} \sum_{J=i+1}^{M} \frac{1}{h_J^2} \qquad (3.40)$$

Equation (3.40) represents the predicted unsaturated hydraulic conductivity of the capillary bundle, calculated from the measured $h(\theta)$ curve. It was first derived by Childs and Collis-George (1950), with later improvements by Marshall (1958) and Millington and Quirk (1959). Equation (3.40) still contains the unknown tortuosity

τ, which must be evaluated by fitting the model to measured $K(\theta)$ data. In the preceding derivation, it was assumed that the length L_c of each of the capillaries was the same, so that the tortuosity factor τ should be a constant. Proper selection of the point of calibration has produced reasonable agreement with the model in certain media (Green and Corey, 1971; Jackson, 1972). However, the model has been able to represent real soil hydraulic conductivities of many soils only when τ is allowed to vary as a function of θ (Fatt and Dykstra, 1951). Since τ can only be evaluated by calibration, this restriction makes the model useful primarily as a conceptual aid in visualizing the relationship between conductivity, pore sizes, and water content.

There have been alternative approaches used to calculate unsaturated hydraulic conductivity from the water characteristic function. Brooks and Corey (1966) developed generalized functional forms for $h(\theta)$ and $K(h)$ in which the model parameters were linked together. This approach has proven to be useful in field measurements of these functions (Russo and Bresler, 1980). Mualem (1976a) has extended the Brooks and Corey (1966) model into a general model for predicting the hydraulic conductivities of different soils. His article includes a large number of examples.

EXAMPLE 3.7: The matric potential–water content function for two hypothetical soils is given in Table 3.3.

Using (3.40), calculate $K(\theta)$ for each soil. Use the measured K_s values for each soil to calculate the tortuosity factor τ.

The factor $\sigma^2 \, \Delta\theta / 2\nu\rho_w g$ has the value 1.14×10^6 cm^3/day for $\Delta\theta = 0.05$. Therefore, (3.40) reduces to

$$K(\theta) = K(\theta_s - L \, \Delta\theta) = 1.14 \times 10^{-6}\tau \sum_{J=L+1}^{M} h_J^{-2} \quad L = 0, \cdots \quad (3.41)$$

To estimate τ for each soil, we first calculate $K(\theta_s)$ using (3.41) and Table 3.3. This produces the value $K(\theta_s) = 9289\tau$ for the sand and 930τ for the loam. Therefore, τ is 0.011 and 0.005 for the sand and loam, respectively.

TABLE 3.3 Matric Potential–Water Content Values for Sandy and Loam Soils

θ	Sand h (cm)	Loam h (cm)
0.50		0
0.45		−37
0.40	0	−113
0.35	−13	−283
0.30	−27	−718
0.25	−47	−2046
0.15	−148	−7316
0.10	−345	

Note: K_s for sand is 100 cm day^{-1} and for loam 5 cm day^{-1}.

TABLE 3.4 **Hydraulic Conductivity Values Predicted from Table 3.3 with Capillary Bundle Model**

	$K(\theta)$	
θ	Sand	Loam
0.5	—	5.0×10^{0}
0.45	—	5.7×10^{-1}
0.40	1.0×10^{2}	8.9×10^{-2}
0.35	2.5×10^{1}	1.3×10^{-2}
0.30	8.2×10^{0}	1.0×10^{-3}
0.25	2.6×10^{0}	1.2×10^{-4}
0.15	1.1×10^{-1}	—
0.10	3.0×10^{-3}	—
0.05	—	—

Inserting these values into (3.41), we obtain the predicted values of K given in Table 3.4.

Figures 3.13 and 3.14 show plots of $h(\theta)$ and $K(\theta)$ for the two soils.

3.3.4 Steady-State Water Flow Problems

When a matric potential difference $\Delta h = h_2 - h_1$ is maintained across a soil column of length $L = z_2 - z_1$, the flow will eventually reach a steady state, at which time the Buckingham–Darcy flux law may be written as

$$J_w = -K(h) \left(\frac{dh}{dz} + 1 \right) \tag{3.42}$$

where now an ordinary derivative is used because h depends only on z.

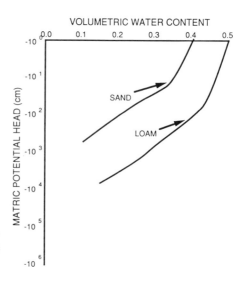

Figure 3.13 Matric potential–water content curves plotted from the data in Table 3.3.

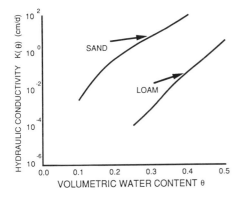

Figure 3.14 Unsaturated hydraulic conductivity–water content function calculated from Figure 3.13 using the capillary bundle model.

Integral Form of Darcy's Law We may calculate the value of the water flux J_w in steady state between any two points z_1 and z_2 in the soil by turning (3.42) into an integral. This is accomplished by moving all factors that depend explicitly on h to the same side of the equation and those that depend on z to the other (see Section 3.5). The series of steps follows:

$$\frac{-J_w}{K(h)} = \frac{dh}{dz} + 1$$

$$-\left(1 + \frac{J_w}{K(h)}\right) = \frac{dh}{dz}$$

$$\frac{dh}{1 + J_w/K(h)} = -dz$$

Now we may integrate this last expression from $h_1 = h(z_1)$ to $h_2 = h(z_2)$:

$$\int_{h_1}^{h_2} \frac{dh}{1 + J_w/K(h)} = -\int_{z_1}^{z_2} dz = z_1 - z_2 \tag{3.43}$$

Equation (3.43) is called the integral form of the Buckingham–Darcy equation. If the unsaturated hydraulic conductivity $K(h)$ has a known functional form, then the integral on the left side may sometimes be integrated exactly to produce an analytic expression for the water flux. This procedure is illustrated in the next section.

Evaporation from a Water Table Although water evaporation in the field is not a steady-state process, a nearly steady upward flow from a water table to a bare soil surface may be established if the daily evaporative demand is reasonably uniform for a long period of time. A classic solution to this problem adapted from Gardner (1958) and Gardner and Fireman (1958) is given in the next example.

EXAMPLE 3.8: Calculate the maximum possible water evaporation rate above a water table located at a distance $z = -L$ below the soil surface ($z = 0$) for a soil that has a hydraulic conductivity–matric potential that may be represented well by the following functional form:

$$K(h) = \frac{K_s}{1 + (h/a)^N} \tag{3.44}$$

where K_s is the saturated hydraulic conductivity and $N > 0$, $a < 0$ are constant parameters representing the shape of the function.

In this problem, $h_1 = 0$ at $z_1 = -L$ (the water table) and $h_2 \to -\infty$ at $z_2 = 0$ representing the maximum capillary attraction to the soil surface and therefore the maximum evaporation rate. Letting $J_w = +E$ represent the evaporation rate, we may rewrite (3.43) as

$$-L = \int_0^{-\infty} \frac{dh}{1 + (E/K_s)(1 + (h/a)^N)} \tag{3.45}$$

Letting $y = h/a > 0$ in (3.45) produces the following change in the integral:

$$-L = a \int_0^{\infty} \frac{dy}{1 + (E/K_s) + (E/K_s)y^N} \approx a \int_0^{\infty} \frac{dy}{1 + (E/K_s)y^N} \tag{3.46}$$

where we have simplified (3.46) by assuming that $E/K_s \ll 1$. Letting $z = y(E/K_s)^{1/N}$ in (3.46) produces the following change in the integral:

$$-L = a \left(\frac{K_s}{E}\right)^{1/N} \int_0^{\infty} \frac{dz}{1 + z^N} \tag{3.47}$$

The integral in (3.47) may be looked up in any standard table of definite integrals, or it may be evaluated by the calculus of residues. It has the value (Abramowitz and Stegun, 1970)

$$\int_0^{\infty} \frac{dz}{1 + z^N} = \frac{\pi}{N \sin (\pi/N)} \tag{3.48}$$

After plugging (3.48) into (3.47), we can solve for the maximum evaporation rate

$$E = K_s \left(\frac{-a\pi}{LN \sin \pi/N}\right)^N \tag{3.49}$$

Equation (3.49) gives the maximum evaporation rate E as a function of the distance L between the soil surface and the water table, and the soil parameters K_s, a, and N are used to model the hydraulic conductivity function (3.44).

Gardner and Fireman (1958) conducted a laboratory experimental test of (3.49) on two soils, a Pachappa fine sandy loam and a Chino clay, by maintaining a hanging water column connected to the bottom of soil columns packed with samples of the two soils. The water column was placed at a specified position below the column to correspond to the water table depth.

Figure 3.15, redrawn from Gardner and Fireman (1958), shows the measured hydraulic conductivity–matric potential functions for the two soils, a Pachappa fine sandy loam and a Chino clay, together with smooth curves from (3.44) fitted to

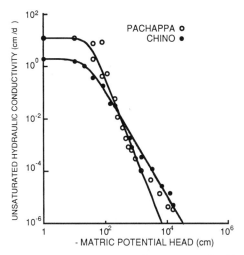

Figure 3.15 The $K(h)$ data (circles) and model fits (curves) using Equation (3.44). (After W. R. Gardner and M. Fireman, Laboratory studies of evaporation from soil columns in the presence of a water table, *Soil Sci.*, Vol. 85, pp. 244–249, © by Williams & Wilkins, 1958.)

the experimental data. Table 3.5 summarizes the values of the parameters K_s, a, and N that were fitted to the two soils.

Figures 3.16 and 3.17 show the predicted relation between evaporation rate and water table depth, using (3.49), together with data points representing the measured evaporation rate from the columns under intense evaporative conditions in the laboratory. The agreement, particularly for large L, is excellent. Note that the Chino clay soil evaporation rate exceeds that from the Pachappa soil for large L.

This simple model calculation demonstrates some fundamental principles that are useful in interpreting water flow behavior in the field environment. Coarse-textured soils, containing mostly large pores that drain at modest suctions, are characterized in the model expression by larger values of N than the finer textured soils, which have a broader distribution of pore sizes. As a consequence, when the distance between the water table and the surface is great enough, the coarse-textured soils offer more resistance to upward water flow than do finer textured soils.

A major problem associated with agriculture in areas overlying shallow ground-

TABLE 3.5 Values of Parameters of Hydraulic Conductivity–Matric Potential Function (3.44) for Two Soils

Soil	K_s (cm day^{-1})	a (cm)	N
Pachappa fine sandy loam	12.3	−29.6	3
Chino clay	1.95	−23.8	2

Source: Gardner and Fireman (1958).

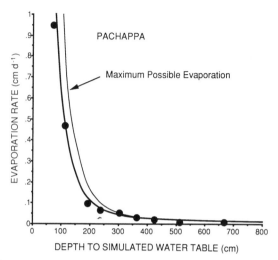

Figure 3.16 Theoretical and experimental rates of evaporation of water from a column of Pachappa sandy loam as a function of depth to simulated water table. (After W. R. Gardner and M. Fireman, Laboratory studies of evaporation from soil columns in the presence of a water table, *Soil Sci.*, Vol. 85, pp. 244–249, © by Williams & Wilkins, 1958.)

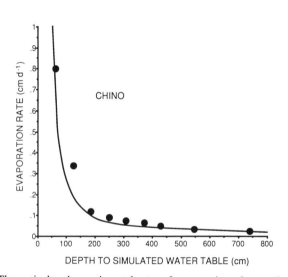

Figure 3.17 Theoretical and experimental rates of evaporation of water from a column of Chino clay as a function of depth to simulated water table. (After W. R. Gardner and M. Fireman, Laboratory studies of evaporation from soil columns in the presence of a water table, *Soil Sci.*, Vol. 85, pp. 244–249, © by Williams & Wilkins, 1958.)

water that is high in salt content is that saline water will move upward to the surface and evaporate, leaving behind salt deposits that can damage crops or deteriorate soil structure. As shown by this model, the upward flow of water in a fine-textured soil can be more significant than in a coarse-textured soil. Thus, to avoid salinization, water tables must be kept lower for fine-textured soils than for coarse-textured ones.

The preceding calculation also illustrates a fundamental difference between saturated and unsaturated flow. In saturated flow, the flux through a finite layer of soil is proportional to the potential difference ΔH across the layer. In unsaturated flow, the conductivity $K(h)$ depends on the matric potential of the soil, so that the flux is not a linear function of ΔH. For example, in the maximum evaporation calculation, an infinite potential difference led to a finite evaporation rate. The reason for this result is that the conductivity in the dry (upper) end of the soil is decreasing rapidly and ultimately approaches zero as $h \rightarrow -\infty$. In fact, as shown by Gardner (1958), the maximum evaporation rate is approached for relatively modest suctions at the surface (i.e., $h \sim -1000$ cm). Further decreases in h at the surface do not increase E.

Steady-State Downward Water Flow   Steady-state downward water flow, while never actually achieved in the field, is nevertheless an approximation of certain flows, such as subsurface drainage in environments under high-frequency irrigation or frequent rainfall. Under these conditions, the Buckingham–Darcy flux equation (3.27) may be replaced by a far simpler expression. This is illustrated in the next example.

EXAMPLE 3.9: Calculate the steady-state downward flux $J_w = -i$ from the surface to a water table at a depth L below the surface. Assume that the soil hydraulic conductivity is described by (3.44) with $N = 2$. With this substitution, the integral form of the Buckingham–Darcy flux law (3.43) may be written as $h_1 = 0$ at $z_1 = -L$ and $J_w = -i < 0$:

$$-(z + L) = \int_0^h \frac{dh}{1 - (i/K_s)\left(1 + (h/a)^2\right)} \qquad (3.50)$$

where the upper limit of the integral has been evaluated at an arbitrary depth z where the matric potential is h.

Letting $y = h/a$ as before, the integral simplifies to

$$-(z + L) = a \int_0^{h/a} \frac{dy}{1 - i/K_s - iy^2/K_s} \qquad (3.51)$$

This integral is a standard form that may be looked up in integral tables (Abramowitz and Stegun, 1970):

$$\int \frac{dy}{\alpha - \beta y^2} = \frac{1}{\sqrt{\alpha\beta}} \tanh^{-1} \frac{y\sqrt{\alpha\beta}}{\alpha} \qquad (3.52)$$

where $\alpha = 1 - i/K_s$ and $\beta = i/K_s$ and $\tanh^{-1}$ is the hyperbolic arctangent.

Thus, applying (3.52) to (3.51),

$$-(z + L) = \frac{a}{\sqrt{(1 - i/K_s)i/K_s}} \tanh^{-1} \frac{h\sqrt{(1 - i/K_s)i/K_s}}{a(1 - i/K_s)} \qquad (3.53)$$

or

$$h = a\sqrt{\frac{K_s}{i} - 1} \tanh\left(\frac{-\sqrt{(1 - i/K_s)i/K_s}}{a}(z + L)\right) \qquad (3.54)$$

Figure 3.18 shows a graph of relative matric potential head h/L versus relative depth z/L for various values of the parameter i/K_s. A significant feature of this curve is that the matric potential approaches a constant value for any flux rate i as the surface is approached, provided that the water table is not too shallow. This important result shows that when water is flowing downward at a constant rate, matric potential gradients dh/dz approach zero and water flows under the influence of gravity alone. Thus, provided that the water table is far below the surface, one may approximate the Buckingham–Darcy law (3.27) for downward flow as

$$J_w \approx -K(h) \qquad (3.55)$$

This approximation is known as *gravity flow*. A general discussion of steady upward and downward flows is given in Raats and Gardner (1975).

Measurement of Unsaturated Hydraulic Conductivity   The unsaturated hydraulic conductivity–matric potential function may be measured in the laboratory in steady-state flow experiments using the principles developed in the preceding. The next example demonstrates how a steady-state evaporation experiment may be used to calculate average values of $K(h)$ in a soil column instrumented with tensiometers.

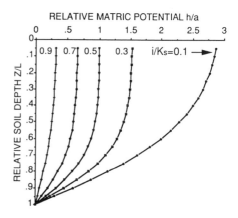

Figure 3.18   Matric potential–depth profiles calculated for steady downward flow with Equation (3.54).

EXAMPLE 3.10: A 50-cm-high soil column is placed above a water source maintaining the bottom of the column ($z = 0$) at saturation ($h = 0$). The top of the column ($z = 50$ cm) is exposed to constant evaporative conditions. Tensiometers are located at z values of 10, 20, 30, and 40 cm within the column. The evaporation rate E in steady state (when the tensiometers and water flux do not change with time) is equal to the flux through the column and is measured by the rate of water entry per unit area into the column from the water source. In steady state the tensiometer readings are as shown in Table 3.6 and the evaporation rate is 0.5 cm day^{-1}.

To calculate $K(h)$, the steady-state form of the Buckingham–Darcy flux law (3.27) is written between two depths z_1 and z_2 where the measured matric potentials are h_1 and h_2, respectively. As an approximation, $K(h)$ between z_1 and z_2 is replaced by its average value $K(\bar{h})$, where $\bar{h} = (h_1 + h_2)/2$. Thus (3.27) becomes

$$J_w = E = -K(h)\frac{dH}{dz} \approx -K(h)\frac{\Delta H}{\Delta z} \tag{3.56}$$

where $\Delta z = z_2 - z_1$ and $\Delta H = H_2 - H_1 = (h_2 + z_2) - (h_1 + z_1)$. Thus

$$K(\bar{h}) = -E\frac{\Delta z}{\Delta H} \tag{3.57}$$

The data in Table 3.6 may be evaluated using (3.57) by following the sequence of steps given in Table 3.7.

Table 3.8 summarizes the calculation of $K(h)$ using this procedure and the data in Table 3.6.

Thus, this experiment yielded four values of the $K(h)$ function, averaged over the range of matric potentials between the successive depth increments.

The resolution of the $K(h)$ measurement could be improved by increasing the density of tensiometers. The $K(h)$ function may also be measured in steady-state

TABLE 3.6 Tensiometer Readings at Steady State in Evaporation Experiment

z (cm)	h (cm)
40	−125
30	−75
20	−40
10	−15
0	0

TABLE 3.7 Sequence of Steps Used to Evaluate $K(h)$ From Steady-State Evaporation Experiment

Step 1	Calculate H at each depth.
Step 2	Calculate ΔH between each depth interval.
Step 3	Calculate $\Delta H/\Delta z$ between each depth interval.
Step 4	Calculate $K(\bar{h})$ between each depth interval with (3.57).
Step 5	Calculate $\bar{h}$ between each depth interval.

TABLE 3.8   Evaluation of $K(h)$ by Procedure in Table 3.7 for Data in Table 3.6

z	h	H	ΔH	$\Delta H/\Delta z$	$K(\bar{h})$	$\bar{h}$
40	−125	−85				
			−40	−4.0	0.125	−100
30	−75	−45				
			−25	−2.5	0.200	−57.5
20	−40	−20				
			−15	−1.5	0.333	−27.5
10	−15	−5				
			−5	−0.5	1.000	−7.5
0	0	0				

downward flow experiments. However, if this method is used, only a small range of matric potentials is covered in a single flow experiment (see Fig. 3.18).

3.3.5   Water Conservation Equation

The steady-state water flow condition discussed in the previous sections describes only a special subset of the possible water transport processes in soil. In general, wetting or drying of the soil will occur as water flows, and the matric potential and water content will be functions of time as well as of space. Such flows are called transient or time dependent and require a more complete mathematical description than steady-state flows. The first step in a complete transient water flow description is to specify the water conservation equation, also called the mass balance or continuity equation.

As in the formulation of the flux laws, we will not attempt to describe water flow at the pore scale but will produce a volume-averaged description of water conservation. We will focus our analysis on a cube of soil occupying a volume $V = \Delta x\, \Delta y\, \Delta z$ at a point (x, y, z) in the flow field (Fig. 3.19). For simplicity, we will assume that water is flowing in the z direction through homogeneous, unsaturated soil.

In the most general case, water may be flowing in and out of the soil volume at different rates, causing the water content of the volume to change. In addition, if plant roots are present inside the volume, some of the water that enters will "disappear" from the volume by entering the plant root and moving toward the leaf within the root xylem. Our statement of water conservation must account for all of these processes.

The water conservation equation is derived by calculating the mass balance for the system during an arbitrarily small time period Δt between t and $t + \Delta t$. Over this time period, the water conservation equation may be stated in words as follows:

Volume of water entering soil volume during Δt

= volume of water leaving soil volume during Δt

+ increase of water volume stored in soil volume during Δt

+ volume of water that has disappeared from the soil volume by plant root uptake during Δt (3.58)

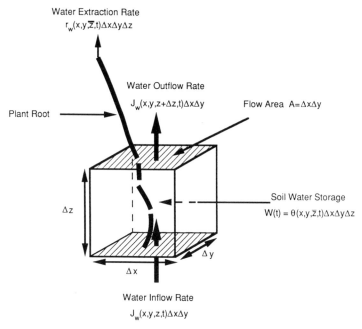

Water Extraction Rate
$r_w(x,y,\overline{z},t)\Delta x \Delta y \Delta z$

Water Outflow Rate

$J_w(x,y,z+\Delta z,t)\Delta x \Delta y$ Flow Area $A = \Delta x \Delta y$

Plant Root

Soil Water Storage

Δz $W(t) = \theta(x,y,\overline{z},t)\Delta x \Delta y \Delta z$

Δy

Δx

Water Inflow Rate
$J_w(x,y,z,t)\Delta x \Delta y$

Figure 3.19 Unit volume of soil used to calculate the water mass balance equation.

For the one-dimensional vertical flow process shown in Fig. 3.19, the volume of water entering the soil volume during Δt is given by

$$\text{Volume of water entering soil volume} = J_w(x, y, z, t + \tfrac{1}{2} \Delta t)A\, \Delta t$$

$$= J_w(x, y, z, t + \tfrac{1}{2} \Delta t)\, \Delta x\, \Delta y\, \Delta t \tag{3.59}$$

where J_w is the water flux in the z direction (rate of flow per unit area) and $A = \Delta x\, \Delta y$ is the cross-sectional area of the inflow surface. The flux is expressed as a function of x, y, z, and t, although it will not be a function of x or y if the flow is in the z direction only. It is assigned an average value[1] at the midpoint $t + \Delta t/2$ of the time interval.

Similarly, at the exit boundary,

$$\text{Volume of water leaving soil volume} = J_w(x, y, z + \Delta z, t + \tfrac{1}{2} \Delta t)A\, \Delta t$$

$$= J_w(x, y, z + \Delta z, t + \tfrac{1}{2} \Delta t)\, \Delta x\, \Delta y\, \Delta t \tag{3.60}$$

[1]Since Δt is arbitrarily small, this will not affect the results.

where now the water flux is evaluated at the point $z + \Delta z$, the location of the outflow surface.

The increase of water volume stored in the soil volume during Δt may be expressed in terms of the volumetric water content[2] θ as

$$\text{Increase of water volume stored in soil volume} = \Delta\theta \, \Delta x \, \Delta y \, \Delta z \quad (3.61)$$

where $\Delta\theta = \theta(x, y, z + \frac{1}{2}\Delta z, t + \Delta t) - \theta(x, y, z + \frac{1}{2}\Delta z, t)$ is the change in θ during the time period Δt. The water content is assigned an average value at the midpoint $z + \frac{1}{2}\Delta z$ of the volume.

Finally, the volume of water removed by plant water uptake may be expressed symbolically as

$$\text{Volume of water disappearing from soil volume} = r_w \, \Delta x \, \Delta y \, \Delta z \, \Delta t \quad (3.62)$$

where r_w is the rate of disappearance of water per unit volume, which may be a function of x, y, z, and t. The function r_w is called a sink term in the conservation equation.

Inserting (3.59)–(3.62) into (3.58) produces

$$J_w(x, y, z, t + \tfrac{1}{2}\Delta t) \, \Delta x \, \Delta y \, \Delta t$$

$$= J_w(x, y, z + \tfrac{1}{2}\Delta z, t + \tfrac{1}{2}\Delta t) \, \Delta x \, \Delta y \, \Delta t + \theta(x, y, z + \tfrac{1}{2}\Delta z, t + \Delta t)$$

$$\cdot \Delta x \, \Delta y \, \Delta z - \theta(x, y, z + \tfrac{1}{2}\Delta z, t) \, \Delta x \, \Delta y \, \Delta z + r_w \, \Delta x \, \Delta y \, \Delta z \, \Delta t \quad (3.63)$$

After dividing by the volume $\Delta x \, \Delta y \, \Delta z$ and rearranging terms, this may be rewritten as

$$\frac{J_w(x, y, z + \tfrac{1}{2}\Delta z, t + \tfrac{1}{2}\Delta z) - J_w(x, y, z, t + \tfrac{1}{2}\Delta t)}{\Delta z} +$$

$$\frac{\theta(x, y, z + \tfrac{1}{2}\Delta z, t + \Delta t) - \theta(x, y, z + \tfrac{1}{2}\Delta z, t)}{\Delta t} + r_w = 0 \quad (3.64)$$

Finally, in the limit as $\Delta z \to 0$ and $\Delta t \to 0$ [recalling definition (3.28) of the partial derivative], (3.64) becomes

$$\frac{\partial J_w}{\partial z} + \frac{\partial \theta}{\partial t} + r_w = 0 \quad (3.65)$$

---

[2]The subscript v on the volumetric water content will be omitted from now on.

Equation (3.65) is called the soil water conservation or continuity equation. If we had allowed the water to flow in an arbitrary direction in Fig. 3.19, the water conservation equation would be written as

$$\frac{\partial J_{w_x}}{\partial x} + \frac{\partial J_{w_y}}{\partial y} + \frac{\partial J_{w_z}}{\partial z} + \frac{\partial \theta}{\partial t} + r_w = 0 \qquad (3.66)$$

where J_{w_x}, J_{w_y}, J_{w_z} are the components of the water flux vector

$$\mathbf{J}_w = J_{w_x}\,\hat{\imath} + J_{w_y}\hat{\jmath} + J_{w_z}\hat{k} \qquad (3.67)$$

and $\hat{\imath}, \hat{\jmath}, \hat{k}$ are unit vectors in the x, y, z directions, respectively.

The water uptake term r_w is included for completeness in (3.65) and (3.66). It is equal to zero when there are no plant roots or other sinks of water present and must be specified with a model when water uptake is occurring. Processes involving water uptake will be discussed later in the book.

3.3.6 Richards Equation for Transient Water Flow

The water conservation equation relates water fluxes, storage changes, and sources or sinks of water. When it is combined with the Buckingham–Darcy flux equation (3.27), an equation may be derived to predict the water content or matric potential in soil during transient flow. We will assume for simplicity that the flow is vertical and that no plant roots are present ($r_w = 0$).

Inserting (3.27) into (3.65) produces

$$\frac{\partial \theta}{\partial t} = \frac{\partial}{\partial z}\left[K(h)\left(\frac{\partial h}{\partial z} + 1\right)\right] \qquad (3.68)$$

Equation (3.68) may not be solved in the form it is in, because it contains two unknowns θ and h and only one equation. This difficulty may be overcome by using the water characteristic or matric potential–water content function $h(\theta)$ to eliminate either θ or h from (3.68). Since either variable may be eliminated, there are two forms of the equation.

Water Content Form of Richards Equation The flux equation (3.27) may be reexpressed as a function of θ alone by the following transformations:

1. Since $K(h)$ is a function of h and $h(\theta)$ is a function of θ, K may be written directly as a function of θ:

$$K\big(h(\theta)\big) \equiv K(\theta) \qquad (3.69)$$

2. The partial derivative $\partial h / \partial z$ may be rewritten by the chain rule of differentiation (Kaplan, 1984) as

$$\frac{\partial h(\theta)}{\partial z} = \frac{dh}{d\theta} \frac{\partial \theta}{\partial z} \tag{3.70}$$

where $dh/d\theta$ is the slope of the matric potential–water content function.

Inserting (3.69) and (3.70) into (3.27) produces

$$J_w = -K(\theta) \frac{dh}{d\theta} \frac{\partial \theta}{\partial z} - K(\theta) \equiv -D_w(\theta) \frac{\partial \theta}{\partial z} - K(\theta) \tag{3.71}$$

where

$$D_w(\theta) = K(\theta) \frac{dh}{d\theta} \tag{3.72}$$

is called the soil water diffusivity.

After this transformation, (3.68) may be written as

$$\frac{\partial \theta}{\partial t} = \frac{\partial}{\partial z} \left(D_w(\theta) \frac{\partial \theta}{\partial z} \right) + \frac{\partial K(\theta)}{\partial z} \tag{3.73}$$

Equation (3.73) is called the water content form of the Richards equation. It is a second-order, nonlinear, partial differential equation called a Fokker–Planck equation and can generally only be solved by numerical methods. Assuming that $D_w(\theta)$ and $K(\theta)$ are known, (3.73) requires two boundary conditions (i.e., the soil surface and deep in the soil) where the behavior of θ or J_w is known as a function of time. It also requires specification of an initial condition describing the water content of the entire profile at $t = 0$.

Modern high-speed computers have made the task of solving (3.73) almost routine when the soil water functions and boundary conditions are known. Use of this equation as a predictive tool is limited primarily by difficulties in measuring D_w and K accurately over the region where the simulation is to be run, particularly when the soil is heterogeneous and each location in the soil has a different $K(\theta)$ and $D_w(\theta)$ function.

Equation (3.73) has also been derived by ignoring hysteresis. The slope $dh/d\theta$ is defined only for a uniform wetting or drying process in which $h(\theta)$ is described uniquely by a single curve (see Fig. 3.12). When repeated wetting and drying cycles are present, (3.73) is invalid and the water diffusivity D_w cannot even be defined. Only a limited amount of study of the influence of hysteresis on water flow has been performed, either theoretically or experimentally, although there are some indications that it may have a substantial influence in certain cases (Curtis and Watson, 1984; Jones and Watson, 1987; Russo et al., 1989).

Matric Potential Form of Richards Equation   The water content time derivative in (3.68) may be rewritten using the chain rule as

$$\frac{\partial \theta}{\partial t} = \frac{d\theta}{dh}\frac{\partial h}{\partial t} \equiv C_w(h)\frac{\partial h}{\partial t} \tag{3.74}$$

where

$$C_w(h) = \frac{d\theta}{dh} \tag{3.75}$$

is called the water capacity function. It is equal to the inverse slope of $h(\theta)$. As with the diffusivity, it is only defined for uniform wetting or drying processes.

Inserting (3.74) into (3.68) produces

$$C_w(h)\frac{\partial h}{\partial t} = \frac{\partial}{\partial z}\left[K(h)\left(\frac{\partial h}{\partial z} + 1\right)\right] \tag{3.76}$$

Equation (3.76) is called the matric potential form of the Richards equation. It may be solved if two boundary conditions and an initial condition are specified, provided that $C_w(h)$ and $K(h)$ are known.

Water Diffusivity Function $D_w(\theta)$ The water diffusivity function may be calculated from $K(\theta)$ and $h(\theta)$ using the definition in (3.73). There are also direct methods for evaluating it in transient flow experiments using theoretical analyses beyond the scope of this book. Useful references for these methods include Bruce and Klute (1956), Whisler et al. (1968), Dirksen (1975, 1979), and Clothier et al. (1983). A review of current methodologies is given by Klute and Dirksen (1986).

Water Capacity Function $C_w(\theta)$ The water capacity function $C_w(\theta)$ defined in (3.75) expresses the increase in matric potential per unit increase in water content. Figure 3.20 shows a typical $h(\theta)$ curve for a sandy soil and the corresponding $C_w(\theta)$ curve. The largest values of C_w are associated with the wet end of the capillary region of the retention curve, where the larger pores are emptying with modest changes in suction. The function is equal to zero at saturation, prior to the onset of desaturation at the air entry suction. It approaches zero at extreme dryness when enormous changes in suction occur with small changes in water content.

3.3.7   Model Functional Forms

In recent years, considerable effort has been spent developing versatile functional forms that have a sufficient number of parameters to represent the range of shapes of the soil water hydraulic functions that might be found in different soils (Mu-

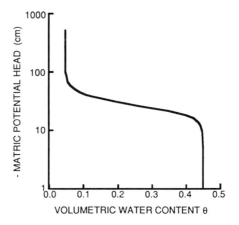

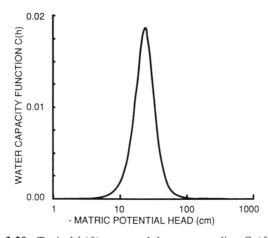

Figure 3.20   Typical $h(\theta)$ curve and the corresponding $C_w(\theta)$ curve.

alem, 1976b; van Genuchten, 1980). Table 3.9 gives the analytic functional forms for $K(\theta)$, $h(\theta)$ prepared by van Genuchten (1980), which have been used extensively in numerical modeling. The expression for $K(\theta)$ [(3.78)] was derived from (3.77) using the theoretical model of Mualem (1976b). A review of general methods for calculating hydraulic conductivity model forms from water characteristic functions is given by van Genuchten and Nielsen (1985).

Equations (3.77) and (3.78) use fitted parameters N, α, K_s, θ_s, and θ_r where the subscript s on K_s, θ_s denotes saturation and θ_r is called the residual water content, referring to the region of $h(\theta)$ where adsorptive forces are dominant and h is decreasing rapidly with little change in θ. The water characteristic function (3.75)

TABLE 3.9 Matric Potential- and Hydraulic Conductivity–Water Content Functional Forms

Function	Mathematical form	
Matric Potential–Water Content	$\tilde{\theta}(h) = [1 + \alpha(-h)^N]^{-M}$	(3.77)
Hydraulic Conductivity–Water Content	$K(\theta) = K_s\tilde{\theta}^{1/2}[1 - (1 - \tilde{\theta}^{1/M})^M]^2$	(3.78)
	where $\tilde{\theta} = (\theta - \theta_r)/(\theta_s - \theta_r)$	
	and $M = 1 - 1/N$	

Source: van Genuchten (1980).

must be compatible with the definitions for $h(\theta)$. This is illustrated in the next example.

> *EXAMPLE 3.11:* Derive a model functional form for $C_w(h)$ using (3.77).
> By definition, $C_w(\theta) = d\theta/dh$. Thus (3.77) is differentiated as follows:

$$\frac{d\theta}{dh} = (\theta_s - \theta_r)\frac{d\tilde{\theta}}{dh} = (\theta_s - \theta_r)\frac{d}{dh}\left[1 + \alpha(-h)^N\right]^{-M}$$

$$= \frac{(\theta_s - \theta_r)(-M)}{\left[1 + \alpha(-h)^N\right]^{1+M}}\frac{d}{dh}\left[1 + \alpha(-h)^N\right]$$

$$= \frac{(\theta_s - \theta_r)(-M)\alpha N(-1)(-h)^{N-1}}{\left[1 + \alpha(-h)^N\right]^{1+M}}$$

or since $M = 1 - 1/N$,

$$C_w(h) = \frac{\alpha(\theta_s - \theta_r)(N - 1)(-h)^{N-1}}{\left[1 + \alpha(-h)^N\right]^{2 - 1/N}} \tag{3.79}$$

3.3.8 Water Flow Calculations in Unsaturated Soil

Steady-State Water Flow through a Crop Root Zone Steady-state water flow rarely occurs in unsaturated soil, particularly in the field, because temperature, evaporation rate, water uptake, and other processes that affect water transport tend to be influenced by the diurnal cycle of radiant energy to the earth's surface. Nevertheless, if the external influences are repeated regularly at approximately the same intensity, the soil water will develop a characteristic response that varies mostly at the surface and smooths out to a steady-state profile beneath the surface. For such a situation, useful information may be obtained by approximating the flow regime with a steady-state model. An illustration of this is given in the next example.

> *EXAMPLE 3.12:* A fully developed crop is irrigated daily at an irrigation rate i_0 centimeters per day. The crop root zone, which extends from $z = 0$ to $z = -L$, has a uniform root density that extracts water at a constant rate everywhere. The total water removed each day is constant and equal to the evapotranspiration (ET) in cen-

timeters per day. ($\text{ET} < i_0$). Calculate the steady-state water flux $J_w(z)$ as a function of the depth in the soil.

The water conservation equation (3.65) in steady state reduces to (since $\partial\theta/\partial t = 0$)

$$\frac{dJ_w}{dz} + r_w = 0 \tag{3.80}$$

where an ordinary derivative is used since J_w is not a function of time. The water uptake per unit volume per unit time, assumed to be constant, is given by

$$r_w = \text{ET}/L \tag{3.81}$$

Thus, after rearranging, (3.80) may be integrated:

$$\int_{-i_0}^{J_w} dJ_w = -\int_0^z \frac{\text{ET}}{L} dz \tag{3.82}$$

so that the flux J_w at depth z is given by

$$J_w(z) = -i_0 - (\text{ET})z/L \quad \text{for } -L < z < 0 \tag{3.83}$$

Thus the drainage flux below the root zone is equal to

$$J_w(-L) = -i_0 + \text{ET} \tag{3.84}$$

Water Flow through Unsaturated Layered Soil   To this point, the discussion of water flow through unsaturated soil has been confined to homogeneous profiles, where the soil water functions ($K(\theta)$, $D_w(\theta)$, $h(\theta)$) are the same everywhere. However, soil is frequently stratified near the surface, containing layers with markedly different water retention and water conducting properties. The mathematical description of water transport through unsaturated layered soil is very complex because of subtle effects that can occur at the interface between layers.

As we saw in the discussion of water flow through saturated layered soil, the layer of least permeability often has a dominant influence on the transport through the system. For example, even though steep hydraulic head gradients are often present, flow through a series of layers of unsaturated soil can be nearly zero under conditions where large and nearly empty pores with small hydraulic conductivities are encountered. Such a condition exists where a wetting front moving through homogeneous soil encounters a layer of coarse sand or gravel. The hydraulic head of the soil just above the wetting front may be of the order of -100 cm of water and that in the dry sand below the front may be as low as -10^3 or -10^4 cm. Thus, the potential gradient at the interface will be large. Despite this, the flow nearly drops to zero as the front reaches the coarse sand layer because there is very little water in fine pores in the sand, and the large pores cannot fill at the low matric potentials present in the upper region. Thus, the cross section for liquid flow is very small. Before any appreciable flow can occur, the hydraulic head in the upper layers of the finer textured soil must rise to a value near zero, at which time some of the pores or channels in the sand will begin to fill, and the hydraulic conductivity of the lower layer will rise. This is illustrated in Fig. 3.21, where a layer of coarse sand in a silt loam soil restricts downward penetration of water. Coarse materials

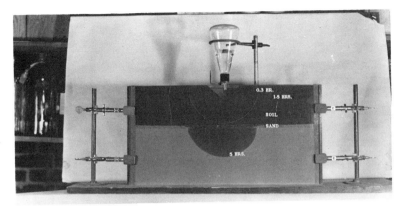

Figure 3.21 Water retention in soil above a sand layer.

such as straw or other organic matter or holes in the soil created by burrowing insects and animals restrict rather than aid flow as long as the hydraulic potentials surrounding them are too low for them to fill with water. For this reason, dry soil often persists through the wet season beneath straw turned under by plowing. Only when large pores and channels connect with the surface where free water can get to them or are beneath the water table do such channels contribute appreciably to liquid flow.

Fine pores in hard pans and clay pans also seriously restrict flow. Such materials become wet rapidly for short distances when first contacted by water because of the high absorptive capacity of fine pores. However, as the distance through which water must move in fine pores increases, the rate of flow decreases; in tight clay it becomes extremely slow. Flow in such materials often is so slow that water tables build up above them. Rapid initial wetting followed by gross restriction of flow is illustrated in Fig. 3.22.

Figure 3.22 Water retention in soil above a clay layer.

Water retention following wetting or redistribution of water is greatly affected by stratification. Clay pans and hard pans often create serious waterlogging because retention is so pronounced above such layers. Coarse layers act much the same as a check valve. Water tables cannot be maintained above a coarse layer; however, since a coarse layer restricts flow at relatively high hydraulic potentials, retention of water above such layers often is appreciably more than it would be in the absence of such a layer. For example, where a coarse gravel layer exists at a depth of about 45 cm in a fine sandy loam, water flow will cease at about -10 cm of hydraulic head. At least three times as much water is retained and available for plant use in the soil above this layer as would be retained if the gravel layer were not present.

3.4 MULTIDIMENSIONAL FLOW

Transient multidimensional water flow problems generally require numerical solution using high-speed computers. There are a few steady-state multidimensional flow problems that may be solved analytically by changing from Cartesian to cylindrical or spherical coordinates. An illustration of this procedure on a well-known saturated flow problem is given in the next example.

EXAMPLE 3.13: A cylindrical well is inserted in a saturated soil aquifer and withdraws water at a uniform volume flow rate Q. Observation wells are placed at radial distances R_1, R_2 from the pumping well, where hydrostatic pressure heads p_1 and p_2 are measured (Fig. 3.23). Calculate the saturated hydraulic conductivity K_s of the aquifer around the well.

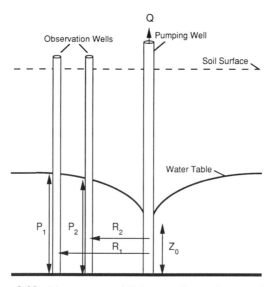

Figure 3.23 Measurement of K_s by a well-pumping experiment.

This water flow system clearly has cylindrical symmetry and will be assumed to have no angular dependence or z dependence. Thus, water will flow radially inward to the pumping well. In this case, the steady-state water conservation equation becomes[3]

$$\frac{1}{r}\frac{d}{dr}(rJ_r) = 0 \tag{3.85}$$

Equation (3.85) may be integrated once, to produce

$$rJ_r = \phi = \text{const} \tag{3.86}$$

At $r = R_0$ (the pumping well radius) the flux $J_r(R_0) = J_{r0}$ may be related to the volume flow rate by multiplying J_{r0} by the surface area $2\pi R_0 z_0$ of the permeable well casing in contact with groundwater. Thus,

$$rJ_r = R_0 J_{r0} = Q/2\pi z_0 \tag{3.87}$$

The radial flux is given by Darcy's law (3.14):

$$J_r = -K_s \frac{dp}{dr} \tag{3.88}$$

where p is the hydrostatic pressure head. If (3.88) is combined with (3.87), the new equation may be written as

$$-K_s \frac{dp}{dr} = \frac{Q}{2\pi z_0 r} \tag{3.89}$$

Equation (3.89) may be integrated after first placing all factors that depend explicitly on r on the same side of the equation. This produces

$$dp = -\frac{Q}{2\pi K_s z_0}\frac{dr}{r} \tag{3.90}$$

Since $p(R_1) = p_1$ and $p(R_2) = p_2$, (3.90) can be integrated:

$$\int_{p_1}^{p_2} dp = -\frac{Q}{2\pi K_s z_0}\int_{R_1}^{R_2}\frac{dr}{r} \tag{3.91}$$

Equation (3.91) can be solved for K_s, with the result

$$K_s = \frac{Q}{2\pi z_0(p_1 - p_2)}\ln\frac{R_2}{R_1} \tag{3.92}$$

Equation (3.92) is the standard equation used for calculating K_s in large aquifers. The measurement represents an average over the volume of the aquifer influenced by the pumping.

---

[3]A good discussion of cylindrical coordinate transformations may be found in Arfken (1985).

3.5 APPENDIX: SOLUTION OF FIRST-ORDER ORDINARY DIFFERENTIAL EQUATIONS

Many of the transport equations of interest in soil physics (notably steady-state equations) can be written as first-order differential equations of the form

$$\frac{dy(x)}{dx} = f(x, y) \qquad (3.93)$$

where $f(x, y)$ is a general expression for a function of x and y, $y(x)$ is the dependent variable (e.g., temperature or pressure), and x is the independent variable (usually position or time). Equation (3.93) requires a starting point or initial condition before it can be solved. We will write this in the form

$$y(x_0) = y_0 \quad \text{at } x = x_0 \qquad (3.94)$$

In general, we can solve (3.93) and (3.94) numerically regardless of the form of $f(x, y)$. In special cases, however, we can integrate (3.93) and derive an analytic relation between y and x. The two major cases where this is possible are discussed in the next two sections.

3.5.1 Method 1: Separation of Variables

If the function $f(x, y)$ in (3.93) can be separated into the product of a function $f_1(x)$ of x and a function $f_2(y)$ of y,

$$f(x, y) = f_1(x)f_2(y) \qquad (3.95)$$

then (3.93) can be integrated after placing all factors depending on x on the dx side of the equation and all factors depending on y on the dy side (constants can go on either side). Thus, after inserting (3.95), we may rewrite (3.93) as

$$\frac{dy}{f_2(y)} = f_1(x) \, dx \qquad (3.96)$$

We now can integrate both sides of (3.96). We can either integrate them indefinitely, adding an unknown constant of integration, or integrate them from (x_0, y_0) to (x, y). In the first case, we get rid of the constant of integration by requiring that the equation obey $y(x_0) = x_0$. Thus, the final solution to (3.93) and (3.94) can be written formally as

$$\int_{y_0}^{y(x)} \frac{dy}{f_2(y)} = \int_{x_0}^{x} f_1(x) \, dx \qquad (3.97)$$

An example of this solution method is now given.

EXAMPLE 3.14:   Solve the ordinary differential equation

$$\frac{dy}{dx} = y^2(x + 1) \tag{3.98}$$

subject to the initial condition

$$y(0) = 1 \quad \text{at } x = 0 \tag{3.99}$$

The right side of (3.98) can be factored into the product $f_1(x) = x + 1$ and $f_2(y)$ $= y^2$. We first rewrite (3.98) by placing it in the form of (3.96):

$$\frac{dy}{y^2} = (x + 1)\, dx \tag{3.100}$$

Now we integrate (3.100) from $y = 1$ to y and $x = 0$ to x.
  These terms are

$$\int_1^y \frac{dy}{y^2} = -\left(\frac{1}{y}\right)\Bigg|_{y=1}^{y=y} = 1 - \frac{1}{y} \tag{3.101}$$

$$\int_0^x (x + 1)\, dx = \left(\frac{x^2}{2} + x\right)\Bigg|_{x=0}^{x=x} = \frac{x^2}{2} + x \tag{3.102}$$

Since these terms are equal, we may write the final solution as

$$y = \frac{1}{1 - x - x^2/2} \tag{3.103}$$

EXAMPLE 3.15:   A saturated soil column of height L has a height d of water ponded on the top of it, while the pressure at the bottom at $z = 0$ is zero. Calculate the water flux J_w through the column and the pressure head $p(z)$ at an arbitrary point z within the column.
  The differential form of Darcy's law (3.13),

$$J_w = -K_s \frac{dH}{dz} = -K_s \left(\frac{dp}{dz} + 1\right) \tag{3.104}$$

may be written as a differential equation in the form of (3.93):

$$\frac{dp}{dz} = -\left(\frac{J_w}{K_s} + 1\right) \tag{3.105}$$

where $p(0) = 0$ at $z = 0$. Since the right side of (3.105) is constant, it may be integrated simply by moving dz to the right. Thus

$$\int_0^p dp = -\left(\frac{J_w}{K_s} + 1\right) \int_0^z dz \tag{3.106}$$

Therefore, the pressure $p(z)$ is simply

$$p(z) = -\left(\frac{J_w}{K_s} + 1\right) z = \phi_1 z \tag{3.107}$$

where ϕ_1 is a constant. Since $p(L) = d$ at the top of the column ($z = L$), then by (3.107)

$$p(L) = d = \phi_1 L \Rightarrow \phi_1 = \frac{d}{L} \Rightarrow p(z) = z\frac{d}{L} \qquad (3.108)$$

Finally, inserting the value of ϕ_1 into (3.107), we can solve for J_w:

$$\phi_1 = \frac{d}{L} = -\left(\frac{J_w}{K_s} + 1\right) \Rightarrow J_w = -K_s\frac{L + d}{L} \qquad (3.109)$$

3.5.2 Method 2: Integrating Factors

There are many functions $f(x, y)$ in differential equations of the form of (3.93) that cannot be factored into the product of a function of x and a function of y. A second type of first-order equation that can always be integrated is one of the form

$$\frac{dy}{dx} - a(x)y = b(x) \qquad (3.110)$$

where $a(x)$ and $b(x)$ are arbitrary functions of x. We can solve this equation as follows. We multiply both sides of (3.110) by the integrating factor,

$$F(x) = \exp\left[-\int_{x_0}^x a(x)\, dx\right] \qquad (3.111)$$

The derivative of $F(x)$ with respect to x is (Kaplan, 1984)

$$\frac{dF}{dx} = -a(x)\exp\left[-\int_{x_0}^x a(x)\, dx\right] = -a(x)F(x) \qquad (3.112)$$

Therefore, after multiplying by $F(x)$ we may write (3.110) as

$$F(x)\frac{dy}{dx} - a(x)F(x)y = F(x)b(x) = F(x)\frac{dy}{dx} + \frac{dF(x)}{dx}y = \frac{dU}{dx} \qquad (3.113)$$

where $U(x) = F(x)y(x)$. Using the initial condition (3.94), we may now restate the differential equation (3.110) in the simple form

$$\frac{dU}{dx} = F(x)b(x) \qquad (3.114)$$

$$U(x_0) = F(x_0)b(x_0) = b(x_0) \quad \text{at } x = x_0 \qquad (3.115)$$

since $F(x_0) = 1$ by (3.112). We may now use the procedure from the previous section to solve (3.114) and (3.115), with the result

$$U(x) = F(x)y(x) = b(x_0) + \int_{x_0}^x F(x')b(x')\, dx' \qquad (3.116)$$

where x' is a dummy variable of integration. Therefore, the solution to (3.110) is

$$y(x) = \frac{b(x_0)}{F(x)} + \frac{1}{F(x)} \int_{x_0}^{x} F(x')b(x') \, dx' \tag{3.117}$$

EXAMPLE 3.16: Solve the differential equation

$$\frac{dy}{dx} = x + y \tag{3.118}$$

$$y(0) = 0 \quad \text{at } x = 0 \tag{3.119}$$

This simple-looking equation cannot be solved by the method in the first section, because $f(x, y) = x + y$ cannot be written as a product of functions of x and y. However, it is of the form of (3.110) with $a(x) = 1$ and $b(x) = x$. Therefore, using (3.111) and (3.119),

$$F(x) = \exp\left[-\int_{0}^{x} dx' \right] = \exp(-x) \tag{3.120}$$

By (3.117), we can write the solution to (3.118) and (3.119):

$$y(x) = \exp[x] \int_{0}^{x} x' \exp[-x'] \, dx'$$

$$\Rightarrow y(x) = \exp[x] - x - 1 \tag{3.121}$$

PROBLEMS

3.1 A long soil column containing tensiometers is in steady-state evaporation, with a water source at the bottom. The area of the column is 100 cm^2. If 150 cm^3 of water must be added to the water source at the bottom each day, calculate the evaporation flux and $K(h)$ from this data. Plot $K(h)$ versus h on semilogarithmic paper.

Z	h
120	-750
100	-300
70	-175
40	-70
20	-30
5	-6
0	0

3.2 A saturated soil column (Fig. 3.24) contains two soil layers, each 10 cm thick, with sand of $K_S = 10 \text{ cm h}^{-1}$ underneath loam of $K_S = 5 \text{ cm h}^{-1}$. The bottom of the soil column is open to the atmosphere ($p = 0$). At $t = 0$, a height of water $d = 10$ cm above the top of the column is ponded on the surface.

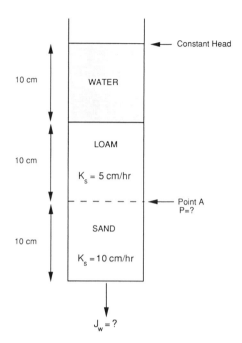

Figure 3.24 Layered soil column.

(a) Assuming steady state, calculate the flux through the column.

(b) Assuming steady state, calculate the water pressure at the sand–loam interface (point *A*).

(c) If this were a falling-head permeameter, what would probably happen as *d* decreased from 10 cm to zero?

3.3 A soil column contains 50 cm of sand over 50 cm of clay (Fig. 3.25). A piezometer at $z = 50$ cm measures the hydrostatic pressure head $p = 50$ cm

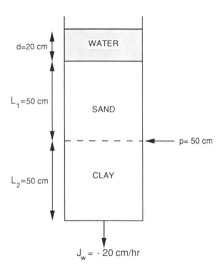

Figure 3.25 Layered soil column.

at the interface. The column is saturated and 20 cm of water are ponded on the top while the bottom is open to the atmosphere ($p = 0$). The steady measured flux rate $J_w = -20$ cm h^{-1}. Calculate the saturated hydraulic conductivity of the sand and the clay and the effective conductivity of the column.

3.4 A cylindrical soil column of 100 cm^2 cross-sectional area and 50 cm height is filled with homogeneous soil that is saturated, and 10 cm of water is continuously ponded on it. The steady-state volume flow rate Q through the soil is 1000 cm^3 h^{-1} (downward).

(a) Calculate the steady-state flux through the column.

(b) Calculate the saturated hydraulic conductivity of the soil. A 1-mm-diameter tube is pushed through the column and hollowed out. In steady state, water flows through the soil in accordance with Darcy's law (3.13) and through the tube in accordance with Poiseulle's law (3.9).

(c) Calculate the volume flow rate through the tube.

(d) Calculate the flux through the column–tube system (Q_{tot}/A_{tot}).

(e) Calculate the effective saturated hydraulic conductivity of the column–tube system.

3.5 A clay lens of $K_c = 0.1$ cm day^{-1} is sandwiched somewhere inside a soil column of height 100 cm that is otherwise filled with sand of $K_s = 200$ cm day^{-1}. A height of water d $= 10$ cm is ponded above the surface.

(a) If the clay lens were not present, what would the steady flux rate be?

(b) If the actual flux rate is $J_w = -15$ cm day^{-1}, how thick is the crust?

(c) Calculate the pressure at the clay–sand interface for two special cases (1) clay on top of the sand and (2) clay on the bottom of the sand.

(d) Discuss the physical implications of your results in part (c).

3.6 A matric potential–water content function is fit to the following theoretical curve:

$$h(\theta) = -30 \left(\frac{0.25}{\theta^2} - 1 \right)^{1/2} \quad \text{for } 0.10 \leqslant \theta \leqslant 0.50$$

(a) Calculate $h(\theta)$ at $\theta = 0.1, 0.2, 0.3, 0.4, 0.5$.

(b) Using the five regions of water content, given in (a) calculate the tube radii and the number of tubes per area of a model porous medium of capillary tubes that would have the same $h(\theta)$ curve ($\sigma = 72$ dyn cm^{-1})

(c) Under a unit hydraulic head gradient, calculate the fraction of the total flow carried by each group of tubes if the model porous medium is saturated.

3.7 (a) Calculate the integral form of the Buckingham–Darcy flux law for steady-state horizontal flow of water from the differential form

$$J_w = -K(h)\frac{dh}{dx}$$

in an analogous manner to the derivation of (3.43).

(b) Calculate an equation relating the maximum horizontal evaporation rate E to the column length L if the left side of the column is kept saturated and

$$K(h) = \frac{K_s}{1 + (h/a)^2}$$

(c) Compare this answer with the corresponding result for the vertical evaporation case [(3.49) with $N = 2$]. Which E is greater for a given L? Why?

3.8 A root zone of thickness L is receiving high-frequency water irrigation at a rate $J_w = -i_0$ at the surface $z = 0$. Assuming that the water uptake distribution is

$$r_w(z) = a(z + L) \qquad -L \leqslant z \leqslant 0$$

and that steady state has developed, calculate the water flux as a function of depth z. Relate the parameter a to the total water loss rate ET (evapotranspiration) in the root zone.

3.9 A vertical soil column is initially saturated to $\theta_s = 0.4$ and rains uniformly [$\theta(t)$ is same at all locations] over its entire height $L = 50$ cm. The soil has a hydraulic conductivity–water content functional form given by

$$K(\theta) = K_{sat} \exp\left[\beta(\theta - \theta_s)\right]$$

where

$$K_{sat} = 100 \text{ cm day}^{-1} \quad \text{and} \quad \beta = 20$$

Use the gravity flow model,

$$J_w \approx -K(\theta)$$

(a) Calculate the average water content of the soil as a function of time.

(b) What is the predicted water content at $t = \infty$.

(c) How long does it take to drain the column to $\theta = 0$?

(d) Discuss the physical significance of (b) and (c). Why did (b) occur?

3.10 A falling-head permeameter is constructed by ponding 20 cm of water at $t = 0$ above a saturated soil column of height $L = 100$ cm. The ponded water, which is held above the column in a chamber of the same area as the column, falls to 0 cm in 1 h. Calculate K_s for this column. Next repeat the analysis approximately by pretending that the water flux through the column was constant and equal to $J_w = 20$ cm h^{-1} and that the height of ponded water was held at 10 cm for the whole experiment. Compare the result calculated by Darcy's law with these approximations to the exact solution and calculate the percentage of error.

3.11 Repeat problem 10 for the case where the soil column is only 5 cm high. Explain the reason for the difference in these two results.

4 The Field Soil Water Regime

Water transport and retention processes in field soils are very complex. The soil surface regime is exposed continually to changing radiation fluxes, which create diurnal cycles of temperature, relative humidity, evaporation, and even water vapor fluxes caused by temperature gradients (see Chapter 6). In addition, water inputs to the soil profile from rainfall or irrigation are often sporadic, causing continual changes in moisture content near the surface.

Despite this complexity, various soil water transport and retention processes such as infiltration, drainage, and evaporation may often be described with conceptual or even quantitative models derived from soil water theory and/or experimentation. This chapter will discuss the principal soil water processes associated with the field regime and introduce conceptual models characterizing their behavior.

4.1 FIELD WATER BALANCE

The general equation describing the water balance at the soil surface may be expressed as

$$P + I - R = ET + D + \Delta W \qquad (4.1)$$

where the terms on the left hand side of (4.1) represent the precipitation P (including dew and frost), applied irrigation water I, and surface runoff R. The sum of these three terms represent the net addition of water to the soil profile over a time period of interest. On the right hand side of (4.1) are evapotranspiration ET, drainage or deep percolation D, and the water storage change ΔW of the soil profile.

Each of the terms in (4.1) represent water flows or storage changes over some arbitrary time interval (e.g., 1 day). All of the terms in (4.1) are positive except for D and ΔW, which may be either positive or negative. A negative value for the drainage term implies that water is flowing upward[1] into the profile where the water balance is conducted, which will occur whenever the hydraulic head gradient is negative. This upward flux could be quite significant in certain situations, such as if a water table is located at a shallow depth below the surface profile, and there is no input of water to the soil for a prolonged time period (See Example 3.8).

---

[1] This sign convention differs from the one used in this book for fluxes, because the word drainage connotes downward flow. Thus, a positive drainage flux is a negative water flux.

The various terms in the water balance equation (4.1) are of considerable importance in disciplines other than soil science. For example, hydrologists must know what fraction of the incoming precipitation will result in direct runoff and in deep percolation to groundwater. Meteorologists are very interested in the estimate of evapotranspiration, since this component of the water balance has a significant influence on the energy budget of the earth. Plant scientists and agronomists are concerned both with evapotranspiration losses by crops and with root zone profile storage changes to determine optimum moisture conditions for plant roots.

In this chapter we will consider each of the terms in the water balance equation using models for these terms derived from the soil water transport equations introduced in the previous chapter wherever possible. Much of the analysis will involve making simplifications to construct approximate models of the flow regime. However, these simplified approaches are based on sound physical principles. Therefore, results of the approximate analysis will provide valuable insight into the behavior of the real system.

4.1.1 Analysis of Field Water Content and Matric Potential Profiles

The last term in (4.1), which represents the water storage change over some time interval, is limited in magnitude by the maximum amount of water (porosity times profile thickness) that can be held by the soil profile. Moreover, if an average water balance is calculated over a long period of time, this term will be less significant than the others. In fact, as long as the input of water from rainfall or irrigation is reasonably regular (i.e., no prolonged drying cycles), the actual soil profile water content in the field will tend to fluctuate about an equilibrium or steady-state value; consequently, over long time periods the storage change may be neglected. Although a true steady-state condition in the field may be more hypothetical than real, changes occurring below the surface zone are frequently very small, and the steady-state model provides a reasonable approximation to the real system.

In practice, it is usually possible to measure P, I, and R with adequate precision in the field. The profile water content and its changes ΔW can also be measured by straightforward means (e.g., a neutron probe). On the other hand, evaporation and drainage are very difficult to measure accurately in the field, and imprecise estimates of these two quantities often lead to erroneous conclusions concerning their relative importance in a given application.

Some idea of the difficulties that may be encountered in the interpretation of field data are demonstrated by Fig. 4.1, taken from Holmes and Colville (1970). In this figure, the soil water content depth profiles under two forested sites are shown at several different times of the year. Fluctuations in soil water content in response to rainfall and evaporation are very pronounced near the soil surface but diminish significantly at greater depths. However, both profiles show some evidence of a storage change deep in the soil during the monitoring period.

By themselves, the water content data do not allow quantitative conclusions to be drawn about the flow processes occurring deep in the soil. However, the water content measurements present a clearer picture of the wetting and drying cycles

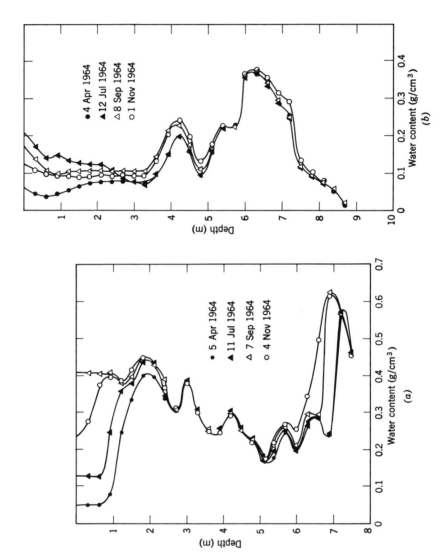

Figure 4.1 Soil water content profiles at various times during the year: (*a*) Young sand, Mount Gambier Forest; (*b*) Kalangadoo sand, Penola Forest. (After Holmes and Colville, 1970.)

when they are combined with water potential measurements in the soil profile. Figure 4.2 shows the matric potential depth distribution at the two sites for the wettest and driest sampling dates. At the Mount Gambier Forest site, it is obvious that there is very little change in matric potential between the 1- and 6-m depths. Hence, the hydraulic head gradient is almost equal to unity (i.e., gravity flow) and the direction of water movement must be downward. There is an upward hydraulic head gradient near the soil surface in April, implying that water is moving toward a dry surface and evaporating. The increase in matric potential between the 7- and 9-m depths probably represents a pulse of water that entered the profile at an earlier time.

The situation in the Penola Forest is significantly different. The hydraulic head gradient is downward between the surface and the 3-m depth and upward between the 5- and the 6-m depths. Furthermore, there is a maximum in the matric potential at about 4 m and separate minima at 5 and 6 m. This profile is probably the consequence of two prolonged drying periods with a period of excessive rainfall between them. This profile is obviously experiencing transient water flow, because the water content and matric potential at a given point vary significantly between the sampling periods. Moreover, water is clearly moving upward through some regions and downward through others. Consequently, some of the profile dried out between May and July, and part of it became wetter. At the Mount Gambier site, on the other hand, the profile appears to be much closer to a steady state, having a net downward flow of water that causes only minor storage changes over time.

4.1.2 Equilibrium and Steady-State Profiles

In Chapter 2, soil water systems at equilibrium, that is, those in which no water flow occurred, were analyzed. In a vertical unsaturated profile at equilibrium, with no solute membranes or air pressure changes, the total potential head h_T is given by

$$h_T = \text{const} = h + z \qquad (4.2)$$

In particular, with a water table located at $z = -L$, the matric potential distribution for an equilibrium soil profile is given by

$$h = -(z + L) \qquad -L \leqslant z \leqslant 0 \qquad (4.3)$$

Thus, the matric potential at any known height above a water table may be used to indicate whether water is flowing upward or downward or the profile is at equilibrium. For example, if a water table is located at a depth $z = -200$ cm and the matric potential at $z = -100$ cm is greater (more positive) than -100 cm, the flow will be downward.

As was shown in Example 3.9, matric potential gradients are close to zero during steady downward water flow, particularly far away from the water table (see Fig. 3.18). In this situation, the Buckingham–Darcy flux equation (3.27) re-

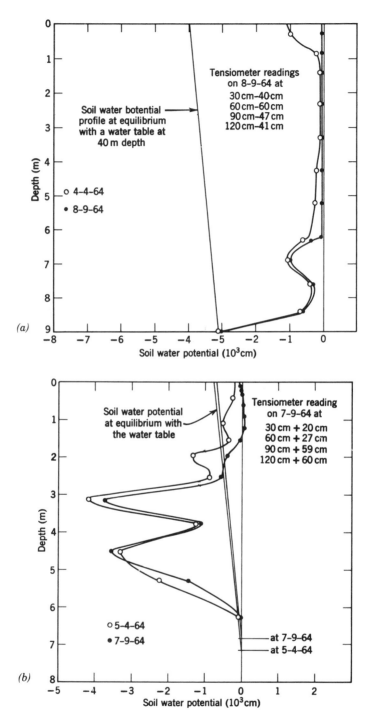

Figure 4.2 Matric potential profiles at the wettest and driest times of sampling for the sites in Fig. 4.1: (a) Mount Gambier Forest; (b) Penola Forest. (After Holmes and Colville, 1970.)

duces to the gravity flow equation

$$J_w = -K(\theta) \tag{4.4}$$

where $K(\theta)$ is the unsaturated hydraulic conductivity. Thus, during prolonged downward flow, the soil water content adjusts to that value necessary to drain the soil under gravity at the imposed rate. This model is consistent with the situation under the Mount Gambier Forest, as shown in Figs. 4.1 and 4.2. Although in the absence of any other evidence one might be tempted to conclude that no water was flowing through the 3–5-m region where the water content was not changing with time, it is far more likely that a steady downward flux of water is continually moving through this region, causing recharge of the dry profile below.

Another situation where gravity flow might approximate drainage reasonably well is in high-frequency irrigation, where water is applied frequently to crops to prevent salt buildup and to allow less water to be used (Rawlins and Raats, 1975). Under these circumstances, profile storage changes below the root zone tend to be minimal and a steady-state gravity flow model can be used to describe flow (Rawlins, 1974).

As shown in the previous chapter, steady-state downward flow through homogeneous soil results in a gravity flow process obeying (4.4) over much of the region above the water table. Hence, the drainage flux must be less than the saturated hydraulic conductivity or the soil will saturate. The situation in layered soils is more complicated. It is possible to apply the Buckingham–Darcy flux equation (3.27) to layered soils to calculate a matric potential–depth distribution, provided that the $K(h)$ relation is known for each layer and both the potential and flux are continuous across the interface between layers. Sometimes the saturated hydraulic conductivity of a given layer will be less than the applied water flux, even though other layers have a higher conductivity. When this happens, a perched water table forms over the impeding layer and positive pressure builds up in the saturated soil region, forcing water to flow at a rate greater than the saturated hydraulic conductivity of the impeding layer. Because of the pressure buildup below it, the layer above the impeding layer will have a hydraulic head gradient less than unity. However, at the same time it will have a higher water content than it would have under gravity flow at the imposed flux rate, so that it will have an unsaturated hydraulic conductivity sufficiently large to transmit the water flow at the reduced gradient. For an example of water flow through layered soil see Srinalta et al. (1969).

In the field, the surface layer of the soil is frequently less permeable than the subsurface horizons, particularly when a surface crust forms. In this case, if a thin layer of water is ponded on the soil surface, the (steady-state) flux through the soil will be less than the saturated hydraulic conductivity of the lower horizons but greater than the saturated conductivity of the surface crust. In the lower horizons, water is driven downward both by gravity and by capillary attraction to the drier lower soil, and the hydraulic head gradient is greater than unity in the region near the surface. If the relation between the unsaturated hydraulic conductivity and the matric potential is known for the soil below the crust, it is possible to calculate the steady-state flux into the soil, provided that the surface crust conductance can be measured or estimated. An analysis of this problem is given by Hillel and Gardner (1969).

A common example of a situation in which a saturated soil region lies on top of an unsaturated one is the profile beneath ponds and canals, where a subsaturated layer of soil may be present between the pond or canal bottom and the saturated region below the water table, even though there is a steady percolation of water from the pond to the water table. A numerical prediction of the water flow lines and water content distribution for a filled canal was obtained by Jeppson and Nelson (1970), who solved the two-dimensional form of the water flow equation for a spatially variable hydraulic conductivity function (Figure 4.3).

4.1.3   Transient Flow Processes in the Field

Because of the many time-dependent water and energy flows moving into the soil through its upper boundary, neither soil water equilibrium nor steady-state water flow ever occur in the field, except approximately below the surface when the boundary influences are periodic or when the surface is exposed to uniform conditions. When all of the time-dependent processes and heterogeneous soil properties of the field soil moisture regime are considered, the flow becomes difficult to describe quantitatively, both because certain processes like hysteresis are difficult to include in the water flow equations and because data characterizing the hydraulic and retention properties of the soil are rarely available at more than a few locations in the soil.

Two different approaches have been taken in developing approximate models of the field water flow regime. The first, the so-called empirical approach, uses mathematical functions that are fitted to experimental observations in the field with only minor concern for the physical principles involved. The second method, which might be called the mechanistic approach, uses models derived from approximate or exact solutions of the water flow equation (3.73) or (3.76). Because the Richards equation is based on conservation and flux laws that are exact in homogeneous media, solutions to practical problems derived from it give insight into the manner in which various soil physical properties influence water movement. Even though it is difficult to apply the flow theory to large heterogeneous areas without considerable simplification, we will adopt the mechanistic approach wherever possible. For the more widely studied processes, some of the more useful empirical expressions will also be discussed.

4.2   INFILTRATION

Infiltration refers to the entry of water into a soil profile from the boundary. Generally, it refers to vertical infiltration, where water moves downward from the soil surface. Since infiltration causes the soil to become wetter with time, water at the leading edge of the wetting pattern advances into the drier soil region ahead of the

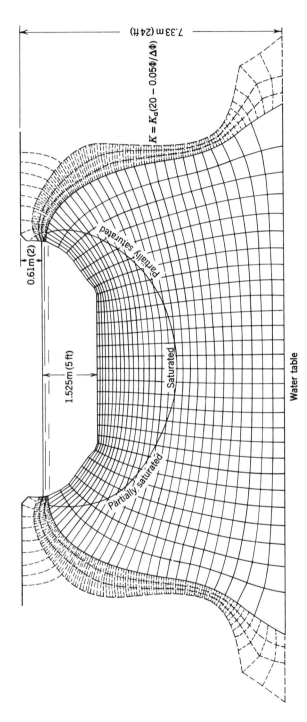

Figure 4.3 Flow net of solution for partially saturated seepage from a canal. The permeability of the soil decreases toward the canal by $K = K_a(2.0 - 0.05\phi/\Delta\phi)$ and consequently is 3.64 times as great at the water table as adjacent to the canal boundary. (After Jeppson and Nelson, 1970.)

front under the influence of matric potential gradients as well as gravity (if infiltration is vertical). During the early stages of infiltration when the wetting front is near the surface, the matric potential gradients predominate over the gravitational force.

The mathematical theory of vertical infiltration, based upon the solution of the Richards equation (3.73), is presented in detail by Philip (1969). Since this mechanistic infiltration model is derived from the physically based water flow equation, it gives considerable insight into the processes governing infiltration and will be presented in some detail in this chapter. However, there are many other vertical infiltration models in use today, the majority of which have been derived empirically from field data. These models all share the common feature that the infiltration rate is highest when water first enters the soil and decreases with time as the wetting front moves away from the surface. The empirical approach to the development of field infiltration equations consists of first finding a mathematical function whose shape as a function of time matches the observed features of the infiltration rate and then attempting a physical explanation of the process. In contrast, the mechanistic approach consists of solving the water flow equation to derive an expression for the infiltration rate. The time dependence of this expression can be interpreted physically as a consequence of the decreasing influence of the matric potential gradient as the wetting front moves farther from the surface during infiltration (Philip, 1969).

4.2.1 Empirical Infiltration Models

Kostiakov Equation   Kostiakov (1932) proposed the infiltration equation

$$I = \gamma t^{\alpha} \tag{4.5}$$

where I is the cumulative amount of infiltration between time zero and t and γ and α are constants. The parameters in (4.5) have no particular physical meaning and are evaluated by fitting the model to experimental data.

By definition, the infiltration rate $i = dI/dt$. Thus, the Kostiakov infiltration rate equation is given by

$$i = \alpha \gamma t^{\alpha - 1} \tag{4.6}$$

Horton Equation   Horton (1933, 1939) was one of the pioneers in the study of infiltration in the field and developed an equation he felt both described the general features of infiltration in different soils and was consistent with his physical concept of the process. The infiltration rate is given in Horton's model by the equation

$$i = i_f + (i_0 - i_f) \exp(-\beta t) \tag{4.7}$$

where i_0 is the initial infiltration rate at $t = 0$, i_f is the final constant infiltration rate that is achieved at large times, and β is a soil parameter that describes the rate of decrease of infiltration. For purposes of comparison with (4.5), (4.7) may be integrated to produce the formula for cumulative infiltration:

$$I = i_f t + \frac{i_0 - i_f}{\beta} \left[1 - \exp\left(-\beta t \right) \right] \tag{4.8}$$

Horton (1940) felt that the reduction in infiltration rate with time after the initiation of infiltration was largely controlled by factors operating at the soil surface. They included swelling of soil colloids and the closing of small cracks that progressively sealed the soil surface. Compaction of the soil surface by raindrop action was also considered important where it was not prevented by crop cover. Horton's field data, similar to those of many other workers, indicated a decreasing infiltration rate for 2 or 3 h after the initiation of storm runoff. The infiltration rate eventually approached a constant value that was often somewhat smaller than the saturated hydraulic conductivity of the soil. Air entrapment and incomplete saturation of the soil were assumed to be responsible for this latter finding. Horton used an exponential function to describe the decreasing infiltration rate since it fit the data reasonably well.

4.2.2 Green–Ampt Infiltration Model

Green and Ampt (1911) derived an approximate mechanistic model of the infiltration process by making several simplifying assumptions about the wetting process during water infiltration. These assumptions were based on approximations of real soil behavior obtained from experience, so that the model they created still has instructive value today.

During an actual infiltration event in which the soil surface is held at a constant matric potential head h_0 with associated water content θ_0 (e.g., by ponding water over it), water enters the soil behind a sharply defined wetting front that moves downward with time (Fig. 4.4a). Green and Ampt replaced this process with one that has a discontinuous change in water content at the wetting front (Fig. 4.4b). In addition, they made the following assumptions: (i) The soil in the wetted region has constant properties (K_0, θ_0, D_0, h_0) and (ii) the matric potential head at the moving front is constant and equal to h_F.

These two assumptions allowed Green and Ampt to solve for the infiltration rate exactly, as shown in the next two examples.

EXAMPLE 4.1: Use the Green–Ampt model to calculate the infiltration rate into a horizontal soil column initially at a uniform water content θ_i that has a water content $\theta_0 > \theta_i$ and an associated matric potential h_0 maintained at the entry surface for all $t > 0$.

From the assumptions of the model, we may replace the wetted soil profile at time t with a uniformly wet region of thickness L. Because the hydraulic conductivity of the entire wetted zone is constant, the infiltration rate i may be calculated using

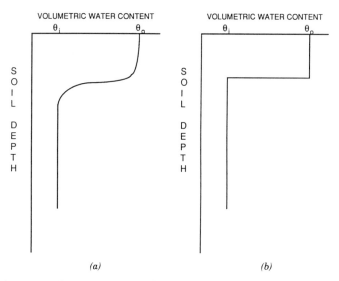

Figure 4.4 (*a*) A soil water content profile during infiltration. (*b*) Corresponding Green–Ampt profile.

Darcy's law (3.14) as

$$i = J_w = -K_0 \frac{h_F - h_0}{L} = K_0 \frac{\Delta h}{L} \tag{4.9}$$

where K_0 is the constant hydraulic conductivity of the "wet" region $0 < x < L$, $\Delta h = h_0 - h_F > 0$, and h_F is the matric potential of the moving front.

Furthermore, the infiltration rate is equal to the time rate of change of water storage in the soil, or

$$i = \frac{d}{dt} \left[(\theta_0 - \theta_i)L \right] = \Delta\theta \frac{dL}{dt} \tag{4.10}$$

where $\Delta\theta = \theta_0 - \theta_i > 0$.

When (4.10) is inserted into (4.9), we obtain a differential equation for the position L of the wetting front:

$$\Delta\theta \frac{dL}{dt} = K_0 \frac{\Delta h}{L} \tag{4.11}$$

This equation may be integrated (see Section 3.5) after all factors that depend explicitly on L are placed on one side. Thus,

$$\int_0^L L \, dL = K_0 \frac{\Delta h}{\Delta\theta} \int_0^t dt \tag{4.12}$$

These integrals are easily evaluated, with the result

$$\frac{L^2}{2} = K_0 \frac{\Delta h}{\Delta\theta} t \equiv D_0 t \tag{4.13}$$

where $D_0 = K_0 \, \Delta h / \Delta \theta$ is the soil water diffusivity of the wet soil region $0 < x < L$.

Finally, the cumulative infiltration $I = L \, \Delta \theta$ is equal to

$$I = \Delta \theta (2 D_0 t)^{1/2} \tag{4.14}$$

and the infiltration rate is

$$i = \frac{dI}{dt} = \Delta \theta (D_0 / 2 t)^{1/2} \tag{4.15}$$

In this model the infiltration rate into the soil is proportional to $t^{-1/2}$.

EXAMPLE 4.2: Repeat Example 4.1 for vertical infiltration. Let $z = 0$ at the soil surface.

The hydraulic head at the surface $z = 0$ is equal to $H_0 = h_0$, whereas at the front, $z = -L$ and $H_L = h_F - L$. Thus, Darcy's law across the wetted region is given by

$$i = -J_w = K_0 \frac{(H_0 - H_F)}{(0 - -L)} = \frac{K_0}{L} (\Delta h + L) \tag{4.16}$$

where, as before, $\Delta h = h_0 - h_F > 0$. Thus,

$$i = -J_w = \Delta \theta \frac{dL}{dt} = \frac{K_0}{L} (\Delta h + L) \tag{4.17}$$

which may be rearranged and integrated as follows:

$$\int_0^L \frac{L \, dL}{\Delta h + L} = \frac{K_0}{\Delta \theta} \int_0^t dt = \frac{K_0 t}{\Delta \theta} \tag{4.18}$$

The integral on the left side may be looked up in standard tables. It is equal to (*Handbook of Chemistry and Physics*, 1987)

$$\int_0^L \frac{L \, dL}{\Delta h + L} = L - \Delta h \ln (1 + L/\Delta h) \tag{4.19}$$

Thus, since $I = \Delta \theta L$, (4.18) may be rewritten as

$$I - \xi \ln \left(1 + \frac{I}{\xi} \right) = K_0 t \tag{4.20}$$

where $\xi = \Delta h \, \Delta \theta$.

For short times, soon after infiltration begins, I is small and the logarithm in (4.20) may be approximated by (*Handbook of Chemistry and Physics*, 1987)

$$\ln (1 + I/\xi) \approx I/\xi - I^2/2\xi^2 \tag{4.21}$$

Therefore, at short times (4.20) reduces to

$$I \approx (2\xi K_0 t)^{1/2} = \Delta \theta (2 D_0 t)^{1/2} \tag{4.22}$$

and

$$i \approx \Delta \theta (D_0 / 2 t)^{1/2} \tag{4.23}$$

which is the same as the horizontal infiltration result (4.15).

At very large times, the rate of change of the second term in (4.20) is small compared with the first and may be neglected. Therefore,

$$i \approx K_0 \qquad (4.24)$$

These results will be of interest in what follows when the exact model for infiltration is discussed.

The Green–Ampt model, because it uses an approximate description of the actual flow regime, has parameters like h_F, which cannot be directly measured. Therefore, it has primarily been used as a conceptual aid in visualizing a complex process, although indirect evaluation of h_F has permitted the Green–Ampt model to be used in certain practical applications (Chong et al., 1982).

4.2.3 Philip Infiltration Model

Although the mathematical theory of water flow was fully developed by the early part of the twentieth century, the Richards equation (3.73) could not be solved without approximations because the equation was nonlinear. Thus, many of the subtleties of unsaturated water flow could not be investigated theoretically. In 1957, J. R. Philip devised a numerical technique for solving the flow equation exactly for a mathematical study of an important practical problem: infiltration into an infinitely deep homogeneous porous medium at a uniform initial water content θ_i that has its boundary (i.e., the soil surface) held at a higher water content $\theta_0 > \theta_i$ (Philip, 1957a–f).

Horizontal Infiltration   The first problem that Philip addressed was that of horizontal infiltration (no gravity). Using the Richards equation (3.73) [with $K(\theta)$ removed] he showed that the infiltration rate i is given exactly by

$$i = \tfrac{1}{2} S t^{-1/2} \qquad (4.25)$$

where $S = S(\theta_0, \theta_i)$ is called the sorptivity and is a function of the boundary and initial water contents θ_0 and θ_i. It is constant during a given experiment in which the inflow end of the uniform horizontal soil column is held at a constant water content. For a given initial water content, S increases as θ_0 increases (Fig. 4.5).

Since S is constant over time, cumulative horizontal infiltration I is given by

$$I = S t^{1/2} \qquad (4.26)$$

Thus, the sorptivity S may be measured simply by determining the slope of I versus $t^{1/2}$.

Vertical Infiltration   The vertical infiltration solution appears as two separate expressions appropriate for short and long times after infiltration commences. The short-time solution is expressed as an infinite series in powers of $t^{1/2}$ as

$$I = S t^{1/2} + A_1 t + A_2 t^{3/2} + \cdots \qquad (4.27)$$

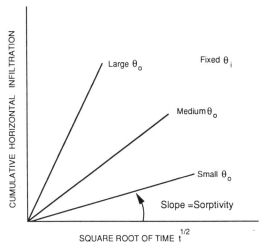

Figure 4.5 Cumulative infiltration into a horizontal soil column of fixed θ_i for various θ_0.

where S is the sorptivity, which appears in the horizontal infiltration solution (4.26), and A_1, A_2, $\cdots$ are constants that depend on the soil properties and θ_0 and θ_i. Philip (1957a) indicated how these constants (and S) could be calculated from $D(\theta)$ and $K(\theta)$.

The Philip model is generally approximated by just the first two terms of (4.27),

$$I = St^{1/2} + At \tag{4.28}$$

with a corresponding infiltration rate

$$i = \tfrac{1}{2} St^{-1/2} + A \tag{4.29}$$

which is valid during the early stages of infiltration.

The long-time infiltration solution calculated by Philip (1957b) has the following properties:

(i) The infiltration rate approaches a constant equal to the value $K(\theta_0)$ of the unsaturated hydraulic conductivity at the surface water content. This is *not* the same as A in (4.29),

$$i \rightarrow K(\theta_0) \tag{4.30}$$

(ii) The wetting front advances without changing its shape.

(iii) The velocity of the moving wetting front approaches a constant value given by

$$V_F = \frac{K(\theta_0) - K(\theta_i)}{\theta_0 - \theta_i} \tag{4.31}$$

Figure 4.6 shows predicted and measured water content profiles from an infiltration experiment reported by Davidson et al. (1963). A zone of almost constant water content extends immediately down from the soil surface. This is sometimes referred to as the *transmission zone*. Both the observed and predicted wetting fronts are extremely steep below the transmission zone and move downward without changing shape. Such a steep wetting front is characteristic of infiltration into relatively dry soil and is easily discernable by eye.

It is obvious from this analysis why the assumption of a uniform water content and a constant hydraulic conductivity above a sharp wetting front was used in the

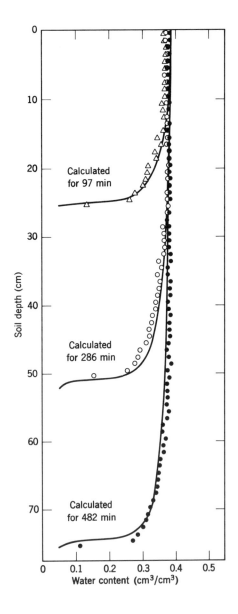

Figure 4.6 Calculated and measured soil water profiles for air-dry Hesperia soil allowed to wet at $\theta_0 = 0.385 \text{ cm}^3 \text{ cm}^{-3}$. (After Davidson et al., 1963.)

approximate infiltration model of Green and Ampt. Because these assumptions maintained a realistic approximate picture of the process, the infiltration equations (4.14) and (4.20) predicted by the Green–Ampt model have properties that are very similar to those calculated by the Philip model (4.26) and (4.28) from an exact analysis.

The rigorous analysis that leads to (4.28) neglects two important factors that cannot always be overlooked in the field. These are the possible entrapment of air in the soil profile during infiltration and the development of surface crusts that decrease the hydraulic conductivity with time. In fact, Horton (1940) attributed the falling infiltration rate to this latter phenomenon. Although the Philip analysis predicts that the infiltration rate must decrease with time due to the decreasing matric potential gradient, even in the absence of a crust, his model (4.28) tends to predict a much slower rate of decrease that continues for a much longer period of time than is usually found with field data. For example, the Horton equation (4.7) generally decreases to a constant value in 2–3 h when used with field-calibrated parameters, whereas the Philip equation (4.28) may continue to decrease for days (Philip, 1969). Green et al. (1970) conducted a theoretical analysis of infiltration that took air entrapment into account and found that this could explain discrepancies between theory and experiment. A comparison of their model predictions to experimentally determined water content profiles during infiltration in the field is given in Fig. 4.7.

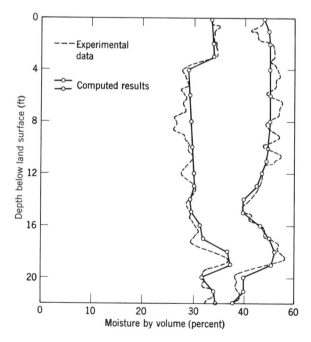

Figure 4.7 Computed and measured soil moisture profiles before and after 51.5 h infiltration using two wetting capillary pressure curves and a variable porosity with depth. (After Green et al., 1970.)

4.2.4   Infiltration into Nonhomogeneous Soil Profiles

One reason that field infiltration data frequently show different characteristics than the models based on theoretical calculations is that field soil profiles are seldom uniform with depth, nor is the water content distribution uniform at the initiation of infiltration. These two effects usually tend to reduce the infiltration rate more rapidly than would be predicted from a model that assumes that the soil was homogeneous.

Another factor sometimes present in the field that is not included in the Philip model is a shallow water table. The potential influence of the water table on water flow and storage changes during infiltration has been investigated in some detail by Freeze (1969). Figure 4.8 shows the matric potential head, hydraulic head, and water content as a function of depth for successive times during infiltration into three different soils in the presence of shallow water tables. These curves were obtained by numerical solution of the water flow equation.

In dealing with nonhomogeneous profiles, it is usually most convenient to divide the profile into layers or horizons, each of which is assumed to be homogeneous. Childs and Bybordi (1969) extended the Green–Ampt approach to stratified soils in which the hydraulic conductivity decreases with depth. Hillel and Gardner (1969, 1970) used a similar approach to study infiltration into soils that have an impeding crust on the soil surface. From this analysis they determined that in the presence of a crust cumulative infiltration increases during the early stages of infiltration at a rate proportional to the square root of time, which also occurs in crust-free soil. At sufficiently large values of time, cumulative infiltration is described approximately by the expression

$$I = K_f t + E \ln (1 + Ft) \qquad (4.32)$$

where I is cumulative infiltration, K_f is the final steady-state infiltration rate, and E and F are parameters whose values depend upon the properties of the soil and of the surface crust.

When a soil layer of different texture and permeability from the surface layer is present in the soil profile, it will reduce the infiltration rate, regardless of whether it is coarser or finer than the surface layer. If the texture is finer, the reduction in infiltration is due directly to its lower permeability. In contrast, a subsurface coarse-textured layer generally has a saturated hydraulic conductivity that is greater than the finer textured layer above it. However, the low matric potential at the wetting front prevents the large, highly conducting pores of the coarse-textured region from filling. The unsaturated conductivity of the resulting partially saturated, coarse-textured region is actually lower than the wetter finer textured region above, and the infiltration rate decreases as the front reaches the interface.

Both situations are illustrated in Fig. 4.9, in which the infiltration rate is plotted as a function of time for layered soil experiments (Miller and Gardner, 1962). In the first case (sand over clay), when the wetting front reaches the buried clay layer, the infiltration rate is immediately reduced and continues to decrease. In the second

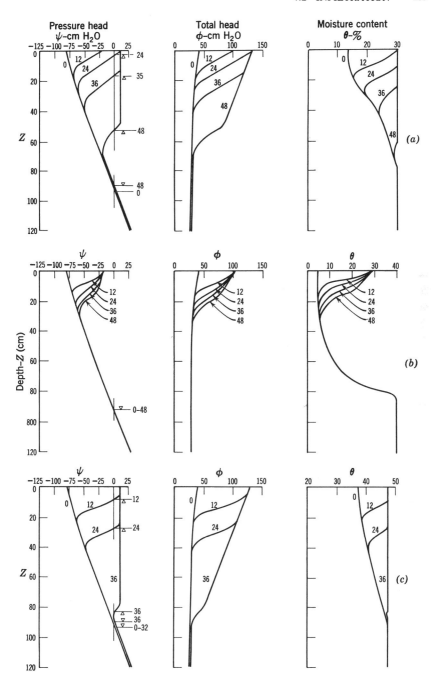

Figure 4.8 Effect of soil type on groundwater recharge and discharge for three soils of different permeability. In each case the infiltration rate $R = 0.1315$ cm min^{-1} is 100 times the recharge rate Q: (a) Del Monte sand; (b) Rehovot sand; (c) Grenville silt loam. (After Freeze, 1969.)

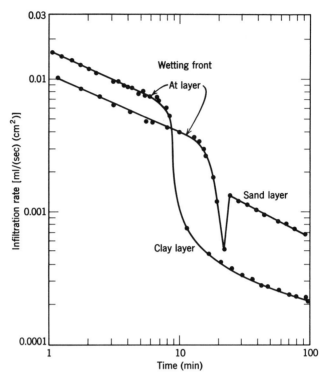

Figure 4.9 Effect of sand and clay layers on infiltration rate into Palouse silt loam as a function of time. (After Miller and Gardner, 1962.)

case (clay over sand), there is an immediate sharp reduction in infiltration rate when the front reaches the sand. However, as the water "piles up" at the interface, the matric potential head at the front increases, which allows larger pores in the sand layer to fill. This raises the conductivity of this layer and allows the infiltration rate to increase. However, the rate does not increase enough to offset the initial decrease, and the new rate is lower than that which would exist in the absence of the sand layer.

Layered soils are also known to induce unstable flow when a fine-textured region lies over a coarse-textured one (Raats, 1973). During infiltration into dry soil layered in this manner, water cannot enter the coarse-textured zone until the pressure has built up sufficiently to wet the larger pores. If this occurs at discrete locations along the wetting front, the new wetted channels in the coarse-textured zone may become conduits for all of the water entering from above. These narrow-flow channels, called fingers, can persist through the entire coarse-textured zone. Since the local flux in these channels can be much higher than the average flux of the continuous front in the fine-textured zone above, the water velocity can be appreciable. A review of flow instabilities in soil is presented in Hillel (1986).

4.2.5   Infiltration When Rainfall Is Limiting

The infiltration models described thus far have all assumed that the surface is maintained at a fixed potential rather than the more usual case where water flux is held at a fixed rate. Under rainfall irrigation, for example, the only time that the surface would be exposed to a fixed potential is if the rainfall rate is so intense that ponding (i.e., positive-pressure head) begins immediately. However, the initial capacity of soil to absorb incoming water is generally very high, and ponding will not commence immediately. In fact, if the rainfall rate never exceeds the final gravity-dominated infiltration rate of a soil, there will never be any runoff.

When the applied rainfall rate is less than the initial infiltration capacity of the soil but greater than the final steady-state, gravity-dominated rate, a transition must occur. The soil will absorb the incoming rainfall until the matric potential gradients near the surface diminish to a point where the water cannot be taken up by the soil profile as fast as it is entering through the surface. At this time, the soil near the surface saturates and ponding develops.

There is no easy way to estimate the time at which runoff begins when a dry soil profile is exposed to prolonged rainfall. Figure 4.10 shows a plot of the predicted infiltration rate as a function of time using a numerical solution of the water flow equation for both ponded surface conditions (dashed curve) and constant-rainfall inputs (solid curves). As shown in this figure, ponding will not occur under constant-rainfall input until some time after the applied rate exceeds the value the infiltration rate would have if the soil were continually ponded. Furthermore, more water will enter the soil prior to ponding when the rainfall rate is lower than when it is higher. Various approximate models have been proposed to determine the ponding time (Kutilek, 1980; Parlange and Smith, 1976; Boulier et al., 1987).

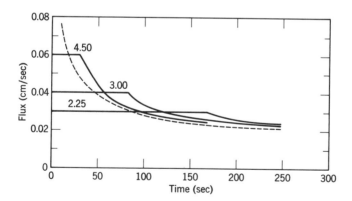

Figure 4.10   Relation between surface flux and time during infiltration into Rehovot sand due to rainfall (solid lines) and flooding (dashed line). The numbers labeling the curve indicate the ratio of the rainfall rate to the saturated hydraulic conductivity. (After Rubin, 1968.)

EXAMPLE 4.3:   Assuming that the ponded (maximum possible) infiltration rate into a soil is given by the Philip model (4.29), calculate (i) the time at which the ponded infiltration rate equals the rainfall rate P and (ii) the amount of time required for a process with a constant rainfall rate R to add the same total amount of infiltration water to the soil as in the ponded process.

The infiltration rate in the Philip model is given by (4.29). Therefore, the time at which this equals the rainfall rate P is given by the solution to

$$\tfrac{1}{2}S/\sqrt{t} + A = P \tag{4.33}$$

Solving for t in (4.33), we obtain the time at which the two rates are equal:

$$t = \frac{S^2}{4(P - A)^2} = t_{min} \tag{4.34}$$

Since this time is less than the ponding time (see Fig. 4.10), it will be called t_{min}.

Cumulative infiltration under ponding, (4.28), equals the cumulative rainfall Pt at a time satisfying

$$S\sqrt{t} + At = Pt \tag{4.35}$$

Solving for t in (4.35), we obtain the time at which the two processes have infiltrated the same amount of water:

$$t = \frac{S^2}{(P - A)^2} = t_{max} \tag{4.36}$$

Because continuous ponding of the soil surface results in the maximum possible cumulative infiltration, ponding will occur under constant-rainfall infiltration prior to this time. Thus, the time given in (4.36) is an overestimate and will be called t_{max}.

The maximum infiltration rate curve shown in Fig. 4.10 may be represented adequately by the Philip model (4.29) with $S = 0.4512$ cm s$^{-1/2}$ and $A = 0.0082$ cm s^{-1}. Table 4.1 summarizes the maximum and minimum ponding time estimates using (4.34) and (4.36) together with the time at which ponding was predicted to occur using the numerical model. This time is intermediate between the two extremes and is somewhat closer to the minimum time than the maximum.

4.2.6   Two- and Three-Dimensional Infiltration

The previous discussion dealt with one-dimensional infiltration in which water is assumed to flow vertically into the soil. Several two- and three-dimensional infiltration models are of interest, since they are more closely related to certain field

TABLE 4.1   Maximum, Minimum, and Calculated Ponding Times for Rehovot Soil Shown In Fig. 4.10.

P	t_{max}	t_{min}	$t_{ponding}$
0.06	73	18	30
0.04	189	47	84
0.03	391	98	170

infiltration measurements and irrigation methods than are the one-dimensional models.

Infiltration from a water-filled semicircular furrow and from a hemispherical cavity were studied in some detail by Philip (1968). Using solutions to the water transport equation, he showed that in both cases the infiltration rate is very high initially and decreases until it approaches a constant rate. This constant rate can be estimated if the dimensions of the cavity or furrow are specified and the hydraulic conductivity of the soil is known. Talsma (1969) found good agreement between field measurements and the predicted rate using Philip's treatment of the problem.

The shape of the wetting front below a local source of water such as a furrow depends very much upon the relative importance of matric and gravitational forces during infiltration. If the matric forces predominate, the wetting front tends to be symmetrical and moves as much laterally as vertically. If gravity is more important, as in the case of very coarse textured soils, then the wetting front is elongated and is more nearly ellipsoid in shape. This behavior is illustrated in Fig. 4.11, which shows lines of equal water content for large times during infiltration from a cylindrical cavity for sandy and clay soils.

There are some fundamental differences between one- and three-dimensional infiltration, which may be illustrated by looking at the large-time infiltration rates predicted by various models. Wooding (1968) derived an approximate expression for the steady rate of infiltration from a circular pond of radius r_0, overlying a soil whose hydraulic conductivity–matric potential function was assumed to be

$$K(h) = K_0 \exp(\alpha h) \tag{4.37}$$

where K_0 and α are constants representing the soil properties. Since $K = K_0$ when $h = 0$, K_0 represents the saturated hydraulic conductivity of a soil obeying (4.37). Using this expression and a simplified form of the three-dimensional water flow equation, Wooding derived the following equation for the steady infiltration flux rate:

$$i_f = K_0 \left(1 + \frac{4}{\pi \alpha r_0^2}\right) \tag{4.38}$$

It is notable that, contrary to one-dimensional flow, the final rate exceeds K_0. This occurs because water may enter and move laterally as well as vertically.

A ring infiltrometer is often used to measure the infiltration capacity of surface soil. This is usually a metal ring 30–60 cm in diameter that is pressed a short distance into the soil and filled with water. The rate of water loss from the ring is taken as an estimate of the one-dimensional infiltration rate. Equation (4.38) gives an estimate of the error involved in this procedure. The constant α will seldom exceed 0.05 cm, and a value of 0.01 cm is more common. A large value for α represents a coarse-textured soil and a small value a soil of finer texture. For a soil in which $\alpha = 0.01$ and a ring diameter $r_0 = 30$ cm, the infiltration rate will be

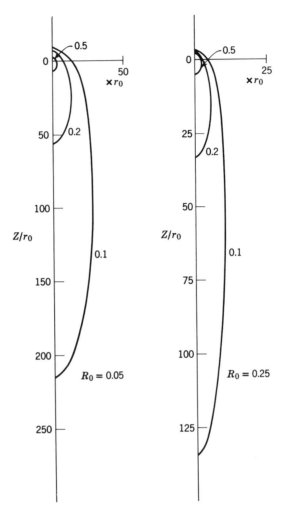

Figure 4.11 The limiting moisture distribution as $t \to \infty$ during infiltration from a cylindrical cavity. Numbers on the curves represent values of $(\theta - \theta_0)/\theta_1 - \theta_0)$. (After Philip, 1968.)

about 40% higher than the infiltration rate for ponded infiltration over a large area. Because of this large error, it is preferable to use a double-ring infiltrometer that has two concentric water-filled rings. The outer ring acts as a border, while the measurement is taken from the inner ring only.

4.3 REDISTRIBUTION

4.3.1 Redistribution of Water in Soil Profiles

The term *redistribution* refers to the continued movement of water through a soil profile after irrigation has ceased at the soil surface. This is a complex process, because the lower part of the profile ahead of the front will increase its water content and the upper part of the profile near the surface will decrease its water content after infiltration ceases. Thus, hysteresis can have an effect on the overall shape and dynamic behavior of the water content profile.

Figure 4.12 shows experimental water content profiles during redistribution stages following three different irrigation cycles. Several general features of the water content profile are apparent from an examination of this figure. First, the water content of the wetted soil profile decreases relatively uniformly over space; there is no abrupt drying of the surface layer. Second, the front of the profile continues to move downward with a sharp boundary between the wet and dry soil regions. The overall change of profile shape is similar to that of a rectangle (whose height is wetted profile thickness and whose width is the difference between wet and dry region water contents) that becomes progressively taller and thinner over

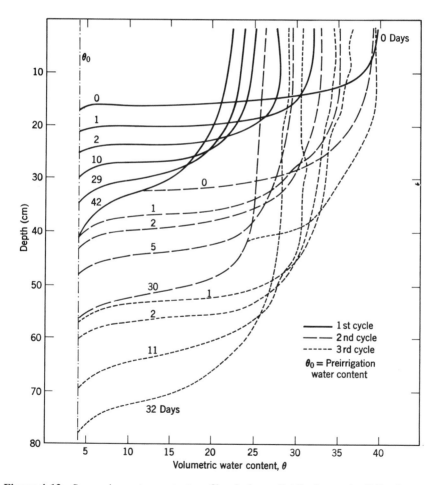

Figure 4.12 Successive water content profiles during redistribution cycles following one, two, and three irrigations of 5 cm each. (After Gardner et al., 1970.)

time while maintaining equal area. Also, because the water content is very uniform within the wetted region, the drainage water across a fixed plane in the profile may be described reasonably well by gravity flow. The "gravity-drained rectangle" model was used by Jury et al. (1976) to describe water redistribution during intermittent irrigation.

Because the water content of the wetted region is continually decreasing, the drainage rate also decreases and the profile slowly approaches equilibrium. Contrary to the equilibrium profile reached after a uniform wetting or drying process, in which water content increases with depth, the water content profile of a redistribution process is more uniform with depth because of hysteresis. As shown in Fig. 4.12, the equilibrium water content for a given matric potential is higher for a drying process than for a wetting process. Thus, the equilibrium profile, where $dH/dz = 0$ or $h = -z$, will have a smaller (more negative) value of matric potential near the surface than deeper in the soil. However, the water content of the surface layer results from a drying process and therefore is larger for a given h than is the water content in the lower region, which results from a wetting process.

Although hysteresis has made exact analysis of redistribution difficult, some approximate models such as the gravity-drained rectangle (Jury et al., 1976) have been developed to produce simple expressions for water content–time relations that have similar features to those of the experimental profiles. For example, it has been observed by a number of workers that the decrease in average water content $\bar{\theta}$ over time in the zone above the wetting front is described accurately by an empirical equation of the form

$$\bar{\theta} = a(t + c)^{-b} \tag{4.39}$$

where a, b, and c are constants. Gardner et al. (1970) have shown that these constants can be related to the soil water diffusivity and permeability if certain simplifying assumptions are made. When the time t in (4.39) is longer than a day or so, the constant c can usually be neglected.

When the factor c may be neglected, (4.39) may be rewritten in the form

$$\log \bar{\theta} = \log a - b \log t \tag{4.40}$$

Hence, values of the constants a and b can be obtained directly from the slope and intercept of a logarithmic graph if the relationship in (4.40) is reasonably accurate.

Values of average water content in various layers of a sandy loam soil profile during drainage after irrigation (Ogata and Richards, 1957) are shown plotted in a log–log graph in Fig. 4.13. The solid lines represent the relation $\bar{\theta} = 0.257t^{-0.128}$ obtained by fitting the data to (4.40).

The rate of water loss from the wetted profile of thickness d, which is equal to the drainage rate, is obtained by differentiating (4.39) and multiplying by d, or

$$dw/dt = -bad(t + c)^{-(b + 1)} \tag{4.41}$$

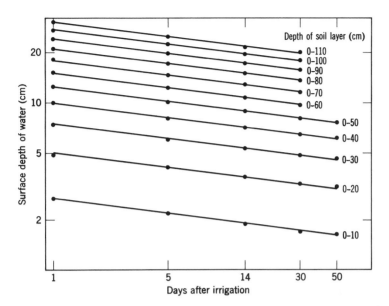

Figure 4.13   Surface depth of water in various layers as related to time (logarithmic scale). (After Ogata and Richards, 1957.)

where $w = \bar{\theta}d$ is the water stored in the draining profile. Since the parameters a and b may be obtained from a logarithmic plot if c is negligible (see (4.39)), the drainage rate may be computed directly.

EXAMPLE 4.4:   Use the gravity-drained rectangle model described in the preceding to derive an expression similar to (4.39), assuming that the unsaturated hydraulic conductivity–water content function is given by

$$K(\theta) = K_s(\theta/\theta_s)^N \qquad (4.42)$$

where K_s and θ_s refer to saturation and $N > 1$ is a constant. Assume initially that the water content profile is saturated to a depth $z = -L$ and that the water content θ_i of the dry layer beneath the wetting front is small ($\theta_i \ll \theta_s$). Thus, the water content–depth distribution at the onset of redistribution is given by

$$\theta = \theta_s \quad \text{if } -L < z < 0$$

$$\theta = \theta_i \approx 0 \quad \text{if } z < -L \qquad (4.43)$$

The gravity-drained rectangle model of redistribution uses two simplifying assumptions. First, the area within the profile "rectangle" is constant. Thus, for $t > 0$ when the water content has reached depth z and has fallen to $\theta < \theta_s$, the water stored in the rectangle is given by

$$L\theta_s = z\theta \qquad (4.44)$$

This is shown in Fig. 4.14. Second, the instantaneous drainage rate at the interface between the wet and dry zones is equal to the gravity flux (4.4). Thus, when the

VOLUMETRIC WATER CONTENT

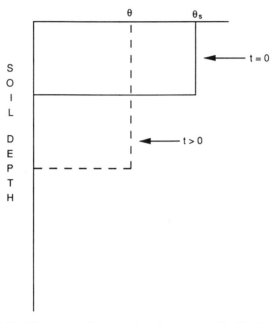

Figure 4.14 The rectangular approximation to the redistribution process.

profile has reached depth z, the mass balance equation for the rectangle, stating that the rate of change of water storage in a fixed soil volume of thickness z is equal to the flux out of the volume at z, is given by

$$z \frac{d\theta}{dt} = \frac{L\theta_s}{\theta} \frac{d\theta}{dt} = -K(\theta) = -K_s \theta^N / \theta_s^N \qquad (4.45)$$

Equation (4.45) may be integrated by placing all factors that depend on θ on the same side (Section 3.8). Thus,

$$\int_{\theta_s}^{\theta} \theta^{-(N+1)} \, d\theta = - \int_0^t \frac{K_s}{L\theta_s^{N+1}} \, dt \qquad (4.46)$$

These integrals may be looked up in any standard table. The result is

$$-\frac{1}{N} \left(\frac{1}{\theta^N} - \frac{1}{\theta_s^N} \right) = -\frac{K_s t}{L\theta_s^{N+1}} \qquad (4.47)$$

We may solve for θ in (4.47) by first rewriting it as

$$\frac{1}{\theta^N} = \frac{NK_s t}{L\theta_s^{N+1}} + \frac{1}{\theta_s^N} \qquad (4.48)$$

and then by inverting (4.48) to

$$\theta(t) = a(t + c)^{-b} \qquad (4.49)$$

where

$$a = (\theta_s^{N+1} L/NK_s)^{1/N} \tag{4.50}$$

$$b = 1/N \tag{4.51}$$

$$c = L\theta_s/NK_s \tag{4.52}$$

The position z of the front is given by (4.44) and (4.49).

This example illustrates how an empirical model (4.39) can be reproduced with a physically based model (gravity flow). It can be shown that for a fixed water table depth even very nonhomogeneous soils tend to drain in such a way that the rate of drainage out of the profile can be related in a simple fashion to the water content of the soil. This provides a useful simplification for the calculation of the field water balance where the assumption of a unique relation between the soil profile water content and the drainage rate is justified. Figure 4.15 shows the drainage rate measured with a weighing lysimeter plotted as a function of the average soil water content of the lysimeter for Plainfield sand (Black et al., 1969). The unsaturated hydraulic conductivity is also shown in the same graph and can be seen to be very nearly equal to the drainage rate, implying that the unit gradient drainage approximation works very well in this soil. Figure 4.16 compares the

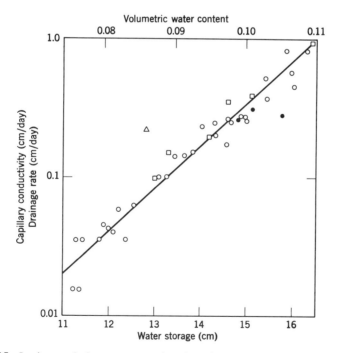

Figure 4.15 Lysimeter drainage rate as a function of water storage (circles) and unsaturated hydraulic (capillary) conductivity of Plainfield sand as a function of soil water content (squares). (After Black et al., 1969.)

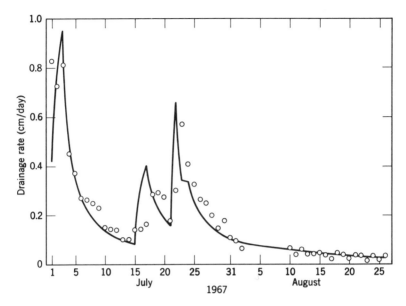

Figure 4.16   Predicted drainage (solid line) from bare Plainfield sand compared with that measured by lysimeter (circles). (After Black et al., 1969.)

drainage rate for the summer season for the soil profile predicted from the line in Fig. 4.15 with the actual drainage rate measured in the lysimeter.

4.3.2   Field Capacity Concept

Field capacity has been defined as the water content of the near surface soil profile (i.e., the root zone) at which drainage becomes negligible. It is an old concept, intended to provide a characteristic index of how much water may be retained from an irrigation or rainfall event after redistribution has ceased. Customarily, field capacity has been arbitrarily defined as the water content of the surface profile after two days of redistribution following infiltration.

In reality, a true field capacity cannot exist, because water will drain continually under gravity as long as no impermeable barrier is present in the soil. However, for many coarse-textured soils, the drainage rate falls to an insignificant level within a few days, after which the water content is changing at such a slow rate that a field capacity concept has practical value. Finer textured soils do not show this abrupt decrease in drainage rate, as evidenced by the sandy loam profile storage as a function of time shown in Fig. 4.13. This soil, as described by (4.39), will have an average water content of 0.235 after 2 days of redistribution. However, the average water content will fall to 0.191 after 10 days and to 0.175 after 20 days, showing that field capacity may not be precisely defined in such a soil. Although in principle one could decide that some specified small drainage rate is low enough to be negligible, in practice what constitutes a negligible drainage rate depends upon the particular problem under consideration. A drainage rate that may be considered unimportant relative to the daily evapotranspiration rate may be significant when summed over an entire year. Because the importance of drainage varies from application to application, no universal criterion may be developed to decide when to neglect downward drainage.

Many attempts have been made to define field capacity in terms of the amount of water retention at a particular matric potential, often to the $-\frac{1}{3}$-bar moisture percentage, or in older literature, to the moisture content of a soil sample wetted and subjected to a force of 1000 g wt in a centrifuge, which corresponds to about $-\frac{1}{3}$ bar. This definition neglects the evidence that water retention in a profile depends on the water transmission properties of the entire profile and on the hydraulic head gradient rather than only on the energy state of water at a particular point in the profile. For example, at drainage rates of a few millimeters per day, water potentials ranging from as high as -0.005 bar in highly stratified soils to as low as -0.6 bar in deep dry land soils have been observed in the surface rooting zone soil profile during redistribution. Furthermore, since many matric potential–water content desorption curves are very flat in this wet range (see Fig. 2.11), huge errors in water retention estimates may be expected from the arbitrary association of field capacity with a particular matric potential value. Furthermore, if one thoroughly irrigates a soil containing a restricting layer such as a hardpan below the surface layer, the surface soil will remain at a high water content and high matric potential for a long period of time after irrigation ceases entirely as a result of the impeding layer. This water content, which would represent the "field capacity" for that soil, would be much higher than the $-\frac{1}{3}$ bar water content of the surface layer.

4.4 FIELD MEASUREMENT OF UNSATURATED HYDRAULIC CONDUCTIVITY

Since it is difficult to create steady-state unsaturated flow in the field, the standard methods for measuring unsaturated hydraulic conductivity in situ involve transient flow experiments. Early field methods (Rose et al., 1965) involved simultaneous measurement of matric potential profiles with tensiometers and water content profiles with neutron probes on field plots undergoing redistribution following initial saturation. The Richards equation (3.68) was written in a finite difference form and solved for the average hydraulic conductivity of discrete layers that provided the best agreement between the observed changes in h and θ.

A problem with the preceding method is that the field soil profile may be heterogeneous, so that using the homogeneous Richards equation (3.68) to interpret the data may lead to inaccurate or even absurd (i.e., negative) values for K in certain locations within the profile. For this reason, $K(\theta)$ is often evaluated with simpler models that are not as sensitive to soil heterogeneity.

As shown in the preceding, experimental observations of redistribution profiles in homogeneous soils show a very uniform water content through the wetted zone.

This suggests that the gravity flow model may be used to describe profile drainage during redistribution without evaporation as

$$L \frac{d\bar{\theta}}{dt} = -J_w(L) \approx -K(\bar{\theta}) \tag{4.53}$$

where $\bar{\theta}$ is the average water content and L is the thickness of the profile. Nielsen et al. (1973) and Libardi et al. (1980) used (4.53) to measure $K(\theta)$ under field conditions. The latter authors also compared this method to the procedure used by Rose et al. (1965).

> *EXAMPLE 4.5:*   A soil profile of $L = 1.0$ m thickness is wetted up to saturation and the surface is covered. Neutron probe readings of water content are taken over time and are used to calculate the quantity of water, $\int_0^L \theta \, dz$, contained in the profile. Using the data given in Table 4.2, calculate the hydraulic conductivity of the soil as a function of water content.
>
> According to (4.53), the hydraulic conductivity is equal to $-L \, d\bar{\theta}/dt$, provided that the gravity flow approximation (4.4) is reasonably valid. The average water content $\bar{\theta}$ is given by
>
> $$\bar{\theta} = \frac{1}{L} \int_0^L \theta \, dz \tag{4.54}$$
>
> Thus, from Table 4.2 we can calculate each of the terms in (4.53) and (4.54), and finally $K(\bar{\theta})$, with the result shown in Table 4.3.

TABLE 4.2   Water Storage Values as Function of Time in 1-m Profile

Water Stored, $\int_0^L \theta \, dz$ (cm)	Time (h)
50	0
45	1
39	5
35	10
32	20
31	30

TABLE 4.3   Evaluation of $K(\Theta)$ Using Equations (4.53) and (4.54) and Table 4.2

$L\bar{\theta}$ (cm)	t (h)	$L \, \Delta\theta$ (cm)	Δt (h)	$\bar{\theta}$	$K(\bar{\theta})$ (cm/h)
50	0	−5	1	0.475	5.0
45	1	−6	4	0.42	1.5
39	5	−4	5	0.37	0.8
35	10	−3	10	0.335	0.3
32	20	−1	10	0.315	0.1
31	30				

Other indirect field methods for measuring unsaturated hydraulic conductivity have been proposed recently. Russo and Bresler (1980) used a modified version of the Green–Ampt (1911) infiltration model to interpret the advance of the wetting front beneath an air entry hydrometer (Bouwer, 1966). By solving the infiltration problem with model functions for $K(h)$ and $h(\theta)$, they were able to estimate the model parameters by forcing agreement between observed and predicted wetting front position.

4.5 WATER FLOW THROUGH STRUCTURAL VOIDS

The water flow equations were derived using the assumption that the soil has a continuous solid matrix that holds unsaturated water in pores and films. Field soil, however, has a number of cracks, root holes, worm channels, etc., whose physical properties differ enormously from the surrounding soil matrix. If filled, these flow channels have the capacity (see problem 3.4) to carry enormous amounts of water at velocities that greatly exceed those in the surrounding matrix (Beven and Germann, 1982).

At the present time there is no complete theory describing water flow through structural voids (sometimes called macropore flow), although a kinematic or shock wave model has shown promising agreement with some laboratory observations (Germann, 1985). There is uncertainty over how important subsurface voids can be in water flow, since if large, they should only fill at matric potentials near saturation. Nonetheless, substantial indirect evidence of flow through structural voids has been obtained by tracer studies. These have shown that a fraction of a chemical application can migrate to substantial depths with only a small amount of water input (Kissel et al., 1974; Richter and Jury, 1986; Roth et al., 1990; Shulin et al., 1987; see review by White, 1986).

Many water flow processes of interest such as groundwater recharge are concerned only with area-averaged water input. Therefore, preferential flow of water through structural voids or by some other means such as unstable flow does not necessarily invalidate the simpler model calculations that assume homogeneous flow. However, preferential flow is of critical importance in solute transport, because it enhances chemical mobility and can increase pollution hazards. This topic will be discussed further in Chapter 7.

4.6 EVAPORATION

The rate of evaporation from a wet, bare soil surface is limited by external meteorological conditions such as wind speed, relative humidity, and the flux of radient energy to the surface (Penman, 1948). In contrast, water loss from a soil with a dry surface layer is regulated primarily by soil water resistances that limit the rate at which water moves upward to the evaporating surface (Philip, 1957h). In this latter case, the water evaporation rate will be less than the maximum potential loss rate dictated by the external conditions.

During transient drying of a soil, control of the evaporation rate can pass from the meteorological factors to the soil resistance. Initially, when the soil surface is wet, evaporation occurs at the potential rate, which is limited by the amount of energy available at the soil surface. If the evaporation rate is intense, the rate of supply of water to the surface from the soil below may not be able to match the rate of loss, causing the surface water constant to become progressively drier. This period, when the evaporation is proceeding at the maximum rate, is called the first stage or *constant-rate* stage of drying. Eventually, the surface approaches a low water content whose vapor pressure is very nearly equal to the atmospheric vapor pressure at the soil boundary. At this time, the surface layer cannot continue to provide water from storage, and the evaporation rate becomes limited by the rate of water movement to the soil surface. This is the so-called second stage or *falling-rate* stage of evaporation, when the evaporation rate decreases continuously with time. It is very difficult to model the evaporation process during the first and second stages of drying, as it requires simultaneous evaluation of water and heat flow (Jury, 1973). However, if one is concerned primarily with average daily evaporation, the solution of the flow equation under isothermal conditions provides a reasonable first approximation to the evaporation rate.

Evaporation losses from initially wetted soil as a function of time were modeled by Ritchie (1972) and in a field study by Shouse et al. (1982) using the following two-stage evaporation model.

Stage 1: Potential Evaporative Loss

During the first stage, the soil surface is wet and the upward flow of water is assumed to be high enough to match the external rate, which is regulated by external conditions at a rate $E = E_p$, the potential loss rate. This lasts for a period of time t_c after irrigation or rainfall ceases. Gradually, gravity-driven drainage and water loss by evaporation deplete the surface layer, and it dries out to a point where regulation of subsequent loss of water shifts to the soil. This triggers the onset of the second stage of drying.

Stage 2: Soil-regulated Evaporative Loss

In the second stage of evaporation, the soil water flow theory predicts that if gravity is neglected, cumulative loss of water will be proportional to the square root of time. [See (4.14) with $\theta_0 < \theta_i$.] This is the model used by Black et al. (1969) and Ritchie (1972).

Thus, the two-stage evaporation model may be described mathematically as

$$E = E_p \quad \text{if } 0 < t < t_c \tag{4.55}$$

$$E_{\text{cum}} = \alpha(t - t_c)^{1/2} \quad \text{if } t > t_c \tag{4.56}$$

where α is a constant. This model was field calibrated in the study of Shouse et al. (1982) on prewetted soil plots that were instrumented with neutron access tubes.

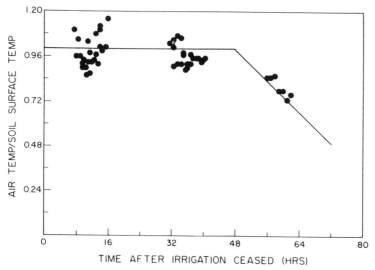

Figure 4.17 Ratio of measured air temperature to measured soil surface temperature as a function of time after irrigation ceased for two bare soil plots. (After Shouse et al., 1982.)

The authors simultaneously monitored the soil surface and air temperatures to determine the time t_c when the soil surface dried and the second stage of drying began.

As shown in Fig. 4.17, the soil temperature of the bare plots increased significantly compared to the air temperature after two days of drying during summer in the sandy loam field where the experiment was conducted. The authors took the value $t_c = 2$ days to be a constant for the field and season. Idso et al. (1974) found

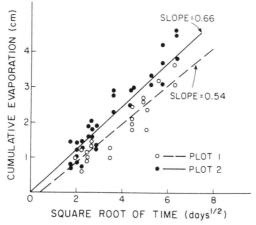

Figure 4.18 Cumulative evaporation versus square root of time after soil surface dries for two bare soil experiments. (After Shouse et al., 1982.)

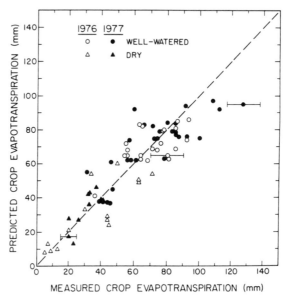

Figure 4.19 Predicted versus measured evapotranspiration (ET) for 2-week intervals of all treatments in 1976 and 1977 ($S_{yx} = 12$ mm). (After Shouse et al., 1982.)

that this time to drying increased significantly in other seasons, to as long as two weeks during winter.

After the second stage of drying began, Shouse et al. (1982) determined the constant α in (4.56) by equating the cumulative evaporative loss to the measured storage change of two soil plots over time. A mean value of α was selected by averaging the replicates (Fig. 4.18).

Shouse et al. (1982) used the evaporation model thus calibrated as part of a model to predict evapotranspiration losses from a growing crop, with good agreement found between predicted and measured evapotranspiration losses over a season (Fig. 4.19).

PROBLEMS[2]

4.1   Use the Green–Ampt model to calculate upward infiltration above a water table. What happens to the cumulative infiltration as $t \to \infty$? Discuss the possible differences between the behavior of infiltration in a Green–Ampt soil and a real soil.

4.2   A soil column of height L is saturated everywhere ($\theta = \theta_s$). At $t = 0$, the bottom is exposed to the air and drainage begins. If the soil has a hydraulic

[2]Problems preceded by a dagger are more difficult.

conductivity–water content relation that fits the model

$$K(\theta) = K_s \exp\left[\beta(\theta - \theta_s)\right]$$

and the column drains according to gravity flow, calculate the average water content and flux $J_w = -K(\theta)$ from the column as a function of time. What is the water content at $t = \infty$? Why did the model produce such an unphysical result?

4.3 The following data were recorded by tensiometers and gamma ray scanning in a 190-cm vertical soil column with steady upward flow from a water table and a steady evaporation rate of 0.5 cm day^{-1}. From the data given, construct and plot (a) $K(h)$, (b) $K(\theta)$, (c) $D(\theta)$, (d) $C(h)$, (e) $h(\theta)$, where K is hydraulic conductivity, D is diffusivity, and C is water capacity $\partial\theta/\partial h$.

Depth (cm)	Water Content	Matric Potential Head (cm)
180	0.11	-780
160	0.13	-480
140	0.15	-340
120	0.17	-240
100	0.20	-180
80	0.25	-125
60	0.30	-75
40	0.40	-45
20	0.48	-22
0	0.50	0

4.4 A weighing lysimeter of surface area 4 m^2 and depth 1.5 m is cropped to wheat and under rainfall irrigation. The following data are recorded by the lysimeter instruments, which record the weight as 0 kg at $t = 0$.

Time (days)	Cumulative Rainfall (cm)	Lysimeter Weight (kg)	Cumulative Drainage (m^3)
0	0	0	0.00
2	2	0	0.04
4	3	0	0.04
6	5	$+40$	0.04
8	7	$+40$	0.08
10	7	$+10$	0.08
12	8	$+30$	0.08
14	9	$+50$	0.08
16	14	$+100$	0.20
18	17	$+80$	0.32
20	18	$+60$	0.36

Calculate the evapotranspiration rate as a function of time, and plot irrigation rate, drainage rate, and ET rate on the same graph in the same units of centimeters per day. What is the leaching fraction of this system averaged over the 20-day period?

4.5   Assuming that the unsaturated hydraulic conductivity of a soil is given by the model function

$$K(\theta) = K_s \exp\left[\beta(\theta - \theta_s)\right]$$

where $K_s = 100$ cm day^{-1}, $\beta = 40$, and $\theta_s = 0.5$, calculate the long-time velocity of the infiltrating water front V_F under continuous ponding with zero head using (4.31) for initial water contents of $\theta_i = 0.0, 0.1, 0.2, 0.3, 0.4, 0.49$. Assuming that this stage began at $t = 0$ (i.e., no capillary attraction), calculate the position of the front after 12 h, and calculate the amount of water that has entered the soil at the different initial water contents.

4.6   When water is infiltrated into the soil while holding the water potential and water content of the surface at a constant value, the infiltration rate changes as a function of time but the water content profile is relatively uniform above the front (see Fig. 4.6). Assuming that the soil is initially very dry, sketch plausible water content profiles for various times during constant-flux infiltration, and discuss the major differences between this case and the case of constant-water-content infiltration.

†4.7   Calculate the minimum and maximum times to ponding under a rainfall rate of 2, 4, and 6 cm h^{-1} as in Example 4.3 for a soil whose maximum infiltration rate obeys the Horton model (4.7), with $i_0 = 10$ cm h^{-1}, $i_f = 1$ cm h^{-1}, and $\beta = 1$ h^{-1}. You will have to calculate the value of t_{max} numerically or graphically.

5 The Soil Thermal Regime

Soil temperature is one of the most critical factors that influence important physical, chemical, and biological processes in soil and plant science. Bacterial growth and plant production both are strongly temperature dependent, as are organic matter decomposition and mineralization. Other important microbiological rate processes, such as biodegradation of pesticides and other organic chemicals, also vary in their intensity with temperature. Many of these rate processes reach maximum levels at some particular range of temperatures and decrease both above and below that point.

Soil temperature affects plant growth first during seed germination. Although seeds of different plants vary in their ability to germinate at low temperatures, all species show a marked decrease in germination rate in soils with low surface temperatures (Russell, 1973). The germination rate will increase significantly with temperature up to a certain point, above which the rate falls off again. Since rapid seed germination ensures an early crop, the temperature of the soil at spring planting thus has a major influence on when the growth stages will occur.

Plant growth after germination is also influenced by soil temperature. Metabolically regulated plant processes, such as water and nutrient uptake, can be diminished below optimum rates at both low and high temperatures, resulting in temperature-dependent growth and yield patterns. For example, corn yields in Iowa were observed to increase almost linearly as a function of soil temperature at 100 cm between the range of 60 and 81.3°F (Allmaras et al., 1964). Above 81.3°F the yields decreased.

This chapter will discuss the major factors influencing soil temperature, beginning with a description of the radiant energy flux to the soil. Subsequently, heat transport equations will be derived to predict the distribution and changes of temperature within the soil as a function both of the external radiation striking the surface and of the principal soil thermal properties.

5.1 ATMOSPHERIC ENERGY BALANCE

5.1.1 Extraterrestrial Radiation

Stefan–Boltzmann Law   The source of all radiant energy for the earth is the sun, which continually emits short-wave electromagnetic radiation as a consequence of its high temperature (≈ 5700 K at the surface). The general formula expressing the energy flux Σ (in watts per square meter) from a body at a temperature T (in kelvin) is given by the Stefan–Boltzmann law (Shortley and Williams, 1965):

$$\Sigma = \epsilon \sigma T^4 \tag{5.1}$$

where σ is the Stefan–Boltzmann constant, which has the value 5.67×10^{-8} W m^{-2} K^{-4}, and ϵ is the emissivity, which equals 1 for a blackbody and has values in the range $0 \leqslant \epsilon \leqslant 1$ for other radiant surfaces.

> EXAMPLE 5.1: Calculate the radiant energy flux from the sun striking the outer edge of the earth's atmosphere assuming that the sun radiates as a blackbody ($\epsilon = 1$). Use the following data:
>
> | | | |
> |---|---|---|
> | Distance from earth to sun | X = | 1.50×10^{11} m |
> | Radius of the sun | R = | 6.97×10^8 m |
> | Surface temperature of sun | T = | 5760 K |

If we assume that the radiant energy from the sun expands radially outward isotropically, the same quantity of energy per unit time will pass through the surface of any sphere enclosing the sun at its center. The total quantity of energy Q in watts emitted is, using (5.1),

$$Q = \Sigma A = (\sigma T^4)(4\pi R^2) \tag{5.2}$$

since $A = 4\pi r^2$ for a sphere.

At the earth's atmosphere, let the radius of the expanding sphere enclosing the sun at its center and having earth at the surface of the sphere be X. The energy flux Σ_E (W m^{-2}) at the earth is therefore equal to, by (5.1) and (5.2),

$$\Sigma_E = Q/4\pi\, X^2 = R^2\, \Sigma/X^2 = R^2\sigma T^4/X^2 \tag{5.3}$$

Plugging in the preceding values, we obtain

$$\Sigma_E = 1.35 \times 10^3 \text{ W m}^{-2} = 1.94 \text{ cal cm}^{-2} \text{ min}^{-1} \tag{5.4}$$

This value is known as the solar constant.

Energy–Wavelength Laws   The radiation that arrives from the sun is not monochromatic; it has a range of wavelengths. The radiant flux density B_λ per unit wavelength for a blackbody is given by Planck's radiation law,

$$B_\lambda = \frac{C_1\, \lambda^{-5}}{\exp\,(C_2/\lambda T)} \tag{5.5}$$

where λ is the wavelength of the radiation in meters, $C_1 = 3.74 \times 10^{-16}$ W m^{-2}, and $C_2 = 1.44 \times 10^{-2}$ m K (Shortley and Williams, 1965).

This distribution has a maximum at a wavelength λ_m that obeys the equation known as Wien's law:

$$\lambda_m T = 2.898 \times 10^{-3} \text{ m K} \tag{5.6}$$

Figure 5.1 (van Wijk, 1963) shows a plot of the radiant flux density (5.5) integrated over a small wavelength interval along with actual measurements of radiation flux density taken at the outer edge of the earth's atmosphere with a spec-

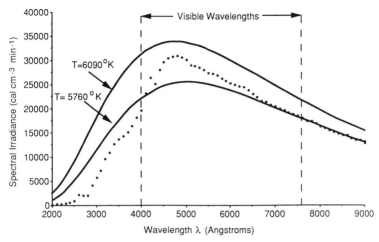

Figure 5.1 Measured (circles) and calculated (lines) using values of radiant energy flux from Equation (5.5) at the outer edge of the earth's atmosphere.

trophotometer. Two values, (5760 and 6090 K) were used for the temperature of the sun in (5.5). The good agreement between the model and the measurement indicates that the sun may be treated to a first approximation as a blackbody. Note that less than half of the total energy flux is in the visible range of from 0.4 to 0.76 μm. However, over 99% of the total radiation is contained in wavelengths of 0.3–4 μm (Chang, 1968). For this reason, solar radiation is known as "short-wave" radiation.

5.1.2 Solar Radiation

Interactions with Atmosphere Much of the short-wave solar radiation that reaches the outer edge of the earth's atmosphere is dissipated before it strikes the soil surface. Some of it is reflected by clouds back into space. A portion of it is absorbed by water vapor, oxygen, ozone, and carbon dioxide molecules in the atmosphere. Part of the radiation is scattered diffusely by molecules and particles in the air; some of this radiation will strike the earth's surface afterward as a scattered beam. The remainder of the radiation reaches the earth's surface directly without striking any atoms in the atmosphere. Thus, the fraction of the radiation from the sun that reaches the earth, called the global solar radiation R_s, is the sum of the direct beam and scattered beam. Figure 5.2 shows a typical distribution of radiation components for a summer day, in which 45% of the extraterrestrial radiation arrives at the surface. The fraction transmitted through the atmosphere is significantly affected by the amount of cloud cover; clouds may reflect anywhere from zero to more than 70% of the incoming extraterrestrial radiation (Van Wijk, 1963).

Net Radiation The solar radiation transmitted through the atmosphere is partitioned further when it strikes the soil or crop surface. A fraction of the solar ra-

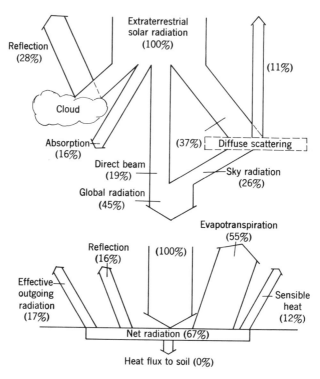

Figure 5.2 Typical partitioning of the extraterrestrial radiation and surface energy balance. (After Geiger, 1965, and Chang, 1961.)

diation, called the albedo a, is reflected at the surface and returns to space. In addition, thermal radiation from the sky, R_{sky} (in the long-wavelength range), strikes the surface, and thermal radiation from the soil or canopy surface, R_{earth}, radiates into space. The net downward radiation component is thus

$$R_N = (1 - a)R_S + R_{nt} \qquad (5.7)$$

where R_{nt} is the net long-wave thermal radiation equal to

$$R_{nt} = R_{sky} - R_{earth} \qquad (5.8)$$

Each of the terms in (5.8) may be calculated with the Stefan–Boltzmann equation (5.1) using an expression for the emissivity ϵ obtained by calibration or a model.

The size of the different terms in the radiation balance (5.7) throughout the year at Hamburg, Germany, is illustrated in Fig. 5.3. The net thermal radiation loss ranged from a low of about 6% in January to a high of about 14% of the extraterrestrial radiation in August. However, it is a dominant component of the net radiation in winter when solar radiation decreases and actually causes R_N to become

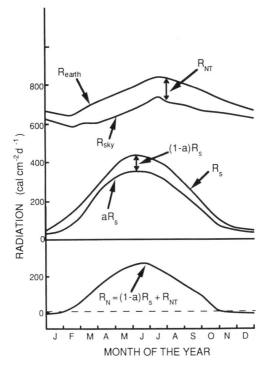

Figure 5.3 Radiation balances throughout the year at Hamburg, Germany. (After Fleischer according to Geiger, 1965.)

negative. The albedo varies from a high of about 0.2 in June to a low of near zero in December.

EXAMPLE 5.2: Calculate the long-wave thermal radiation energy flux R_{earth} from the earth assuming $T = 300$ K and $\epsilon = 0.95$. Estimate the wavelength at which the maximum energy flux occurs.

The energy flux density is calculated from the Stefan–Boltzmann equation (5.1):

$$R_{earth} = \epsilon\sigma T^4 = 436 \text{ W m}^{-2} = 0.63 \text{ cal cm}^{-2} \text{ min}^{-1}$$

This is about 32% of the value of the solar constant.

The maximum radiation wavelength is calculated from Wien's law (5.6):

$$\lambda_m = 9.66 \times 10^{-6} \text{ m} = 966 \text{ Å}$$

This wavelength is in the infrared region of the electromagnetic spectrum. Thus, we do not see the earth glowing as it radiates.

The fraction of the solar radiation that remains as net radiation at the surface varies significantly with climate, latitude, and surface cover. Table 5.1 summarizes data for different climates on the fraction of global radiation that is retained as net radiation (Chang, 1968). Since the global radiation decreases significantly

TABLE 5.1   Ratio of Net Radiation to Global Radiation during 24-h Periods

Location	Type of Cover	Global Radiation (langleys/day)		Ratio of Net Radiation to Global Radiation
Copenhagen, Denmark, 55° 44'	Grass	463	(July)	0.51
Rothamsted, England, 51° 58' N	Grass, tall crops	550		0.41
		550		0.46
Davis, CA, 38° 30' N	Grass	750	(June)	0.56
		175	(December)	0.33
Aspendale, Australia, 38° 2' S	Grass	181	(July)	0.18
		689	(January)	0.69
Hawaii, 21° 18' N	Sugar cane	725	(Summer)	0.69
		400	(Winter	0.65
	Pineapple	710	(May–June)	0.66
		500	(December)	0.53

Source: After Chang (1968).

during the winter months away from the equator, there is a greater seasonal variation in this ratio in northern and southern latitudes than in equatorial regions. In fact, as seen in Fig. 5.3 and illustrated again in Fig. 5.4, net radiation may even be negative in winter months.

5.1.3   Physical Factors Affecting Solar Radiation

Albedo   The fraction or percentage of incoming solar radiation that is reflected at the crop or soil surface is called the albedo. It is dependent upon the nature of the surface, the angle of the sun, and latitude. The albedo increases significantly with distance from the equator, as shown in Fig. 5.5, where the percentage of incoming solar radiation that is reflected ranges from a low of 70% near the equator to a high of 56% near the north pole (Houghton, 1954). The albedo coefficients of various surfaces are given in Table 5.2. In general, water surfaces have a lower albedo than cropped or soil surfaces, reflecting less than 10% of the radiation striking them. Canopy surfaces have albedos that vary between 5 and 25%. The lower values are characteristic of forest canopies and the higher values are more representative of nonequatorial crops at full ground cover. Crops at full cover in the tropics, however, have much lower albedos than at higher latitudes. The high position of the sun in the sky in tropical climates causes the radiation to strike the surface of the earth normally, which causes less reflection than when the sunlight arrives at an angle (Chang, 1968). For this reason, the albedo is higher in the early morning and later afternoon because of the lower angle of the sun (Chang, 1961).

The principal factors influencing the albedo of a bare ground surface are the soil type and the moisture level of the surface soil, mainly because of their effect on soil color. For example, dry, light-colored desert soils (serozems) have twice

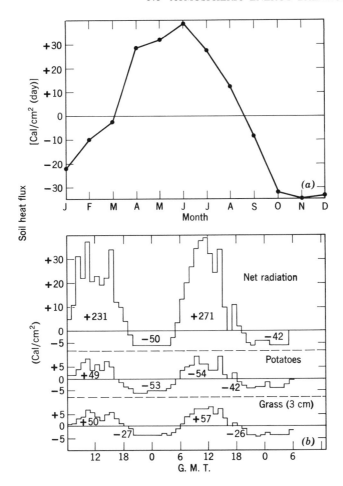

Figure 5.4 Annual variation in net radiation. (After Monteith, 1958.)

the albedo of dry, dark-colored chernozems (Table 5.2). Therefore, adding water to light soils, which darkens the surface, can decrease the solar reflection by up to 50%. Other alterations of soil surface color will also affect the albedo. Stanhill (1965) showed that application of white $MgCo_3$ to the surface of a bare soil doubled its albedo, causing a subsequent 10°C decrease in soil temperature.

Latitude The angle at which the sun's rays meet the earth greatly influences the amount of radiation received per unit area for two reasons. First, the radiation approaching the earth at an angle moves through more of the atmosphere and is subjected to greater scattering, reflection, and adsorption than radiation moving directly through the atmosphere. Second, radiation striking the earth at an angle to the vertical has a higher albedo than radiation coming in normally. The amount of radiation per unit area reaching the earth is proportional to the cosine of the

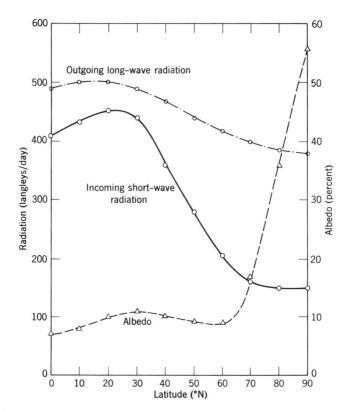

Figure 5.5 Radiation and albedo distributions in the Northern Hemisphere in relation to latitude. (After Houghton, 1954.)

angle made between the perpendicular to the surface and the direction of the radiation as it strikes the surface.

The radiation distribution in the Northern Hemisphere in relation to latitude is illustrated in Fig. 5.5 (Houghton, 1954). Both the incoming global and outgoing long-wave daily radiation are highest at the lower latitudes. The global radiation decreases rapidly at latitudes above 30° and reaches a constant amount above 60°. The albedo is lowest (<10%) in tropical regions, increases slightly in the middle latitudes, and rises sharply in the polar areas. The relatively constant daily global radiation and the tremendous increase in the percentage of reflection at the higher latitudes are the result of the lower solar elevation. Thus, in the tropical latitudes solar radiation is high and albedo low, producing extreme warming conditions for all 12 months of the year.

The outgoing long-wave radiation shown in Fig. 5.5 is highest at low latitude and decreases at high latitude, principally because the surface temperatures in these latitudes follow the same pattern.

TABLE 5.2 Relation of Nature of Surface on Reflection of Short-Wave Radiation

Nature of Surface	Albedo (%)
Fresh snow cover[a]	75–95
Light sand dunes[a]	30–60
Meadows and fields[a]	12–30
Forests[a]	5–20
Water surfaces[a]	3–10
Soils[b]	
Serozem	
Dry	25–30
Wet	10–12
Clay	
Dry	23
Wet	16
Chernozem	
Dry	14
Wet	8
Crops[c]	
Corn (New York)	23.5
Sugar Cane (Hawaii)	5–18
Pineapple (Hawaii)	5–8
Potatoes (USSR)	15–25

[a] From Geiger (1965).
[b] From Chudnovskii (1966).
[c] From Chang (1968).

Exposure   The effects of latitude may be simulated on a small scale within certain latitudes by changes in the direction of exposure and the degree of slope of the land. For example, the angle at which the rays of the sun strike a steep south slope is entirely different from that on a steep north slope; the southern slope receives more solar radiation per unit area. These differences in exposure have great ecological and agricultural significance, inasmuch as the temperature of the soil is always higher on southern-facing slopes than on northern ones. Wollney (1878) pioneered the study of the effect of exposure on the temperature of surface soils. He found that the southern exposures were always several degrees warmer than the north slopes. Because this increased warmth was due entirely to the direct action of the sun's rays, greater temperature variations from night to day were observed on the south exposure. He also found that the temperature differences between exposures increased as the slopes became steeper. The direction of exposure, however, was of greater significance than the degree of slope.

Exposure has very little influence on the surface heat balance in the tropics because of the high elevation of the sun. It also has little influence in polar climates, because much of the radiation striking the surface is diffuse sky radiation, which reaches north and south slopes at the same intensity. Consequently, exposure is of greatest consequence in climates at intermediate latitudes, where the solar elevation is significant and diffuse sky radiation is not the dominant form of solar energy.

Distribution of Land and Water   In general, island climates are less variable than continental climates. The presence of large bodies of water tends to stabilize the temperature because of the high specific heat of water, which is responsible for the absorption of large amounts of heat. In addition, the atmosphere over land masses surrounded by bodies of water is highly saturated with water vapor, which reduces the amount of radiant energy reaching the earth.

In contrast, continental climates can undergo greater seasonal extremes in temperature. For example, the central plains region and corn belt of the United States are characterized by hot summers and cold winters. Even the nights are warm in summer, because of the large amount of thermal radiation from the earth.

The climate of the coastal regions bordering large land masses is influenced by both the land and water. Characteristic ocean currents in the area can have a major influence on any land mass that comes in contact with them. The gulf current along the coast of Great Britain and the Japanese current along the shores of the northwestern United States are examples of warm currents that significantly affect the climate of the land they touch.

Vegetation   Vegetation is an important factor governing soil temperature because of the insulating properties of plant cover. Bare soil is unprotected from the direct rays of the sun and becomes very warm during the hottest part of the day. When cold seasons arrive, exposed soil rapidly loses its heat to the atmosphere. On the other hand, a good vegetative cover intercepts a considerable portion of the sun's radiant energy, which prevents the soil beneath from becoming as warm as bare soil during the summer. By preventing heat loss at night and shielding the surface during the day, vegetative cover reduces the daily variations in soil temperature relative to those in bare soil. In winter, the vegetation acts as an insulator to reduce the rate of heat loss from the soil. For this reason, frost penetration is more rapid and the depth of freezing is greater in bare soils than in those under a vegetative cover.

The major ways in which vegetation affects the soil energy balance are by (i) altering the albedo of the surface (Table 5.2), (ii) decreasing the depth of penetration of global radiation through the canopy (Chang et al., 1965), (iii) increasing the removal of latent heat by evapotranspiration, and (iv) decreasing the rate of heat loss from the soil by insulating the surface.

Mulches can affect the thermal regime of the soil in several ways. Light-colored plastic mulches transmit short-wave solar energy to the soil but prevent the loss of long-wave thermal radiation, thereby producing a greenhouse effect that warms the soil underneath. Mulches that have low thermal conductivities will decrease the conduction of heat into and out of the soil, causing mulch-covered soil to be cooler during the day and warmer during the night than bare soil. For this reason, mulches are often used to protect the soil from excessive cooling during the winter and to prevent the soil from warming up early in the spring.

5.2 SOIL SURFACE ENERGY BALANCE

5.2.1 Energy Balance Equation

The net radiation (5.7) arriving at the soil or crop canopy surface will distribute its energy in a number of different ways depending upon the surface and atmospheric conditions. If we disregard lateral inputs of heat to the surface and neglect transient energy changes caused by heating or cooling the surface or canopy, we may write a one-dimensional, steady-state heat energy balance at the surface as follows:

Net heat energy arriving at surface = net heat energy leaving surface   (5.9)

Components of Energy Balance   There are three major transport processes carrying heat away from the soil surface. The first, called the sensible or convective heat flux S, represents the vertical transport of warm air from the surface zone to the atmosphere above the surface. This occurs predominantly by turbulent convection (bulk flow) of air. The second, called the soil heat flux J_H, represents the vertical transport of heat into the soil. The third process is the conversion of heat by evaporation and the subsequent transport of water vapor from the surface zone to the atmosphere above the surface. This is called the latent heat flux $H_v \cdot \mathrm{ET}$, where ET is the evapotranspiration of water vapor flux to the atmosphere (evaporation + plant transpiration) and H_v is the latent heat of vaporization. The vapor transport away from the surface also occurs predominantly by convection. Thus, the steady-state heat balance equation is given by

$$R_N = S + H_v \cdot \mathrm{ET} + J_H \tag{5.10}$$

Each of the terms on the right side of (5.10) are positive when they move away from the surface (Fig. 5.6).

   At first glance it may seem strange to have a water vapor flux in the heat balance equation. However, it requires energy (585 cal g^{-1} at 20°C) to evaporate liquid water, and therefore the water vapor carries away a portion of the energy that arrives at the surface. Thus, the presence or absence of evapotranspiration at a surface can significantly influence the size of the other heat flow terms on the right side of (5.10). Table 5.3 represents two extremes in the partitioning of the components of (5.10) and clearly illustrates the importance of evapotranspiration on the partitioning of radiation.

   Figure 5.6 and Table 5.3 apply to daily averages or to daylight periods when radiation is incident on the surface. The surface energy balance is quite different

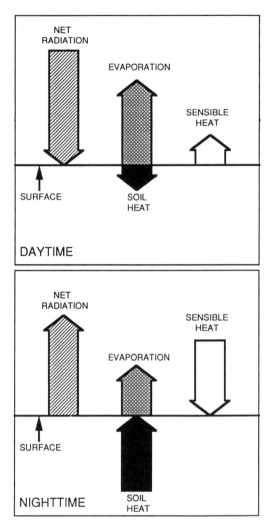

Figure 5.6 Schematic illustration of the components of the surface energy balance. (After Tanner, 1968.)

TABLE 5.3 Representative Relative Daytime Values of Components of Surface Energy Balance Equation (5.10) for Bare, Dry Soil and Well-watered Crop under Full Cover

Radiation Component	Bare, Dry Soil	Watered, Fully Developed Crop
S/R_N	.45	0.30
$H_v \cdot ET/R_N$	≈ 0	0.70
J_H/R_N	.55	≈ 0

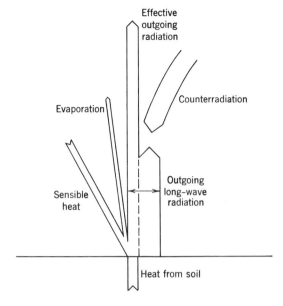

Figure 5.7   Nocturnal heat exchange at earth's surface. (Adapted from Rose, 1966.)

at night, when the net transfer of energy is usually away from the soil to the air
(Fig. 5.7).

5.2.2   Measurement of Evapotranspiration

There are numerous direct and indirect methods for measuring evapotranspiration
from bare or cropped soil. One of the most useful methods for estimating evapotranspiration from surfaces that are sufficiently moist that the rate of loss is limited
by the external or potential conditions is called the Penman equation (Penman,
1948). This equation is derived by combining the energy balance equation (5.10)
with aerodynamic formulas describing the turbulent transport of water vapor and
heat to the atmosphere above the canopy surface.

Aerodynamic Transport Equations   Water vapor and heat are carried away from
the immediate vicinity of the canopy surface by turbulent bulk flow of parcels of
air containing quantities of heat and vapor. These parcels are carried upward in a
chaotic manner, exchanging their contents with other parcels, which in turn are
carried away from the surface. The net result is a transport of heat and water vapor
between two heights, Z_0 (the canopy "surface") and some arbitrary elevation Z_1.
This turbulent flux is proportional to the vapor concentration or temperature difference between Z_0 and Z_1. Thus, the turbulent flux of sensible heat away from
the surface may be written as

$$S = h_H \left[\rho_a C_a T_0 - \rho_a C_a T_1 \right] = \rho_a C_a h_H (T_0 - T_1) \qquad (5.11)$$

where ρ_a is the density of the air, C_a is the specific heat of the air, T is temperature, and h_H is the turbulent transfer coefficient for heat between Z_0 and Z_1. The transfer coefficient depends primarily on wind speed and surface roughness characteristics. The quantity $\rho_a C_a T$ is the heat concentration or heat content per unit volume. Similarly, the water vapor transport equation may be written as

$$ET = h_v \left(\rho_{v0}^* - \rho_{v1} \right) \tag{5.12}$$

where ρ_v is water vapor density and h_v is the turbulent transfer coefficient for water vapor between Z_0 and Z_1. The asterisk denotes vapor saturation, since the surface is assumed to be moist. Equation (5.12) may be rewritten in terms of the vapor pressure P_v using the ideal gas law

$$\rho_v = \frac{M_v P_v}{RT} = \frac{M_v}{M_a} \frac{P_v}{P_a} \tag{5.13}$$

where M_v is the molecular weight of water vapor, R is the universal gas constant, M_a is the molecular weight of air, and P_a is the air pressure. Thus, (5.12) may be written in terms of the vapor pressure difference:

$$ET = \frac{M_v h_v}{M_a P_a} \left(P_{v0}^* - P_{v1} \right) \tag{5.14}$$

Penman Combination Equation   Equations (5.10), (5.11), and (5.14) are the basis for the Penman equation, which is derived by making the following approximations:

(i) The transfer coefficients for heat and water vapor are assumed to be equal:

$$h_v = h_H = h \tag{5.15}$$

This is called the similarity hypothesis, which is reasonably accurate under most conditions at the surface.

(ii) The slope y of the function $P_v^*(T)$ is approximated by

$$y = \frac{dP^*v}{dT} \approx \frac{P_{v0}^* - P_{v1}^*}{T_0 - T_1} \tag{5.16}$$

The first step in the derivation is to rewrite (5.14) as

$$ET = \frac{M_v h_v}{M_a P_a} \left(P_{v0}^* - P_{v1}^* + P_{v1}^* - P_{v1}^* \right) \tag{5.17}$$

Equation (5.16) is then inserted into (5.17):

$$\text{ET} = \frac{M_v h_v}{M_a P_a} \left[y(T_0 - T_1) + \Delta P_{v1} \right] \tag{5.18}$$

where $\Delta P_{v1} = P_{v1}^* - P_{v1}$ is the vapor pressure saturation deficit at Z_1. Equation (5.18) is then combined with (5.11) to produce

$$H_v \cdot \text{ET} = \frac{H_v M_v h_v}{M_a P_a} \left(\frac{yS}{\rho_a C_a h_H} + \Delta P_{v1} \right) \tag{5.19}$$

If we set $\rho_a C_a P_a M_a / M_v H_v = \gamma$, called the *psychrometer constant*, and let $h_v = h_H$, (5.19) may be solved for the sensible heat flux S in terms of ET:

$$S = \frac{y}{\gamma} H_v \cdot \text{ET} - \frac{\rho_a C_a h}{y} \Delta P_{v1} \tag{5.20}$$

Finally, (5.20) is inserted into the energy balance equation (5.10). When this equation is solved for the latent heat flux, the result is

$$H_v \cdot \text{ET} = \frac{y}{y + \gamma} \left(R_N - J_H + \frac{\rho_a C_a h}{y} \Delta P_{v1} \right) \tag{5.21}$$

Equation (5.21) is Penman's equation for potential water loss from a crop or canopy. Usually, the group of terms multiplying the saturation deficit is correlated with the wind speed,

$$\frac{\rho_a C_a h}{y} = f(u) = au + b \tag{5.22}$$

where a and b are constants and u is horizontal wind speed. Doorenbos and Pruitt (1976) discuss ways of selecting values of a and b to represent a given climate. They also provide tables for calculating net radiation from solar radiation, temperature, and other climate factors.

5.3 HEAT FLOW IN SOIL

5.3.1 Heat Flux Equation

Heat energy may be transported through soil by a number of different mechanisms, including conduction, radiation, convection of heat by flowing liquid water, convection of heat by moving air, and convection of latent heat. The two most important processes of heat transport in soil under normal conditions are conduction and convection of latent heat.

Conduction refers to the transport of heat by molecular collisions. For a pure solid substance, the conductive heat flux J_{H_c} in one dimension is described by Fourier's law as

$$J_{H_c} = -\lambda \frac{dT}{dz} \tag{5.23}$$

where T is temperature and λ is a constant called thermal conductivity. This equation describes the heat flow in rigid bodies whose composition remains unchanged during the transport of heat.

Convection of latent heat refers to the transport of the latent heat energy (energy per mass required for vaporization) in water vapor. Since any internal evaporation–condensation phase change in soil will liberate or consume heat energy, transport of water vapor from one location to another constitutes a transport of heat in a latent form. This is illustrated schematically in Fig. 5.8.

Water is evaporated at point A in the system by the addition of heat. The resultant vapor is moved laterally to point B by wind from a fan. At point B it condenses upon hitting a cold wall, causing the liberation of heat. The net result is a lateral transport of heat from A to B, traveling in latent form as the vapor moves. The equation for the latent heat convection J_{H_L} is simply

$$J_{H_L} = H_v J_v \tag{5.24}$$

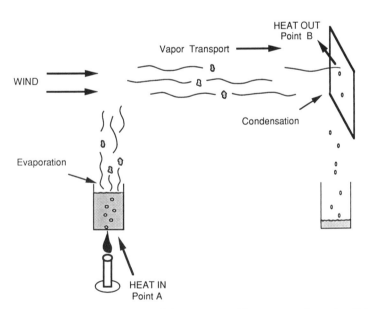

Figure 5.8 Hypothetical illustration of the transport of heat energy from A to B by convection of latent heat carried by water vapor.

where J_v (g cm^{-2} s^{-1}) is the water vapor mass flux and H_v (erg g^{-1}) is the latent heat of vaporization.

The expression for the net flux of soil heat may thus be written as

$$J_H = -\lambda^* \frac{dI}{dz} + H_v J_v \tag{5.25}$$

where λ^* is the instantaneous value of the thermal conductivity of the moist porous medium.

Equation (5.25) is largely formal because, as pointed out by de Vries (1958), imposition of a temperature gradient on a moist soil sample will cause liquid water and water vapor to move as well as heat, thus changing the value of λ^*. Therefore, unlike in a rigid solid body, λ^* cannot be measured directly in soil. As shown in Chapter 6, the water vapor flux may be written approximately in one dimension as

$$J_v \sim -D_{T_v} \frac{dT}{dz} \tag{5.26}$$

where D_{T_v} is the thermal vapor diffusivity. Equation (5.26) is accurate except in very dry soil where the relative humidity drops below unity.

Inserting (5.25) into (5.26), we obtain

$$J_H = -(\lambda^* + D_{T_v} H_v) \frac{dT}{dz} \equiv -\lambda_e \frac{dT}{dz} \tag{5.27}$$

where λ_e is the effective thermal conductivity of the porous medium, including the effects of conduction and convection of latent heat. The parameter λ_e is measurable by any method that measures λ in a solid, provided that (5.25) and (5.26) are valid.

EXAMPLE 5.3: A steady-state experiment is conducted on a soil sample to measure its thermal conductivity (Fig. 5.9). The sample, of thickness L, is placed between two glass plates of thickness d. The outside edge of each plate is connected to constant-temperature reservoirs. Thermocouples are attached to the inside edge of each plate. The following data are collected:

Position	Temperature
$z = 0$	T_0
$z = d$	T_A
$z = d + L$	T_B
$z = 2d + L$	T_1

Calculate λ_e for the soil layer as a function of the measured temperatures and the known thermal conductivity λ_g of the glass plates.

If lateral movement of heat is negligible, then we may assume that the heat flux through each of the glass plates is equal to the heat flux through the soil. Thus,

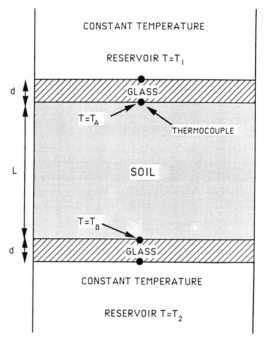

CONSTANT TEMPERATURE

RESERVOIR T=T₁

d GLASS

T=T_A

THERMOCOUPLE

L SOIL

T=T_B

d GLASS

CONSTANT TEMPERATURE

RESERVOIR T=T₂

Figure 5.9 A device for the steady-state measurement of soil thermal conductivity.

writing Fourier's law (5.23) through the upper plate, we obtain

$$-J_H = -\lambda_g(T_A - T_1)/d \tag{5.28}$$

which may be calculated if λ_g is known and T_A, T_0, and d are measured. Once J_H is known, Fourier's law may be written across the soil layer between points A and B (assuming that the glass is in good contact with the soil):

$$-J_H = -\lambda_e(T_B - T_A)/L \tag{5.29}$$

After equating (5.28) and (5.29), we solve for λ_e:

$$\lambda_e = \frac{LJ_H}{(T_B - T_A)} = \frac{(\lambda_g L/d)(T_A - T_1)}{T_B - T_A} \tag{5.30}$$

Notice how similar the approach to problem solving is for heat flow and for saturated water flow (see Section 3.2). This is because Darcy's law (3.13) and Fourier's law (5.23) have the same mathematical form.

5.3.2 Heat Conservation Equation

The heat conservation equation is derived in exactly the same manner as the water conservation equation in Chapter 3, by conducting a heat energy balance on a small cubic soil volume (see Fig. 3.19). Assuming for simplicity that heat is flowing in the vertical z direction, we record the heat balance during a short time interval Δt between t and $t + \Delta t$. In words, the heat balance equation is written as

Amount of heat energy flowing into soil volume during Δt

    = amount of heat energy flowing out of soil volume during Δt

    + increase in heat energy stored in soil volume during Δt

    + amount of heat energy that has disappeared from soil volume

    during Δt by reactions   (5.31)

For one-dimensional vertical heat flow, the first term in (5.31) may be written as

Amount of heat flowing into volume during Δt

$$= J_H \left(x, y, z, t + \tfrac{1}{2} \Delta t \right) \Delta x \, \Delta y \, \Delta t \qquad (5.32)$$

where J_H is the soil heat flux evaluated at the average time $t + \tfrac{1}{2}\Delta t$ and $\Delta x \, \Delta y$ is the cross-sectional area of the soil volume element in Fig. 3.18.

Similarly, the second term may be written as

Amount of heat flowing out of volume during Δt

$$= J_H(x, y, z, + \Delta z, t + \tfrac{1}{2} \Delta t) \Delta x \, \Delta y \, \Delta t \qquad (5.33)$$

The increase in heat energy stored in the soil volume may be expressed in terms of the heat content per soil volume H as

Increase in heat energy stored in volume

$$= \left[H(x, y, z + \tfrac{1}{2} \Delta z, t + \Delta t) - H(x, y, z + \tfrac{1}{2} \Delta z, t) \right] \Delta x \, \Delta y \, \Delta z \qquad (5.34)$$

where H is evaluated at the midpoint $z + \tfrac{1}{2} \Delta z$ of the volume element.

The amount of heat disappearing from the volume by reactions is expressed symbolically as

Amount of heat disappearing from volume during $\Delta t = r_H \, \Delta x \, \Delta y \, \Delta z \, \Delta t$ (5.35)

where r_H is the rate of loss of heat per soil volume by reactions.

After inserting (5.32)–(5.35) into the heat balance equation (5.31) and rearranging terms, we obtain

$$\frac{J_H\left(x, y, z + \Delta z, t + \tfrac{1}{2} \Delta t\right) - J_H\left(x, y, z, t + \tfrac{1}{2} \Delta t\right)}{\Delta z}$$

$$+ \; \frac{H\left(x, y, z + \tfrac{1}{2} \Delta z, t + \Delta t\right) - H\left(x, y, z + \tfrac{1}{2} \Delta z, t\right)}{\Delta t}$$

$$+ \; r_H = 0 \qquad (5.36)$$

As the size of the volume shrinks ($\Delta z \to 0$) and the time interval $\Delta t \to 0$, (5.36) reduces to the differential heat conservation equation

$$\frac{\partial H}{\partial t} + \frac{\partial J_H}{\partial z} + r_H = 0 \tag{5.37}$$

In three dimensions, (5.37) is written as

$$\frac{\partial H}{\partial t} + \frac{\partial J_{Hx}}{\partial x} + \frac{\partial J_{Hy}}{\partial y} + \frac{\partial J_{Hz}}{\partial z} + r_H = 0 \tag{5.38}$$

where

$$J_H = J_{Hx}\,\hat{\imath} + J_{Hy}\,\hat{\jmath} + J_{Hz}\,\hat{k} \tag{5.39}$$

is the heat flux vector.

The heat sink term r_H should be included in the heat balance whenever a source or sink of heat (e.g., radioactive soil, chemical reactions) generates or consumes nonnegligible quantities of energy. Since we have already included the conversion of latent heat by evaporation in the heat flux term, it should not be placed in the equation as a sink term. Normally we will assume that $r_H = 0$.

The heat content per unit volume H may be written as

$$H = C_{\text{soil}}(T - T_{\text{ref}}) \tag{5.40}$$

where C_{soil} is the soil volumetric heat capacity and T_{ref} is an arbitrary reference temperature at which $H = 0$.

When the heat flux (5.27) and the heat content (5.40) are inserted into the heat conservation equation (5.37), we obtain, (assuming $C_{\text{soil}} = \text{const}$),

$$C_{\text{soil}}\frac{\partial T}{\partial t} = \frac{\partial}{\partial z}\left(\lambda_e \frac{\partial T}{\partial z}\right) \tag{5.41}$$

If the z dependence of λ_e is neglected, (5.41) reduces to

$$\frac{\partial T}{\partial t} = K_T \frac{\partial^2 T}{\partial z^2} \tag{5.42}$$

where $K_T = \lambda_e/C_{\text{soil}}$ is the soil thermal diffusivity. Equation (5.42) is called the heat flow equation.

5.3.3   Thermal Properties of Soil

Heat Capacity   The heat capacity per unit volume of a substance is defined as the quantity of heat required to raise a unit volume of the substance one degree of temperature. For a mixture of materials such as soil, the heat capacity per volume of the composite material is the sum of the heat capacities of the constituents weighted by their volume fractions. Thus, we may express the soil heat capacity as

$$C_{\text{soil}} = X_a C_a + X_w C_w + \sum_{j=1}^{N} X_{s_i} C_{s_i} \qquad (5.43)$$

where X refers to volume fraction, C to heat capacity per volume, and the subscripts a, w, s_i to air, water, and solid constituent i (out of a total of N different solid materials in the soil). The heat capacity per volume C of a pure substance may also be expressed as

$$C = \rho c \qquad (5.44)$$

where ρ is the density of the substance and c is the specific heat, or the heat capacity per unit mass. Table 5.4 summarizes specific heat values for the principal solid materials found in soil. A study of Table 5.4 reveals that the principal soil minerals differ little in their heat capacity values and that the specific heat of organic matter is higher than that of soil minerals. For this reason, de Vries (1963) recommended using average values of 0.46 and 0.60 cal cm^{-3} °C^{-1} for the heat

TABLE 5.4   The Specific Heat of Various Soil Constituents

	Specific Heat (cal g^{-1} °C^{-1})			
Material	Lang (1878)	Ulrich (1894)	Kersten (1949)	Bowers and Hanks (1962)
Coarse quartz sand	0.198	0.191	0.190	0.19
Fine quartz sand	0.194	0.192	0.197	—
Quartz powder	0.209	0.189	—	—
Kaolin	0.233	0.244	—	—
Feldspar	—	0.194–0.205	0–0.190	0.210–0.220
Mica	—	0.206–0.208	—	—
Apatite	—	0.183	—	0.22
Dolomite	—	0.222	—	0.23
Al$_2$O$_3$	0.217	—	—	—
Fe$_2$O$_3$	0.163	0.165	—	—
Humus	0.477	0.443	—	—
Calcareous sandy soil	0.249	—	—	—
Humus calcareous sandy soil	0.257	—	—	—
Garden soil	0.267	—	—	—
Clay	—	—	—	0.27
Silty clay	—	—	—	0.26
Silt loam	—	—	0.164–0.194	—

capacity per volume of soil minerals and organic matter, respectively, and the value 1 cal cm^{-3} °C^{-1} for water. The heat capacity of air is small and may be neglected. With these average values, (5.43) may be expressed as

$$C_{\text{soil}} = \theta + 0.46(1 - \phi - X_0) + 0.60X_0 \qquad (5.45)$$

where X_0 is the volume fraction of organic matter, ϕ is porosity, θ is volumetric water content, and $\phi - X_0$ is the volume fraction of all of the soil minerals. The soil heat capacity can also be measured directly by calorimetry. Equation (5.45) has been shown to correspond well with values of C_{soil} measured directly by calorimetry (de Vries, 1963).

Thermal Conductivity   Since the soil is a granular medium consisting of solid, liquid, and gaseous phases, the thermal conductivity will depend upon the volumetric proportions of these components, the size and arrangement of the solid particles, and the interfacial contact between the solid and liquid phases. The thermal conductivity of quartz is 26.3 mcal cm^{-1} s^{-1} °C^{-1} when measured parallel to the axis of the crystal and 16.0 mcal cm^{-1} s^{-1} °C^{-1} when determined perpendicular to this axis. The corresponding thermal conductivity values for water and air are 2.4 and 0.06 mcal cm^{-1} s^{-1} °C^{-1}, respectively. Thus, the ratio of the thermal conductivities for quartz, water, and air is 333:23:1.

Because of this huge difference, it is obvious that the thermal conductivity of a granular soil will depend upon the intimacy of the contact of the solid particles and the extent to which air is displaced by water in the pore spaces between the particles.

The results of measurements of λ_e by several investigators are given in Table 5.5. The relative size of the thermal conductivity of different soils follows the order sand > loam > clay > peat. Since the thermal conductivities of the mineral components of the solid phase are all of the same order of magnitude (Smith and Byers, 1938), the thermal conductivity differences shown in Table 5.5 are related to the degree of packing and porosity of the system. Thermal conductivity diminishes with decreasing particle size due to reduced surface contact between the particles through which heat will readily flow. Patten (1909) observed that the thermal conductivity of carborundum was lowered about 70% as the size of particles decreased from 450 to 6 nm. He also found that the conductivity of quartz particles was only one-fifteenth to one-twentieth that of a solid quartz block.

Increasing the bulk density of soils lowers the porosity and improves the thermal contact between the solid particles as well as reduces the volume of low-conducting air. The impact of soil porosity on thermal conductivity is shown in Figs. 5.10 and 5.11. Van Rooyen and Winterkorn (1959) evaluated the thermal conductivity measurements reported in Russian investigations on a chernozem soil (Fig. 5.10). As the bulk density of the dry soil increased from 1.1 to 1.5 (porosity decrease from about 59 to 43%), the thermal conductivity increased from 1.0 to 2.1 mcal cm^{-1} s^{-1} °C^{-1}. The data of van Duin (1963) in Fig. 5.11a indicate similar relationships, showing that a 50% decrease in the porosity of sand (as well as clay)

TABLE 5.5   Thermal Properties of Different Soils

Type of soil	Thermal Conductivity, $\times 10^{-3}$ (cal cm^{-1} s^{-1} °C^{-1})				Thermal Diffusivity, $\times 10^{-3}$ (cm^2 s^{-1})		
	Relative,[a] von Schwarz (1879)	Geiger (1965)	Nakshabandi and Kohnke (1965)	van Duin (1963)	Geiger (1965)	Nakshabandi and Kohnke (1965)	van Duin (1963)
Sand							
Wet	100	4.00	4.35[b]	3.70[c]	7.0	12.6[b]	4.4[c]
Dry	85.5	0.55	0.35	0.37	3.5	1.5	1.7
Clay							
Wet	90.3	3.50	1.40[d]	3.18[c]	11.0	3.2[d]	3.7[c]
Dry	74.3	0.17	0.25	0.37	1.2	1.5	1.8
Loam							
Moist	99.3	—	4.25[e]	—	—	6.0[e]	—
Dry	83.3	—	0.45	—	—	1.8	—
Peat							
Wet	58.8	0.85	—	0.82[f]	1.2	—	1.2[f]
Dry	51.5	0.20	—	0.11	2.0	—	1.3

[a] Moist sand = 100.
[b] 23% saturation with water.
[c] 100% saturation with water.
[d] 38% saturation with water.
[e] 75% saturation with water.
[f] $66\frac{2}{3}$% saturation with water.

caused a doubling of the thermal conductivity. Similarly, van Duin's data in Fig. 5.11b show that a 50% decrease in porosity caused a large increase in the thermal diffusivity $K_T = \lambda_e / C_{soil}$ of sand over the entire range of moisture content.

The increase in thermal conductivity as a result of raising the soil bulk density within normal ranges is small compared with the impact of adding water to the soil. The presence of water films at the points of contact between particles not only improves the thermal contact between particles but also replaces air in the soil pore space with water, which has about 20 times the thermal conductivity of air. The curves in Figs. 5.10 and 5.11 point out quite clearly the rapid increases in thermal conductivity and diffusivity as the percentage of water in the soil pore space rises. The greatest rate of increase in conductivity occurs at the lower moisture contents. If the thermal conductivity of the dry soil (Fig. 5.10) with a bulk density of 1.1 g cm^{-3} is taken as the reference value, the dry soil thermal conductivities at bulk densities of 1.2, 1.3, and 1.5 increase by a factor of 1.3, 1.65, and 2.1, respectively, with respect to the reference value. In contrast, when the pore spaces are 25% filled with water, the thermal conductivities increase by a factor of greater than 4 relative to the reference value. When the water saturation values rise to 50%, the ratios range from 6.7 to 8.6, showing that the effect of bulk density on thermal conductivity is greater at the higher moisture contents. These same relationships hold in the curves of van Duin in Fig. 5.11.

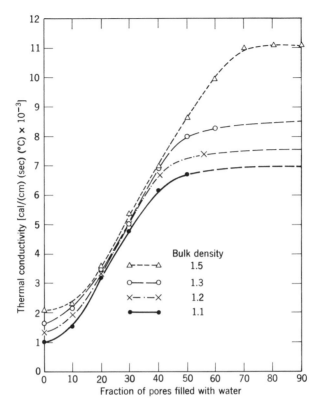

Figure 5.10 Soil thermal conductivity–water content relation as affected by bulk density. (Adapted from data of van Rooyen and Winterkorn, 1957.)

In order to visualize the impact of water on thermal conductivity, consider two dry quartz spheres that are in contact with each other. Most of the heat conduction takes place through a relatively small cross-sectional area at the point of contact. However, with the addition of a small amount of water in the wedges at the point of contact, the surface through which heat is conducted increases greatly because the water flow pathway is small and water conducts heat far more readily than air does. This causes the rapid rise of thermal conductivity at low water contents. As more water is added, the films become thicker, and the effect is less pronounced per unit of water added; thus, the rate of rise of thermal conductivity is slower in this region.

Thermal diffusivity (K_T) (Fig. 5.11b) first increases rapidly with increasing water content to a maximum and then decreases (Patten, 1909). This is explained by the fact that heat capacity rises linearly with water content, whereas thermal conductivity experiences its most rapid rise at low moisture contents, causing the ratio λ/C to have an internal maximum as a function of θ. Despite the change in K_T with θ in the dry range, the thermal diffusivity is relatively constant over a large

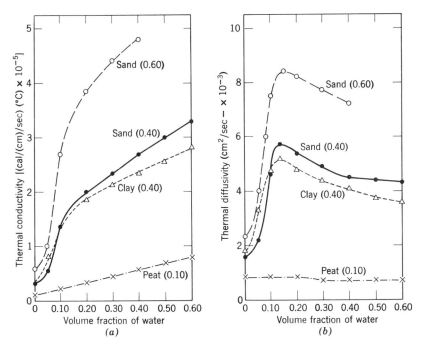

Figure 5.11 Soil thermal conductivity (*a*) and thermal diffusivity (*b*) as a function of water content for various soil types. Numbers refer to porosity. (After van Duin, 1963.)

range of water content. This means that the heat flow equation (5.42) may be treated as though it were linear as a first approximation. This can greatly simplify calculations, as shown in Section 5.3.4.

Measurement of Thermal Conductivity   The steady-state method for determining the thermal conductivity of moist soils (as illustrated by Example 5.3) has two major weaknesses. First, water will redistribute under the influence of a steady-state temperature gradient (Jury and Miller, 1974), creating a nonuniform profile within the column. Second, this method is strictly a laboratory technique and cannot be used in situ. The transient-state cylindrical probe (Jackson and Taylor, 1965; de Vries and Peck, 1968) overcomes these difficulties, although there is some temperature-induced moisture flow. The method consists of a thin metal wire that is heated electrically to serve as the heat source and a thermocouple to measure the temperature rise. These are placed inside a cylindrical tube, which is inserted into the soil. When the wire is connected to a voltage source, the wire heats up, causing heat to flow radially. The temperature of the thermocouple probe in contact with the soil is given by the equation

$$T = T_0 = \frac{q}{4\pi\lambda}\left[d + \ln\left(t + t_0\right)\right] \tag{5.46}$$

where T_0 is the temperature at time t_0, $T - T_0$ is the temperature rise, q is the heat flowing per unit time and unit length of wire, d is a constant that depends on the location of the thermocouple, and t_0 is a correction constant that depends upon the dimensions of the probe as well as the thermal properties of both the probe and the soil.

Equation (5.46) is obtained by solving the heat flow equation in cylindrical coordinates for the appropriate initial and boundary conditions (Carslaw and Jaeger, 1959). If $T - T_0$ is plotted against $\ln(t)$, a straight line is obtained for times $t \gg t_0$. The thermal conductivity is then calculated by the revised equation

$$\lambda = \frac{q}{4\pi S} \qquad (5.47)$$

where S is the measured slope of T versus $\ln(t)$. The value of the power dissipation per unit length q is calculated from the current i applied to the wire and the measured resistance r per unit length of the wire.

5.3.4 Applications of Heat Flow Equation

Steady-State Heat Flow Problems   The steady-state heat flow equation (5.27) is formally identical to Darcy's law (3.13) and (3.104) for saturated water flow. In heat flow, the driving force is a temperature gradient instead of a hydraulic head gradient, and the coefficient of proportionality between driving force and flux rate is the thermal conductivity rather than the hydraulic conductivity. Thus, all of the principles developed for solving flow problems in saturated soil are applicable to steady-state heat flow problems as well. This is illustrated in the next example

EXAMPLE 5.4:   A soil column contains 50 cm of dry quartz sand ($\lambda_s = 0.5$ mcal cm^{-1} s^{-1} °C^{-1}) over 25 cm of dry loam ($\lambda_L = 0.25$ mcal^{-1} s^{-1} °C^{-1}). The top of the column is held at $T = 30$°C and the bottom at $T = 5$°C. Calculate the steady-state heat flux through the two layers and the temperature at the sand–loam interface.

The first step is to find the equivalent thermal conductivity λ_{eq} of the two layers. By analogy with (3.25), we may write

$$\frac{L_1 + L_2}{\lambda_{eq}} = \frac{L_1}{\lambda_1} + \frac{L_2}{\lambda_2} \qquad (5.48)$$

$$\frac{75}{\lambda_{eq}} = \frac{50}{0.5} + \frac{25}{0.25} = 200 \qquad (5.49)$$

where $\lambda_{eq} = 75/200 = 0.375$ m cal cm^{-1} s^{-1} °C^{-1}.

Now that the equivalent conductivity has been calculated, the heat flux equation may be written across the entire column using Fourier's law (5.27) expressed between $z_1 = 0$ and $z_2 = L_1 + L_2 = 75$ cm:

$$J_H = -\frac{\lambda_{eq}}{L_1 + L_2}(T_2 - T_1) = -0.125 \text{ cal cm}^{-2}\text{ s}^{-1}$$

Since the heat flux is known, we now can write Fourier's law across the lower half of the column between $z_1 = 0$ and $z_3 = L_1 = 25$ cm:

$$J_H = -0.125 = -\frac{\lambda_L}{25}(T - 5) = -0.01(T - 5)$$

or $T = 17.5°C$.

Annual Temperature Changes in Soil   In the field, temperature is constantly changing, so that steady-state models are rarely useful near the surface. Instead, the time-dependent heat flow equation (5.42) must be used to solve for soil temperature. There are a few cases where the time-dependent equation is very easy to use.

Figure 5.12 is a graph of the long-term monthly average soil temperature near the surface measured at a research station in Davis, California (NOAA, 1986).

It is clear from Fig. 5.12 that we could represent this graph very well by an equation of the form

$$T(t) = T_A + A \sin \omega t \tag{5.50}$$

where T_A is the annual average temperature, A is the amplitude of the surface fluctuations, and $\omega = 2\pi/\tau$ is the angular frequency, where τ is the period of the wave. Equation (5.50) expresses one boundary condition for the heat flow equation (5.42). We also must specify what happens deep in the soil as $z \rightarrow -\infty$ in order to solve (5.42) for the temperature. Since the source of heat at the surface is periodic, it is reasonable to assume that far below the surface the soil will stay at the average temperature,

$$\lim_{z \rightarrow -\infty} T(z, t) = T_A \tag{5.51}$$

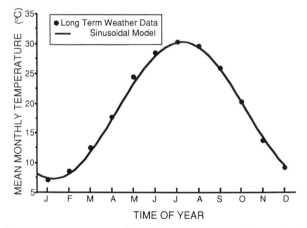

Figure 5.12 Long-term average monthly soil temperatures at 10 cm depth measured at Davis, California. (Data from NOAA, 1986.)

The solution of Equation (5.42) with the boundary conditions (5.50) and (5.51) may be obtained by several techniques (such as Laplace transforms). However, this is a well-known problem whose solution has been published (Carslaw and Jaeger, 1959):

$$T(z, t) = T_A + A \exp(z/d) \sin(\omega t + z/d) \qquad -\infty < z < 0 \quad (5.52)$$

where

$$d = \sqrt{2K_T/\omega} = \sqrt{K_T \tau / \pi} \qquad (5.53)$$

Equation (5.52) is a sine wave whose amplitude decreases with depth and that is phase shifted by an amount that increases with depth (Fig. 5.13). The constant d given in (5.53) is called the damping depth. It depends on the physical properties of the soil as well as on the angular frequency ω of the surface change.

Inspection of (5.52) and Fig. 5.13 reveals several properties of the solution.

1. The temperature at any depth is a periodic sine wave with amplitude $A \exp(-z/d)$.
2. The phase of the wave is retarded with respect to the surface by a time lag $\Delta t = z/\omega d$.
3. All depths have the same period $\tau = 2\pi/\omega$.
4. All depths have the same annual average temperature T_A.

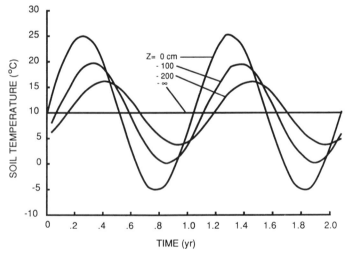

Figure 5.13 Graph of the temperature at the surface and three depths as a function of time using Equation (5.52).

TABLE 5.6 Thermal Profiles for Plainfield Sand

x (cm)	Amplitude (°C)	Time Shift (days)	T_{min} (°C)
0	17.0	0	−10.0
50	13.6	13	−6.4
100	10.9	26	−3.9
200	6.8	52	0.2
300	4.6	78	+2.4
500	1.8	131	+5.2
1000	0.2	262	+6.8

Note: $\lambda = 0.002$ cal cm^{-1} °C^{-1} s^{-1}; $C = 0.40$ cal cm^{-3} °C^{-1}; $K_T = 430$ cm^2 day^{-1}; $d = 224$ cm; $T_A = 7$°C; $A = 17$°C.

We can insert some realistic values for the soil thermal properties to estimate the size of the predicted phase lag and amplitude attenuation. Table 5.6 summarizes the properties of Plainfield sand at Wisconsin (Jury, 1973) and gives the amplitude and phase lag at several depths, assuming that the soil surface amplitude and average temperature are equal to $A = 17$°C and $T_A = 7$°C, respectively.

One striking feature of the data in Table 5.6 is the large phase lag at various depths. If the warmest temperatures at the surface occur in July, they will occur in September at 200 cm and in December at 500 cm. Further, we notice that below 200 cm the minimum temperature is above freezing, which is useful to know if we have to bury a water pipe, for example.

Notice that the damping depth completely characterizes the penetration of the thermal wave. Suppose we were sending down waves with a period of one day instead of one year. (We have to be more careful here because the daily temperature wave is not as well described by a sine wave, and it would vary more from day to day.) Then $\omega = 2\pi/(1$ day$)$ and the daily damping depth $d = 11.8$ cm. This means that the wave will damp out before getting very deep at all. For example, a wave of amplitude A at the surface will drop to $0.08A$ at $z = 30$ cm. This effect may be summarized in the following general principle: *Rapidly changing surface temperatures will not penetrate as deeply into the soil as slowly changing temperatures of the same amplitude.*

Not all climates have soil temperature that varies in a sinusoidal manner. Also, soil surface temperature rather than air temperature should be used as the upper boundary condition because the soil temperature is determined to a great extent by evaporation and soil cover and may differ considerably from the air temperature above it.

The annual wave model provides a means of measuring the average thermal diffusivity of the soil. The procedure is illustrated in the next two examples.

EXAMPLE 5.5: Two max–min thermometers are buried in the soil for one year and then excavated. The recorded temperatures (°C) are given in the following. Calculate the average thermal diffusivity and damping depth for this soil.

Depth (cm)	T_{max}	T_{min}	$T_{max} - T_{min}$
−100	25.0	7.0	18
−200	21.5	10.5	11

Since the sine function in (5.52) has maximum and minimum values of ± 1, respectively, we may write equations for T_{max} and T_{min} at any z:

$$T_{max}(z) = T_A + A \exp(z/d) \tag{5.54}$$

$$T_{min}(z) = T_A - A \exp(z/d) \tag{5.55}$$

Subtracting the second equation from the first, we obtain

$$\Delta T(z) = T_{max} - T_{min} = 2A \exp(z/d) \tag{5.56}$$

Since the max–min temperature difference has been measured at two depths, both unknowns A and d may be solved for in (5.56). Thus, we have

$$\frac{\Delta T(z_1)}{\Delta T(z_2)} = \frac{A \exp(z_1/d)}{A \exp(z_2/d)} = \exp\left(\frac{z_1 - z_2}{d}\right) \tag{5.57}$$

After taking the natural logarithm of both sides of (5.57), we may solve for d:

$$d = \frac{z_1 - z_2}{\ln\left[\Delta T(z_1)/\Delta T(z_2)\right]} \tag{5.58}$$

Inserting the values in the table, we obtain $d = 203$ cm. Finally, solving for K_T in (5.53), we obtain (with $\tau = 365$ days)

$$K_T = \pi d^2/\tau = 355 \text{ cm}^2 \text{ day}^{-1}$$

EXAMPLE 5.6: A similar study is conducted in a different soil, this time using daily thermocouple temperature measurements at two locations to determine the time at which the maximum temperature occurs. From the data that follows, calculate d and K_T for this soil.

Depth (cm)	Day of maximum T
−100 cm	August 1
−200 cm	August 31

The maximum value of the sine function occurs when its argument has the value $\pi/2$. Therefore, the equation describing the maximum value of the temperature reduces to

$$\omega t + z/d = \pi/2 \tag{5.59}$$

Thus, the two data points in the table each satisfy (5.59). Subtracting one equation from the other, we obtain

$$\omega(t_2 - t_1) = \frac{z_1 - z_2}{d} \tag{5.60}$$

Solving (5.60) for d, we obtain

$$d = \frac{z_1 - z_2}{\omega(t_2 - t_1)} = \frac{\tau(z_1 - z_2)}{2\pi(t_2 - t_1)} \tag{5.61}$$

Inserting the values in the table, we obtain $d = 194$ cm, and from (5.53), we obtain $K_T = 324$ cm^2 day^{-1}.

5.3.5 Soil Temperature Observations

Diurnal Variations   Early investigations by Wollney (1883) and Bouyoucos (1913) studied the diurnal variations in soils as affected by the nature of the soil, type of surface cover, and incoming radiation. Figure 5.14 from Yakuwa (1945) shows diurnal soil temperature at four depths in a loam during the summer. In the morning before sunrise, the minimum temperature of the soil was lowest at the surface and increased with depth. For example, at about 4:30 A.M., when the surface temperature was approximately 13°C, the temperature at 20 cm was 24°C. Thus, heat is leaving the soil at this time. Because of the time lag associated with soil heat flow when the surface temperature is changing, the lower depths continue to cool for a period of time.

Also evident from Fig. 5.14 is the decreasing amplitude of the diurnal wave at lower depths, which is consistent with the damping depth concept discussed in the previous section.

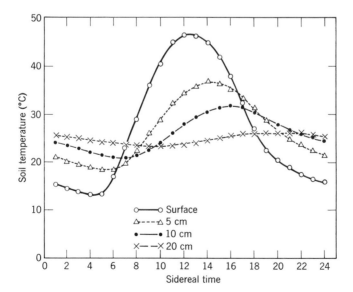

Figure 5.14 Diurnal variations in temperature measured at different depths in a loam soil. (After Yakuwa, 1945.)

EXAMPLE 5.7:   Although the sine wave model of soil surface temperature does not quantitatively describe the shape of the daily soil surface temperature in most cases, the diurnal damping depth is still useful in analyzing the penetration of waves with a daily frequency. Assuming that the thermal diffusivity of the loam soil of Yakuwa has a constant value of about 4.5×10^{-3} cm^2 s^{-1} (see Fig. 5.11), analyze and interpret the data in Fig. 5.14 using the damping depth concept.

The daily wave has a period of 1 day. Using (5.53) and the value of $K_T = 4.5 \times 10^{-3}$ cm^2 s$^{-1} = 389$ cm^2 day^{-1}, we obtain a daily damping depth value $d = 11$ cm. Assuming that the mean temperature in Fig. 5.14 is 25°C, the surface amplitude is approximately 20°C. The predicted amplitude at 5 cm is $e^{-5/d} = 0.63$ as large as at the surface, or 12.7°C. Thus, the maximum temperature at this depth should be about 38°C. Similarly, at 10 and 20 cm, the amplitude should be 12.1 and 3.3, and the maximum temperatures should be 33 and 28°C, which compare reasonably well with the figure.

Annual Variations   The seasonal variations in soil temperature with depth are similar in character to the diurnal changes. The summer months (June and July in the Northern Hemisphere), like midday, represent the peak of the global radiation and the maximum temperatures. The winter months have an effect similar to nocturnal daily temperatures. The California data of Smith (1932) in Fig. 5.15 are typical of the variations observed.

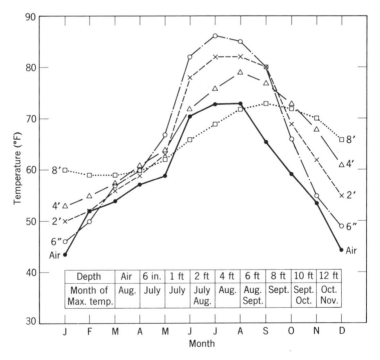

Figure 5.15   Monthly variation of soil temperature in relation to depth. (From data of Smith, 1932.)

From the middle of May until the first of August, the soil from 6 in. to 8 ft (15–240 cm) was warmer than the air, except for the 8-ft depth. The deeper layers were warmer during the winter months than the 6-in. depth; therefore, the heat flux was upward during this period. The temperature gradient reversed from about May 1 to the middle of September, so that heat flow was downward during this period. Maximum temperatures at 6 in. and 2, 4, and 8 ft occurred approximately on July 1, July 15, August 1, and September 1, respectively. These seasonal differences were associated with the incoming global radiation and the thermal properties of the soil profile as related to changes in moisture content and temperature gradients.

EXAMPLE 5.8: The curves in Fig. 5.15 for the soil temperatures yield the following values for $\Delta T = T_{max} - T_{min}$ and $t^* = $ time of maximum temperature.

Soil Depth (cm)	$T_{max} - T_{min}$ (°F)	t^* (assuming July 1 = 0) (days)
15	40	0
60	32	14
120	25.5	30
240	14	61

Plot $\ln \Delta T$ and t^* versus z, and use linear regression to estimate d and K_T for this soil.

Figure 5.16 shows the plots and the linear regression lines, both of which describe the data well.

From (5.56), $\ln \Delta T = \ln (2A) + z/d$. Therefore, the inverse slope of the line in Fig. 5.16a is d, which is equal to 217 cm.

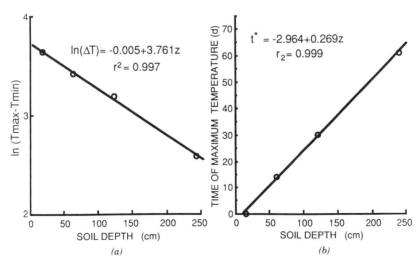

Figure 5.16 Graphical analysis of data in Fig. 5.15 using (a) Equation (5.56) and (b) Equation (5.59).

Also, from (5.59), the time of maximum temperature obeys the equation

$$t^* = \frac{\pi}{2\omega} - \frac{z}{\omega d} = \frac{\tau}{4} - \frac{z\tau}{2\pi d} \tag{5.62}$$

Therefore, the slope of the line in Fig. 5.16b is equal to $\tau/2\pi d$, from which we calculate $d = 216$ cm.

Obviously, the soil temperature model for the annual wave is very consistent with the data of Smith (1932) in Fig. 5.15.

PROBLEMS[1]

5.1   Two thermocouples are buried in the soil at $z = -75$ cm and $z = -150$ cm. During the year they record maximum and minimum temperatures as follows:

z	T_{max}	T_{min}
-75	20.75	-0.75
-150	17.70	2.30

Calculate the damping depth for the annual wave, the soil thermal diffusivity K_T, the annual average temperature, and the amplitude of the wave at the surface. What will be T_{max} and T_{min} at $z = -225$ cm?

5.2   A temperature difference of 20°C (0 and 20°C) is established across a soil column containing oven-dry quartz sand of thermal conductivity $\lambda_Q = 0.020$ cal cm^{-1} s^{-1} °C^{-1} and oven-dry silt $\lambda_s = 0.010$ cal cm^{-1} s^{-1} °C^{-1}. Each layer is 50 cm thick. Calculate the soil heat flux in cal cm^{-2} s^{-1} and the temperature at the interface in the middle of the column. See Fig. 5.17.

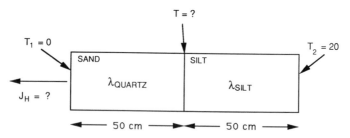

Figure 5.17   Illustration of experimental system.

5.3   The climate in Palm Springs, California, is characterized by hot summers and mild winters. A reasonable function to use to represent the annual soil surface temperature is

$$T(t) = 25 + 15 \sin(\omega t)$$

$$\omega = 2\pi \text{ per year}$$

[1]Problems preceded by a dagger are more difficult.

An individual builds a subterranean home in the desert, with a layer of 2.5 m of soil over the roof extending back to the soil surface. The soil has a thermal diffusivity $K_T = 400 \text{ cm}^2 \text{ day}^{-1} = 146{,}000 \text{ cm}^2 \text{ yr}^{-1}$.

Calculate the maximum and minimum annual temperatures of the roof of the house and determine how much later the maximum temperature will occur compared to the surface.

5.4 The thermal conductivity–water content data for a soil were fitted to the empirical function

$$\lambda(\theta) = a - be^{-c\theta}$$

and the following values were obtained:

$$a = 4.5 \times 10^{-3} \text{ cal cm}^{-1} \text{ °C}^{-1} \text{ s}^{-1}$$

$$b = 4 \times 10^{-3} \text{ cal cm}^{-1} \text{ °C}^{-1} \text{ s}^{-1}$$

$$c = 4$$

Use the following relation for soil volumetric heat capacity (in cal cm^{-3} °C^{-1}),

$$C(\theta) = 0.46(1 - \phi) + \theta$$

where ϕ is porosity. Calculate and plot $\lambda(\theta)$, $C(\theta)$, and $K_T = \lambda/C$ between oven dry and saturation ($\theta_s = 0.5$). Calculate the average value of K_T between 0.1 and 0.4 approximately. What is the maximum percentage of deviation from the average value in this range?

5.5 An experiment is conducted in which λ_e is measured to be $\lambda_e = 0.003$ cal cm^{-1} s^{-1} °C^{-1} at $T = 20$°C. From a second experiment, the water vapor mass flux J_v is fitted to the following model:

$$J_v = -0.015 \exp\left(\frac{T}{20}\right)\frac{dT}{dZ} \quad (\text{g cm}^{-2} \text{ day}^{-1})$$

where T is in degrees Celsius. Assume $H_v = 585$ cal g^{-1} is constant and λ^* is constant.

(a) Calculate λ^* from the information given.

(b) Calculate λ_e at $T = 30, 40, 50, 60$°C.

(c) Calculate the fraction of the total λ_e attributable to latent heat flow at each temperature.

5.6 Assume that the soil described in problem 5.1 is placed in a soil column of length $L = 50$ cm and that a temperature $T = 10$°C is placed at $z = 0$ and $T = 60$°C is placed at $z = 50$ cm. Calculate (a) the steady-state heat flux and (b) the temperature $T(z)$ within the column. (*Hint*: λ_e is a function of T.)

5.7   The surface temperature of a soil with $K_T = 300$ cm^2 day^{-1} is measured as a function of time for one day, when partial cloud cover causes changes to occur about every half h, as shown in Fig. 5.18.

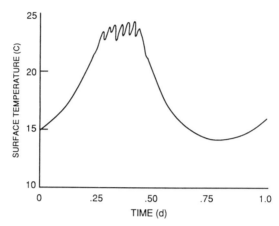

Figure 5.18   Surface temperature as a function of time.

Sketch a reasonable plot of $T(t)$ at $z = 15$ cm. Quantitatively justify why you sketched it in the way you did.

5.8   Two thick copper plates are used to sandwich a thin styrofoam plate as shown in Fig. 5.19. The left plate outer edge is held at $T = 5°C$ and the

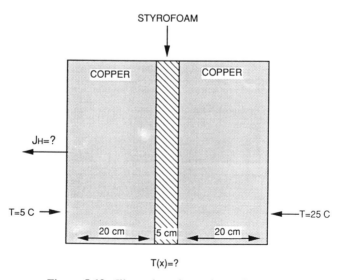

Figure 5.19   Illustration of experimental system.

right plate outer edge at $T = 25°C$. Calculate the heat flux through the layers and the temperature distribution everywhere in the interior ($\lambda_{copper} \sim 1$ cal cm^{-1} s^{-1} °C^{-1}; $\lambda_{styro} = 10^{-4}$ cal cm^{-1} s^{-1} °C^{-1}).

†5.9   The Fourier series formula for the square-wave surface temperature may be expressed as

$$T(0, t) = \begin{cases} T_0 & 0 < t < \tau/2 \\ 0 & \tau/2 < t < \tau \\ T_0 & \tau < t < 3\tau/2 \\ 0 & 3\tau/2 < t < 2\tau \quad \text{etc.} \end{cases}$$

where τ is the period of the square wave as given by

$$T(0, t) = \frac{T_0}{2} + \frac{2T_0}{\pi} \sum_{N=1}^{\infty} \frac{1}{2N - 1} \sin \frac{(4 N - 2)\pi t}{\tau}$$

Using the property that the sum of two solutions in the linear heat flow equation (5.42) is a solution of the equation, calculate the soil temperature $T(z, t)$ resulting from such a heat load at the surface. Plot the relative temperature $T(z, t)/T_0$ as a function of time at $z = 10, 20, 30$ cm assuming that $K_T = 400$ cm^2 day^{-1} and $\tau = 1$ day. [*Hint*: Each term in the series has a different period $\tau_N = \tau/(2N - 1)$.]

6 Soil Aeration

Soil aeration refers to the transport of gases through the soil air space and to the exchange of gases between the soil and the atmosphere. Two important gases in the soil air are carbon dioxide, which is produced as a byproduct of plant root respiration and biological activity, and oxygen, which is consumed in soil by the same processes.

Plant roots require oxygen to function normally. For most plant species, translocation of oxygen from the leaves to the roots is inadequate to supply O_2 at the required rate, and the roots must supplement their supply from the soil air. As the soil O_2 is depleted, it must be replaced by O_2 moving from the atmosphere above the surface into the soil. This transport occurs primarily by gaseous diffusion.

6.1 COMPOSITION OF SOIL AIR

The composition of the gases comprising the soil air depends upon the rate of respiration of microorganisms and plant roots, on the solubility of CO_2 and O_2 in water, and on the rate of gaseous exchange with the atmosphere.

In general, the CO_2 concentration in soil air will always exceed the small atmospheric value of 0.03%, because CO_2 is produced as a byproduct of plant root respiration and microbial breakdown of carbon-based organic compounds in the soil. The extent of buildup of CO_2 gas depends not only on the intensity of these processes but also on the ease with which the gas can escape to the atmosphere. In poorly aerated soils with substantial respiratory and microbial activity, it is not uncommon to observe CO_2 concentrations that are hundreds of times higher than atmospheric levels (Russell, 1973).

The major processes that produce CO_2 in soil also decrease O_2 concentrations. Thus, it is common to observe complementary profiles of these two gases, with CO_2 rising and O_2 falling with distance below the soil surface. The concentrations of both gases are likely to vary at a given location at different times of the year (Fig. 6.1). These seasonal differences reflect not only changes in respiratory and microbial activity but also varying diffusive resistance in the soil caused by differences in water content.

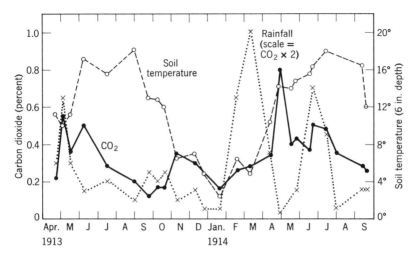

Figure 6.1 Seasonal variation in the CO_2 content of soil air.

6.2 GAS REACTIONS IN SOIL

6.2.1 CO_2 Production in Soil

Production of CO_2 gas in soil will be highest when microbial and plant activity is at a maximum, which will depend on the crop and climate. The metabolic activity of microorganisms tends to increase with temperature up to a maximum and then decrease as temperature is raised further. In addition, CO_2 concentrations will build up when the soil is moist and transport of gas to the atmosphere is impeded. Figure 6.2 shows O_2 and CO_2 profiles over time for a well-drained sandy loam soil and a poorly drained silty clay from the same geographical area. The clay soil shows much more pronounced O_2 depletion and CO_2 buildup than the well-aerated soil (Russell, 1973).

Wide ranges of CO_2 evolution rates in soil have been observed under different conditions. Monteith et al. (1964) reported CO_2 fluxes of 1.5 g m^{-2} day^{-1} in winter and 6.7 g m^{-2} day^{-1} in summer for a clay soil bare of vegetation. Currie (1970) measured values of 1.2 g m^{-2} day^{-1} in winter and 16 g m^{-2} day^{-1} in summer in bare soil and corresponding values of 3.0 and 35 g m^{-2} day^{-1} in soil cropped with kale.

6.2.2 O_2 Consumption in Soil

When the soil is aerobic, the volumes of CO_2 evolved and O_2 depleted tend to be comparable. However, O_2 depletion has frequently been found to be less than the commensurate CO_2 evolution, which implies that anaerobic respiration is occurring in part of the soil volume. Clark and Kemper (1967) reported mean values of

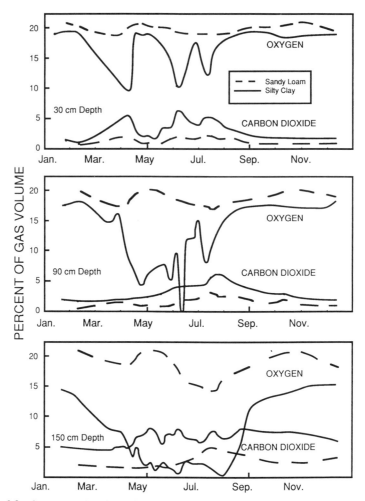

Figure 6.2   Oxygen and carbon dioxide content of the soil air at three depths in a sandy loam and a silty clay apple orchard. (After Russell, 1973. Reprinted with permission of Longman.)

O_2 depletion between 2.5 and 5.0 g m^{-2} day^{-1} in bare soil and values approximately twice as high under cropped conditions. Greenwood (1971) measured a mean consumption rate of about 8 g m^{-3} day^{-1} in a crop at full cover. In the study by Currie (1970), the O_2 consumption rates were between 60 and 75% of the CO_2 production rates, reaching a maximum of 24 g m^{-2} day^{-1} under cropped kale in the summer.

Even in bare soil, the rate of oxygen consumption by native microorganisms would rapidly remove all oxygen from the soil air if not replaced from the soil atmosphere. This is illustrated in Example 6.1.

EXAMPLE 6.1: Assuming that the soil air is at atmospheric O_2 concentration, calculate how long it would take to deplete the O_2 level of the top 0.5 m of a soil that is at 0.20 volumetric air content if the O_2 consumption rate by microorganisms is 5.0 g m^{-2} day^{-1}. Assume that no gas enters the soil from the atmosphere.

The atmospheric O_2 concentration is approximately 300 g m^{-3}. The top 0.5 m of soil contains 0.1 m^3 m^{-2} of soil air. Thus, there is an oxygen supply of about 30 g m^{-2} in the top 0.5 m, which would last about six days if not replenished from above.

As the previous example illustrates, biological activity in the soil requires frequent replenishment of oxygen from the atmosphere above the surface. Because it is evolved as oxygen is consumed, carbon dioxide gas similarly would reach very high levels in soil if the gas did not escape to the atmosphere. Romell (1922) estimated that the CO_2 concentration at 20 cm in the soil could double in 1.5 h and increase by a factor of 10 in 14 h just from bacterial production if gaseous exchange with the atmosphere was prevented.

6.3 GAS TRANSPORT THROUGH SOIL

6.3.1 Gas Conservation Equation

In this section, the mathematical framework for describing gas transport through soil will be developed. The starting point for the analysis, just as it was with water and heat flow, is a statement of conservation for the mass species (in this case a gas) which is subject to transport and transformation processes in the soil.

To simplify the analysis, we will assume that the gas is moving in the z direction only and that the gas is insoluble. Later in the book we will analyze transport of chemicals that have a substantial solubility as well as vapor density in soil.

As we did with water and heat flow in soil, we imagine that there is a small cubical volume of dimensions Δx, Δy, Δz located inside the soil (see Fig. 3.19). The chemical species we wish to characterize is flowing as a gas into and out of the volume in the z direction, reacting in the soil, and changing its mass concentration within the soil air space. Thus, the statement of mass conservation for the chemical in the volume $V = \Delta x\, \Delta y\, \Delta z$ during a short period of time from t to $t + \Delta t$ is

Mass of chemical entering volume through area $\Delta x\, \Delta y$ at z during Δt

$=$ mass of chemical leaving volume through area $\Delta x\, \Delta y$

at $z + \Delta z$ during Δt

$+$ increase of chemical mass stored in volume during Δt

$+$ loss of chemical mass from volume during Δt by biological

or chemical reactions $\hspace{3cm}$ (6.1)

We can develop a differential equation to express this conservation statement mathematically by defining some functions to represent the processes described in (6.1).

The mass of chemical per unit soil volume, C_T, is called the total chemical or solute concentration. When the chemical is insoluble, this mass is all stored in the vapor space. The flow of gas per unit area per unit time, J_g, is called the soil gas flux. The loss of mass per unit soil volume per unit time, r_g, is called the gas reaction loss rate. It is sometimes called a sink term.

Equation (6.1) is a special case of the more general solute mass balance equation. It will be generalized in Chapter 7 to allow dissolved and adsorbed solute phases to be present as well.

The relationship between the mathematical functions, which are assumed to depend on z and t but not on x and y, and the conservation statement in (6.1) is as follows:

Mass of gas entering V at z during Δt = flux $\times$ area $\times$ time

$$= J_g(z, t + \tfrac{1}{2} \Delta t)(\Delta x \, \Delta y) \, \Delta t \tag{6.2}$$

Mass of gas leaving V at $z + \Delta z$ during $\Delta t = J_g(z + \Delta z, t + \tfrac{1}{2} \Delta t) \, \Delta x \, \Delta y \, \Delta t$

Increase of mass stored in volume during Δt $\qquad\qquad\qquad$ (6.3)

$$= \left(\frac{\text{final mass}}{\text{volume}} - \frac{\text{initial mass}}{\text{volume}} \right) \times \text{volume}$$

$$= \left[C_T(z + \tfrac{1}{2} \Delta z, t + \Delta t) - C_T(z + \tfrac{1}{2} \Delta z, t) \right] \Delta x \, \Delta y \, \Delta z \tag{6.4}$$

Loss of mass from V during Δt by reactions $= \dfrac{\text{loss}}{\text{volume/time}} \times \text{volume} \times \text{time}$

$$= r_g(z + \tfrac{1}{2} \Delta z, t + \tfrac{1}{2} \Delta t)(\Delta x \, \Delta y \, \Delta z) \, \Delta t \tag{6.5}$$

After plugging (6.2)–(6.5) into (6.1) and dividing by $\Delta x \, \Delta y \, \Delta z \, \Delta t$, we obtain

$$\frac{C_T(\bar{z}, t + \Delta t) - C_T(\bar{z}, t)}{\Delta t} + \frac{J_g(z + \Delta z, \bar{t}) - J_s(z, \bar{t})}{\Delta z} + r_g(\bar{z}, \bar{t}) \tag{6.6}$$

where $\bar{z}$ and $\bar{t}$ are the average values of z and t. In the limit $\Delta z \to 0$, $\Delta t \to 0$, (6.6) reduces to the differential form of the gas conservation equation:

$$\frac{\partial C_T}{\partial t} + \frac{\partial J_g}{\partial z} + r_g = 0 \tag{6.7}$$

Notice the similarity of this equation to the water conservation and heat conservation equations (3.65) and (5.37), respectively.

For an insoluble gas, all of the solute mass resides in the soil air space, and the total chemical concentration C_T reduces to

$$C_T = aC_g \tag{6.8}$$

where a is the volumetric air content (volume of soil air per volume of soil) and C_g is the soil gas concentration (mass of gas per volume of soil air). With (6.8), the gas conservation equation (6.7) reduces to

$$\frac{\partial}{\partial t}(aC_g) + \frac{\partial J_g}{\partial z} + r_s = 0 \tag{6.9}$$

where the volumetric air content may be removed from the time derivative if it does not change with time. Equation (6.9) is called the gas conservation equation for an insoluble gas.

Before (6.9) can be applied to solve gas transport problems, a gas flux law must be specified, relating J_g to the gas concentration C_g.

6.3.2 Gas Convection in Soil

In principle, gases in soil can be displaced by bulk movement of the soil air phase in response to differences in the total air pressure. Within the air space of the soil, these pressure changes could be induced by soil temperature changes, barometric pressure fluctuations in the atmosphere above the surface, wind blowing over the soil surface, and infiltration. These effects are described in what follows.

Temperature Effects Temperature may influence the renewal of soil air in two ways. First, there may be temperature differences between various parts of the soil. It is possible that the contraction and expansion of the air within the pore spaces as well as the tendency for warm air to move upward may cause some exchange between soil regions with different temperatures and perhaps with the atmosphere. Second, the soil and the atmosphere usually have different temperatures. This temperature differential could cause an exchange of air volume between the atmosphere and the soil air in the immediate surface layer.

It is difficult to estimate the significance of temperature effects on gas exchange in soils. Romell (1922) suggested that the contribution to normal soil aeration caused by daily variations of temperature within the soil was less than 0.13% and that caused by temperature differences between soil and atmosphere could be responsible for no more than 0.5%. Thus, it appears that temperature is a minor factor in soil aeration.

Barometric Pressure Effects Theoretically, according to Boyle's law, any increase in the barometric pressure of the atmosphere should cause a compression and subsequent decrease in the volume of the soil air, thereby allowing air from the atmosphere to penetrate the soil pores. On the other hand, any decrease in barometric pressure should produce an expansion of the soil air, allowing part of it to enter the atmosphere above the soil. Thus, gas exchange between soil and atmosphere caused by pressure fluctuations should occur from time to time, provided that changes in the barometric pressure of the atmosphere are not damped out before they influence the air pressure within the soil pores.

Buckingham (1904) calculated the magnitude of the soil–atmosphere gas exchange due to barometric pressure changes and showed that atmospheric air would penetrate a permeable soil column 3 m deep only to a depth of about 3.0–5.6 mm, depending upon the magnitude of the barometric pressure change. Thus, it is evident that fluctuations in atmospheric pressure have little influence upon soil aeration, even where there is good contact between the soil air and atmosphere. Romell (1922) estimated that no more than about 1% of the normal aeration of soils can be attributed to variations in barometric pressure.

Wind Effects   One might expect that the pressure and suction effects of high winds at the soil surface would exert some influence upon the renewal of the soil air. Romell (1922) also analyzed this contribution to soil aeration and concluded that wind action could not be responsible for more than about 0.1% of the normal aeration on vegetated soils. More recent experiments on the effects of air turbulence on the transfer of vapor in the soil suggest that bulk flow of air in response to pressure fluctuations may not be negligible in porous, bare soils (Farrell et al., 1966). For example, air blowing across a bare surface with a wind speed of 15 mph can penetrate coarse sand and mulches to a depth of several centimeters. Even though there is no net mass flow of soil gases, the fluctuations in air pressure at the soil surface result in a mixing of air within the surface that enhances gas transport beyond that due to diffusion alone (Scotter et al., 1967).

Rainfall Effects   The infiltration of rainfall into the soil may cause a renewal of the soil air in two ways. First, as the water infiltrates, it displaces soil air from the pores. During the subsequent redistribution of the infiltrating front, the pores are refilled with air from above. Second, soil air is enriched by dissolved O_2 that is carried into the soil by infiltrating water and exchanged later with the soil air phase. We might at first expect that a large rainfall could completely renew the soil air, especially if the water is able to displace the major portion of the air within the pores. In many instances, however, a considerable amount of entrapped air remains, which is not forced out of the soil by infiltered water. Romell (1922) estimated that rainfall accounts for only about 7–9% of the normal aeration, depending on the soil type and climate.

Mass Flow of Gases into Buildings   Concern over high levels of radon gas in certain homes has prompted intense research recently on the mechanisms by which this gas moves from the soil. Radon gas is a vapor phase reaction product formed during the radioactive decay of radium, a natural constituent of many soils. Recent studies (Nero and Nazeroff, 1985) have verified that substantial air pressure differences exist between the soil air and the interior of houses in contact with the soil air. The lower pressure of the air within the house is a consequence of ventilation, circulation of air in the heating and cooling systems, and the geometry of the home and can permit substantial entry of soil gases into the house by mass flow. Once in the home, the radon gas can pose a substantial health hazard since it is a potent carcinogen. It appears to be possible to minimize the air entry by altering the points of contact between the soil and the house (Nazeroff et al., 1987).

6.3.3 Gas Diffusion

Since convection is usually negligible except under the preceding special circumstances, the major mechanism of gas transport is diffusion of vapor within the soil air space.

In free air, the gas diffusion flux J_g (mass per area per time) is expressed by Fick's law of diffusion:

$$J_g = -D_g^a \, \partial C_g / \partial z \tag{6.10}$$

where D_g^a (in square centimeters per second) is the binary gaseous diffusion coefficient in free air. Its value depends on the chemical diffusing in the air and on the temperature and air pressure (Bird et al., 1960). At atmospheric pressure and $T = 25°C$, it varies between about 0.15 and 0.25 cm^2 s^{-1} for gases of low molecular weight and decreases with increasing size of the diffusing molecule. Jury et al. (1983a) recommended using an average value of about 0.05 cm^2 s^{-1} for pesticides of intermediate molecular weight (e.g., 100–300 g/mol). For small molecules like CO_2, the value is higher ($D = 0.14$ cm^2 s^{-1}, *Handbook of Chemistry and Physics*, 1987).

Equation (6.10) will overestimate the flux of gas through soil, both because gas must diffuse through a longer path length to get from one point to the other and because the cross-sectional area available for flow is reduced by solid and liquid barriers. Consequently, the gas flux in soil is calculated by modifying the diffusion flux in air (6.10) by a gas tortuosity factor $\xi_g < 1$, producing the new flux equation

$$J_g = -\xi_g D_g^a \frac{\partial C_g}{\partial z} = -D_g^s \frac{\partial C_g}{\partial z} \tag{6.11}$$

where $D_g^s = \xi_g D_g^a$ is the soil gas diffusion coefficient.

A number of studies have been conducted to determine the influence of the porous medium on gas diffusion through soil. In early work, such studies were confined to porous media with very high air content a, so that the principal factor affecting the tortuosity ξ_g was the solid matrix. Buckingham (1904) proposed an equation of the form $\xi_g = \epsilon a$, where ϵ is a constant. Penman (1940a,b) studied the diffusion of carbon dioxide through packed soil cores with a range of $0.195 < a < 0.676$ and recommended 0.66 as an average value for ϵ. Flegg (1953), working with soil aggregates in the range of $0.35 < a < 0.75$, obtained values of ϵ

between 0.53 and 0.89. Using soil with a porosity of 0.355, van Bavel (1952) measured a value of ϵ of 0.58.

When soil is not air dry, the relation between ξ_g and air content becomes more complex. Currie (1965) found a nonlinear relation between ξ_g and a for various porous media as a function of relative saturation. He suggested that a power law of the form $\xi_g = \alpha a^\beta$ be used in moist soil. A theoretically based model for ξ_g was derived by Millington (1959) and Millington and Quirk (1961) in which

$$\xi_g = a^{10/3}/\phi^2 \tag{6.12}$$

where ϕ is the porosity.

This model has been shown to be in good agreement with data over a wide range of soil water content (Sallam et al., 1984). A summary of various tortuosity models proposed in the literature is given in Table 6.1. A plot of these models as a function of a taken from Sallam et al. (1984) demonstrates that the various models agree closely in the air-dry region, where most of the early measurements were taken (Fig. 6.3).

TABLE 6.1   Models for Gas Tortuosity Factor ξ_g as Function of Volumetric Air Content a

Model	ξ_g
Penman (1940a,b)	$0.66a$
Marshall (1959)	$a^{1.5}$
Currie (1960)	γa^μ
Burger (1919)	$a/[k - a(k - 1)]$
Millington and Quirk (1961)	$a^{10/3}/\phi^2$

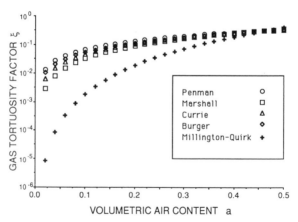

Figure 6.3 Theoretical ratios of gas diffusion coefficient in soil to gas diffusion coefficient in air as a function of volumetric air content proposed by various scientists. (After Sallam et al., 1984.)

Measurement of Gas Diffusion Coefficients in Soil   The standard method for measuring gas diffusion coefficients in soil in the laboratory is the diffusion chamber method (Rolston, 1986a). In this method, a soil column that is devoid of gas and is open at one end is suddenly brought into contact with a closed air chamber containing a concentration of the gas of interest. Over time, the gas diffuses from the chamber into the soil at a rate that depends on the soil gas diffusion coefficient. The concentration in the air chamber can be related theoretically to the diffusion coefficient, and therefore the latter can be selected so as to optimize the agreement between theory and experimental observation.

Measurement of Gas Flux in Field   Soil gas flux is difficult to measure in the field because the gas diffusion coefficient is normally not known. Therefore, (6.11) cannot be used directly to evaluate J_g. The most common methods for estimating gas flux involve flux chambers at the soil surface (Rolston, 1986b). These chambers either allow gas to accumulate in a box placed over the surface (Matthias et al., 1980) or trap the gas at the surface while maintaining a low concentration (Ryden et al., 1978). In either method, only the gas escaping at the surface is measured, which might be very different from the rate of production (Jury et al., 1982a). Errors involved in the operation and interpretation of these devices are discussed in Rolston (1986b).

6.3.4 Gas Transport Equation

The gas transport equation is formed by inserting the gas flux equation (6.11) into the gas conservation equation (6.9). If we assume D_g^s and a are constant, we obtain

$$a \frac{\partial C_g}{\partial t} = D_g^s \frac{\partial^2 C_g}{\partial z^2} + r_g \qquad (6.13)$$

This equation describes the changes in gas concentration as a function of space and time in soil.

6.4. GAS TRANSPORT MODELING IN SOIL

The gas transport equation (6.13) is mathematically identical to the heat transport equation (5.41) when the gas reaction term r_g is absent. For this reason, the heat transport examples discussed in the previous chapter equally well apply to inert gas transport problems with the same boundary conditions. This is illustrated in the next example.

EXAMPLE 6.2: An inert gas ($r_g = 0$) diffuses through a soil column whose inlet end $z = 0$ is at a uniform concentration C_0 while the outlet end $z = L$ of the column is maintained free of gas, $C_L = 0$. Calculate the steady-state gas concentration within the column.

In steady state, $\partial C_g / \partial t = 0$. Thus, the gas transport equation (6.13) reduces to

$$\frac{d^2 C_g}{dz^2} = 0 \tag{6.14}$$

This may be integrated once over z, so that

$$\frac{dC_g}{dz} = \phi_1 \tag{6.15}$$

where ϕ_1 is an unknown constant of integration.

Equation (6.15) may be rearranged to the form

$$dC_g = \phi_1 \, dz \tag{6.16}$$

and integrated a second time, with the result

$$C_g = \phi_1 z + \phi_2 \tag{6.17}$$

where ϕ_2 is a constant of integration. The two constants may be evaluated using the boundary conditions

$$C_g = C_0 = \phi_2 \quad \text{at } z = 0 \tag{6.18a}$$

$$C_g = 0 = \phi_1 L + C_0 \quad \text{at } z = L \tag{6.18b}$$

Thus, $\phi_1 = -C_0 / L$, and the gas concentration profile is

$$C_g(z) = C_0 (1 - z/L) \tag{6.19}$$

6.4.1 Steady-State O_2 Transport and Consumption

Most transport processes occurring in the natural environment are time dependent. However, steady-state solutions to certain problems whose parameters do not change appreciably over time may provide reasonable approximations to the actual situation. One such problem is oxygen consumption in the surface zone of soil. In a fully developed crop root zone, it is reasonable to treat the oxygen consumption rate as constant in space and time. Since the O_2 concentration in the atmosphere is uniform, a steady-state flow of gas from the atmosphere to the soil is possible, provided that gas cannot escape below the root zone. In steady state, the gas conservation equation (6.13) reduces to

$$\frac{dJ_g}{dz} + r_g = 0 \tag{6.20}$$

If we assume that no gas flows below the root zone at $z = -L$, then the lower boundary condition may be expressed as

$$J_g(-L) = 0 \tag{6.21}$$

Equation (6.20) may be rearranged to the form

$$dJ_g = -r_g \, dz \tag{6.22}$$

which may be integrated from $z = -L$ to z:

$$\int_0^{J_g} dJ_g = -r_g \int_{-L}^z dz \qquad (6.23)$$

This trivial integration results in

$$J_g(z) = -r_g(z + L) \qquad (6.24)$$

Thus, the gas flux decreases linearly from a maximum value of $-r_g L$ at the surface to a low of zero at the bottom of the root zone. It is negative because the direction of flow is downward.

The gas concentration in the profile may be calculated by substituting the gas flux equation (6.11) into (6.24):

$$-r_g(z + L) = J_g = -D_g^s \, dC_g/dz \qquad (6.25)$$

Equation (6.25) is a first-order differential equation, which may be integrated after all factors depending explicitly on z are placed on the same side (see Section 3.5):

$$dC_g = \frac{r_g}{D_g^s} (z + L) \, dz \qquad (6.26)$$

Equation (6.26) may now be integrated from $z = 0$ (where $C_g = C_0$) to z:

$$\int_{C_0}^{C_g} dC_g = \frac{r_g}{D_g^s} \int_0^z (z + L) \, dz \qquad (6.27)$$

The integrals in (6.27) are straightforward, and we obtain

$$C_g(z) = C_0 + \frac{r_g}{D_g^s} \left(\frac{z^2}{2} + Lz \right) \qquad (6.28)$$

Equation (6.28) is plotted in Fig. 6.4 for various values of $r_g L^2/D_g^s$. Note that for large values of this parameter the gas concentration drops to zero within the root zone. This behavior is explored further in the next example.

EXAMPLE 6.3: Calculate the maximum value of soil air content a below which some part of the root zone will be completely depleted of oxygen. Assume that the Millington–Quirk model of tortuosity (6.12) is valid.

Since the gas concentration decreases uniformly with depth, we can define the condition at which the soil begins to suffer from oxygen depletion by setting the concentration (6.28) equal to zero at the bottom of the root zone:

$$C_g(-L) = 0 = C_0 - \frac{r_g L^2}{2D_g^s} \qquad (6.29)$$

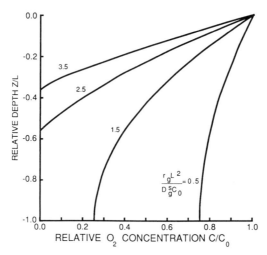

Figure 6.4 The O_2 profiles in a root zone as a function of the dimensionless parameter $\xi = r_g L^2 / D_g^s C_0$.

Thus, the critical value of the soil gas diffusion coefficient that satisfies (6.29) is

$$D_g^s = \xi_g(a)D_g^a = \frac{r_g L^2}{2 C_0} \qquad (6.30)$$

If we substitute the Millington–Quirk tortuosity model (6.12) into (6.30), we may solve for the value of air content that satisfies (6.29):

$$a = \left(\frac{r_g L^2 \phi^2}{2 C_0 D_g^a}\right)^{3/10} \qquad (6.31)$$

This is the smallest value of air content for which the entire root zone receives oxygen. Above this value, we can expect symptoms of oxygen deficiency to develop.

6.4.2   Steady-State and Transient CO_2 Transport and Evolution

The steady-state model used for O_2 depletion in a root zone is less applicable to the problem of CO_2 evolution because the evolving gas can diffuse downward as well as upward unless a barrier to downward diffusion, such as a water table, is present near the surface. This special case is discussed in the next example.

EXAMPLE 6.4:   Calculate the steady-state CO_2 profile in soil, assuming that the gas is produced at a constant rate $r_g = -R$ (negative because of production rather than consumption of gas) and that the CO_2 flux is zero at $z = -L$.
   The gas conservation equation (6.9) reduces to

$$\frac{dJ_g}{dz} - R = 0 \qquad (6.32)$$

for the case of steady flow with uniform evolution of CO_2. Equation (6.32) may be easily integrated from $z = -L$ to z, producing

$$J_g = R(z + L) \tag{6.33}$$

Thus, the flux has a maximum value of RL at the surface. When the gas flux equation (6.11) is inserted into (6.33), the result is a first-order differential equation for the concentration:

$$J_g = -D_g^s \, dC_g/dz = R(z + L) \tag{6.34}$$

Equation (6.34) may be integrated after placing all factors that depend on z on one side:

$$dC_g = -\frac{R}{D_g^s}(z + L) \, dz \tag{6.35}$$

Equation (6.35) may now be integrated from $z = 0$ (where $C = C_0$) to z, producing the concentration profile

$$C_g(z) = C_0 - \frac{R}{D_g^s}\left(\frac{z^2}{2} + Lz\right) \tag{6.36}$$

Equation (6.36) is plotted in Fig. 6.5 as a function of $RL^2/C_0 D_g^s$. Large values of this parameter, such as when the air content is low and diffusion resistance is high, result in substantial CO_2 evolution in the soil.

It is interesting to contrast the steady-state solution in Fig. 6.5 with the solution to the time-dependent problem described by (6.13) for the case where there is no barrier to downward gas diffusion in the soil. We may solve this problem by assuming that the gas production term is equal to $r_g = -R$ for $0 > z > -L$ and r_g

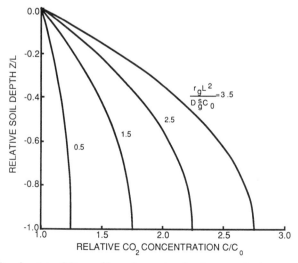

Figure 6.5 Steady state CO_2 profiles when a barrier is present at $z = -L$ for various values of the dimensionless parameter $\xi = r_g L^2/D_g^s C_0$.

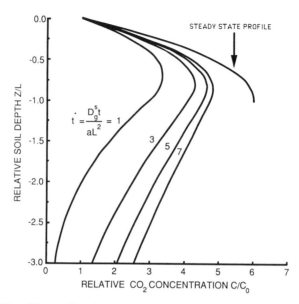

Figure 6.6 The CO_2 profiles in the soil as a function of dimensionless time $t^* = D_g^s t/aL^2$ when no barrier is present at $z = -L$.

$= 0$ for $z < -L$ and that the air content is uniform. Equation (6.13) with this value of the source term r_g may be solved by the method of Laplace transforms (Arfken, 1985). The solution is shown in Fig. 6.6 for the case $RL^2/2D_g^2C_0 = 1$. Note that concentration always lies below the steady concentration calculated for the case of no flow below $z = -L$ because now gas may diffuse downward as well as upward. Steady state can only be reached when the soil air below the root zone reaches the steady-state value at $z = -L$. This situation is unlikely to ever occur in soils with deep water tables.

6.4.3 O_2 Depletion at the Plant Root Interface

The O_2 concentration in the soil air space at a given depth in the soil is regulated by the diffusive resistance of the soil and by the rate of demand of the plant roots and soil biota. For a given O_2 level, the rate of entry of gas to the root is regulated further by diffusion from the air space into the root (Currie, 1962; Lemon, 1962). Thus, the amount of O_2 that reaches the root surface is controlled both by the rate of gaseous exchange between the soil air and the atmosphere and by the rate of transfer of O_2 from the soil pores to the root. The latter process takes place through water films that exist around the plant roots and the soil particles; therefore, the final O_2 diffusion to the root must take place through the liquid phase. Wiegand and Lemon (1958) proposed the following equation to explain the O_2 supply to plant roots:

$$C_R = C_p - \frac{qR^2}{2D_l^s} \ln \frac{R_e}{R} \tag{6.37}$$

where C_R is the oxygen concentration dissolved in water at the root surface (g cm^{-3}), C_p the dissolved O_2 concentration at the liquid–gas interface (g cm^{-3}), q the rate of O_2 consumption within the root (g cm^{-3} s^{-1}), R the root radius (cm), R_e the radius of the root plus the moisture film (cm), and D_l^s the diffusion coefficient in the liquid–solid soil matrix around the root (cm^2 s^{-1}). Equation (6.37) is derived by solving the steady-state diffusion equation in cylindrical coordinates.

The liquid diffusion coefficient in soil, D_l^s, may be represented by a tortuosity model

$$D_l^s = \xi_l(\theta)D_l^w \tag{6.38}$$

where D_l^w is the diffusion coefficient of O_2 in pure water and $\xi_l(\theta)$ is the liquid diffusion tortuosity factor.

The water film around the root represents an enormous barrier to diffusion. The ratio of D_l^w to D_g^a is approximately 10^{-4} for gases that also dissolve. This means that a 1-mm water film offers about the same diffusion flow resistance as 10 m of air with the same concentration difference across it. As a result, even relatively high values of O_2 in the soil air may not be sufficient to prevent aeration problems in plant roots surrounded by thick water films (Kristensen and Lemon, 1964).

6.5 FLOW OF WATER VAPOR THROUGH SOIL

6.5.1 Water Vapor Flux Equation

Because Fick's law of diffusion in soil [Eq. (6.11)] describes well the transport of gases of low solubility in soil, the first researchers studying water vapor movement assumed that it would behave similarly to other gases. Thus, they wrote the vapor flux equation as

$$J_{wv} = -\xi_g D_v^a \frac{\partial \rho_v}{\partial z} \tag{6.39}$$

where ξ_v is the gas tortuosity factor, which might be represented by either the Penman or the Millington–Quirk model (see Table 6.1). The water vapor density or vapor concentration ρ_v can be expressed in terms of the temperature and matric potential head h as

$$\rho_v = \rho_v^* \cdot \text{RH} = \rho_v^* \exp \frac{hM}{\rho_w RT} \tag{6.40}$$

where RH is the relative humidity and ρ_v^* is the saturated vapor density. Unless the soil is very dry (i.e., $h < -20,000$ cm) the vapor density may be approximated by its saturated value, which is a function only of temperature. Thus, we may rewrite (6.40) in the region where the vapor approximation is valid as

$$J_v = -\xi_g D_v^a \frac{d\rho_v^*}{dT} \frac{\partial T}{\partial x} \quad (\text{RH} \sim 1) \tag{6.41}$$

Note that the slope of the saturated vapor density–temperature curve is a known property of water and may be looked up in any handbook. It was thought that (6.41) should be able to describe water vapor movement in soil. However, in laboratory tests of (6.41), when a temperature gradient was placed across a soil column, the measured vapor flux was up to 10 times larger than the predicted value (Gurr et al., 1952). The reason for this discrepancy eluded researchers for many years but was finally resolved satisfactorily by a theoretical paper written by Philip and de Vries in 1957.

The model of Philip and de Vries contains two theoretical assumptions causing it to differ from the simple theory described by (6.41). Their first observation was that water vapor may condense on one side of a liquid barrier and evaporate across the other side, thereby short circuiting part of the liquid-filled pore space (Fig. 6.7).

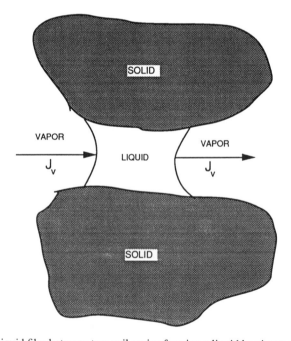

Figure 6.7   Liquid film between two soil grains forming a liquid barrier to vapor transport.

Thus, until the porous medium is very wet and the liquid-filled pore space is continuous, the available pore space for diffusion to be used in the tortuosity factor in (6.41) is not the air content a but $a + \theta = \phi$.

The second enhancement mechanism proposed by Philip and de Vries (1957) relates to the use of an average temperature gradient in (6.41). The thermal conductivity of the solid phase is very much greater than that of the liquid phase, which in turn is very much greater than the thermal conductivity of the air phase. Since water vapor moves primarily through the air spaces, the mean temperature gradient measured by sensors that average over all three phases should be much smaller than the temperature gradient across the vapor-filled space, which drives the water vapor. Thus, (6.41) should be written as

$$J_{wv} = -\xi' D_v^a \frac{d\rho_v^*}{dT} \omega \frac{dT}{dx}$$ (6.42)

where

$$\xi' = 0.66\phi \quad \text{or} \quad \phi^{4/3}$$ (6.43)

and

$$\omega = \left(\frac{\partial T}{\partial x}\right)_v \bigg/ \left(\frac{\partial T}{\partial x}\right)_{\text{soil}}$$ (6.44)

By using a theoretical model of the soil thermal conductivity (de Vries, 1963), the authors were able to calculate values for ω that ranged from 2 to 3. With the two enhancement mechanisms, the modified theory of Philip and de Vries given by Equations (6.42)–(6.44) removed most of the discrepancies between theory and experiment.

6.5.2 Approximate Water Vapor Flux Law

A detailed analysis of a modified form of the Philip and de Vries (1957) model showed that a very simple approximate flux law may be used to describe water vapor movement in soil. This flux law is given by

$$J_v \approx -2F(T) \, dT/dx$$ (6.45)

where $F(T)$ is a known function of the temperature that may be calculated without doing any experiments. Details of this approximate flux law are given in Jury and Letey (1979). Equation (6.45) was shown to be in reasonable agreement with existing experimental data, while having the advantage of circumventing the extremely difficult problem of measuring water vapor movement directly.

6.6 MEASUREMENT OF O_2 DIFFUSION AND CONSUMPTION IN SOIL

One of the earliest methods used to determine the O_2 diffusion coefficient in the soil pore space was the technique described by Raney (1949). It consisted of allowing O_2 to diffuse into a chamber that was inserted into the soil. From analyses of the gas concentration in the chamber made at 10-min intervals, the O_2 diffusion rate was estimated from a solution to the gas flow equation (6.13) applied to the geometry of the chamber.

Willey and Tanner (1963) determined the O_2 concentration in the soil air with a polarographic electrode, which is enclosed in a plastic membrane that is permeable to O_2. When a small voltage is applied between the polarographic cathode and the nonpolar anode, almost all of the O_2 that diffuses to the cathode is reduced and the concentration of O_2 at the electrode surface is very close to zero. The resulting current is proportional to the rate of reduction of O_2, which in turn is limited by the rate of diffusion of O_2 to the electrode. The measurement is applicable to O_2 diffusion in situ.

The platinum electrode was introduced by Lemon and Erickson (1952) to measure the oxygen diffusion rate (ODR) in soils. The method is based upon the principle that the electric current resulting from the reduction of O_2 at the platinum surface is governed (within limits) solely by the rate at which O_2 diffuses to the electrode surface. By placing a known voltage between the platinum electrode and a saturated calomel cell, the authors were able to calculate the O_2 diffusion rate from the observed current. A recent review of the platinum electrode and its use in soils is presented by Phene (1986).

According to Fick's law of diffusion (6.11), the measured ODR for an electrode placed in soil should depend on both the O_2 concentration gradient at the electrode surface and the O_2 diffusion coefficient in the water film around the electrode. Thus, the measured value of ODR represents the integrated effects of the O_2 concentration in solution, the O_2 diffusion coefficient, the thickness of the moisture film around the electrode, the length of the diffusion path to the electrode, and the electrode radius (Letey and Stolzy, 1964; Stolzy and Letey, 1964). The method fails in relatively dry soils, since the electrode must be wetted so that the whole electrode surface is covered by a moisture film. For this reason, it works better in finer textured soils whose pores remain filled at lower matric potentials than in coarse-textured soils. The ODR decreases as the moisture content increases, because the water film around the electrode becomes thicker, increasing the gas diffusion resistance.

Temperature affects the ODR by increasing the rate of plant respiration and the diffusion coefficient of O_2 in water. Also, the solubility of O_2 in the liquid phase is decreased. These combined effects result in about a 1.8% increase in the ODR per degree Celsius (Stolzy and Letey, 1964).

Because so many factors can affect the ODR measurement, it is only qualitatively related to the aeration status of soil (McIntyre, 1971). It is important therefore that standard techniques be used so that results from different investigators will be comparable.

Bertrand and Kohnke (1957) obtained a good correlation between the ODR values measured at the 20-cm depth with the platinum electrode and the gas tortuosity factor $\xi_g = D_g^s / D_g^a$ determined at the same depth with an oxygen analyzer. However, the correlation was not as good at the 40-cm depth. Wiersum (1960) found that visual observations of the depth of root penetration correlated very well with measurements by the platinum electrode.

Because of the similarity between the electrode shape and a plant root, ODR measurements can give relative estimates of soil aeration status when they are taken at different depths. Figure 6.8 shows measured ODR values with depth as a function of O_2 concentration at the soil surface. The values decrease with depth and with O_2 concentration, showing the need to take measurements at several locations to obtain a representative value for the root zone. Absolute values of ODR have meaning only when correlated with observed plant performance. Stolzy and Letey (1964) indicate that root initiation is reduced or stopped in many plants at ODR values between 18 and 23 g cm^{-2} min^{-1}.

The ODR for optimum top growth appears to be greater than 40×10^{-8} g cm^{-2} min^{-1}. The reader is referred to the excellent review of Stolzy and Letey (1964) for detailed evaluations of the impact of the ODR on plant responses.

The importance of drainage on the ODR is illustrated in Fig. 6.9. These data are self-explanatory and emphasize the necessity of lowering the water table to

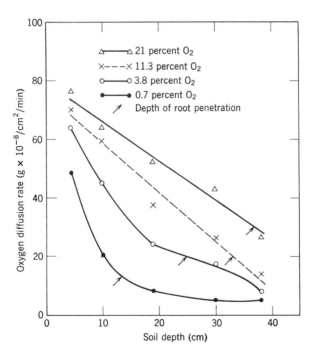

Figure 6.8 Oxygen diffusion rate as a function of depth for various oxygen concentrations at the soil surface. (After Stolzy and Letey, 1964.)

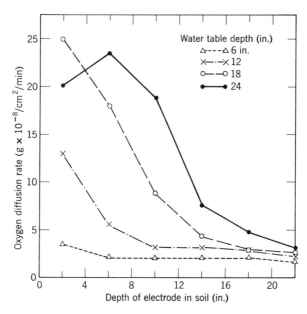

Figure 6.9 Oxygen diffusion rates at a given soil depth as a function of depth of water table. (After Williamson and van Schilfgaarde, 1965.)

provide the proper aeration status for plant roots. The data indicate that aeration falls below optimum levels 20–25 cm above the water table irrespective of its depth.

PROBLEMS

6.1 The steady-state gas concentration in spherical coordinates may be written for radially symmetric flow as

$$\frac{1}{r^2} \frac{d}{dr} (r^2 J_g) + \Omega_g = 0$$

where J_g is the gas flux in the radial direction and Ω_g is the gas consumption rate.

(a) For the special case of constant gas consumption rate calculate the gas flux $J_g(r)$ inside a spherical aggregate of radius r. Assume that the flux vanishes at $r = 0$.

(b) Using the gas flux equation

$$J_g = -D_g^s \frac{dC_g}{dr}$$

$$J_s[x, y, z, t + \tfrac{1}{2}\Delta t]\, \Delta x\, \Delta y\, \Delta t$$

$$= J_s[x, y, z + \Delta z, t + \tfrac{1}{2}\Delta t]\, \Delta x\, \Delta y\, \Delta t$$

$$+ \left(C_T[x, y, z + \tfrac{1}{2}\Delta z, t + \Delta t] - C_T[x, y, z + \tfrac{1}{2}\Delta z, t]\right)\Delta x\, \Delta y\, \Delta z$$

$$+ r_s[x, y, z + \tfrac{1}{2}\Delta z, t + \tfrac{1}{2}\Delta t]\, \Delta x\, \Delta y\, \Delta t \tag{7.2}$$

where J_s is the total solute flux (mass flowing per area per time), C_T is the total solute concentration (mass of solute per volume of soil), and r_s is the reaction rate per volume (loss of solute per soil volume per time).

Equation (7.2) is next divided by $\Delta x\, \Delta y\, \Delta z\, \Delta t$ and rearranged, producing

$$\frac{J_s[x, y, z + \Delta z, \bar{t}] - J_s[x, y, z, \bar{t}]}{\Delta z}$$

$$+ \frac{C_T[x, y, \bar{z}, t + \Delta t] - C_T[x, y, \bar{z}, t]}{\Delta t} + r_s(\bar{z}, \bar{t}) = 0 \tag{7.3}$$

where $\bar{z}, \bar{t}$ are the average values of z and t, respectively. Taking the limit of (7.3) as $\Delta x, \Delta y, \Delta z, \Delta t \to 0$, we obtain

$$\partial C_T/\partial t + \partial J_s/\partial z + r_s = 0 \tag{7.4}$$

Equation (7.4) is called the solute conservation equation in one dimension. For the special case where the solute is insoluble, it reduces to the gas conservation equation (6.9).

7.1.1 Solute Storage in Soil

In general, chemicals that are soluble in water and have a nonnegligible vapor pressure can exist in three phases in soil: as a gas in the soil air, as a dissolved solute in the soil water, and as a stationary phase adsorbed to the soil organic matter or charged clay mineral surfaces. Mathematically, we may express the total solute concentration C_T in terms of the phase contributions as

$$C_T = \rho_b C_a + \theta C_l + a C_g \tag{7.5}$$

where C_a is adsorbed solute concentration, expressed as mass of sorbant per mass of dry soil; C_l is dissolved solute concentration, expressed as mass of solute per volume of soil solution; and C_g is gaseous solute concentration, expressed as mass of chemical vapor per volume of soil air. The units in which the solute phases are expressed correspond to the way in which they are measured. For example, if a soil air sample of volume V_a is withdrawn and a mass M_g of chemical vapor is detected in it, then $C_g = M_g/V_a$. The bulk density ρ_b, volumetric water content θ, and volumetric air content a convert each of these phase concentrations to mass per *soil* volume, which is the unit in which C_T is expressed.

7.1.2 Solute Flux through Soil

When chemical is present in more than one phase, it can move through soil either as a vapor or as a dissolved constituent (the sorbed phase is assumed to be stationary):

$$J_s = J_l + J_g \tag{7.6}$$

where J_l is the flux of dissolved solute and J_g is the flux of solute vapor. As discussed in Chapter 6, gas is transported primarily by diffusion, so that the gas flux is given by Fick's law (6.11), repeated here as

$$J_g = -\xi_g(a)D_g^a \frac{\partial C_g}{\partial z} = -D_g^s \frac{\partial C_g}{\partial z} \tag{7.7}$$

where D_g^s is the soil gas diffusion coefficient.

The flux of dissolved solute consists of two terms: the bulk transport or convection of dissolved chemical moving with flowing soil solution and the diffusive flux of dissolved solute moving by molecular diffusion within the soil solution.

The bulk flow or convection of dissolved solute J_{lc} may be written formally as

$$J_{lc} = J_w C_l \tag{7.8}$$

where J_w is the water flux. In porous media, (7.8) does not represent all of the transport by flowing solution because the soil water flux J_w is an approximate quantity that has been averaged over many soil pores; it does not represent the actual water flow paths, which must curve around solid obstacles and air space. Consequently, (7.8) does not describe the solute convection completely; it fails to describe the tortuous convective motion of solute relative to the average convective motion. This extra motion is called hydrodynamic dispersion (Bear, 1972), and the total solute convection is described by

$$\text{Total convection} = J_w C_l + J_{lh} \tag{7.9}$$

where J_{lh} is called the hydrodynamic dispersion flux.

The other term contributing to solute transport in the dissolved phase is called the liquid diffusion flux J_{ld}, which is written formally as

$$J_{ld} = -\xi_l(\theta)D_l^w\frac{\partial C_l}{\partial z}$$

$$= -D_l^s\frac{\partial C_l}{\partial z} \tag{7.10}$$

where D_l^w is the binary diffusion coefficient of the solute in water, D_l^s is the soil liquid diffusion coefficient, and $\xi_l(\theta)$ is the liquid tortuosity factor to account for the increased pathlength and decreased cross-sectional area of the diffusing solute in soil. Since air as well as solid particles form barriers to liquid diffusion, $\xi_l(\theta)$ is a strongly increasing function of water content θ. Since the liquid tortuosity modifies liquid diffusion in the same way that the gas tortuosity modifies gas diffusion, the Millington–Quirk (1961) tortuosity relation (6.12) may be modified by substituting θ for a to represent $\xi_l(\theta)$:

$$\xi_l(\theta) = \theta^{10/3}/\phi^2 \tag{7.11}$$

7.1.3 Convection–Dispersion Model of Hydrodynamic Dispersion

The erratic transport of solute by hydrodynamic dispersion has been studied theoretically and experimentally for many years. A special case of all possible flows, called convective–dispersive transport, occurs in porous media under the following two conditions: (i) the medium is homogeneous through the volume in which solute transport occurs and (ii) the time required for solutes in stream tubes of different velocity to mix (by diffusion or transverse dispersion) along a direction normal to the direction of mean convection is short compared to the time required for solutes to move through the volume by mean convection (Taylor, 1953). For example, in a cylindrical soil column this condition requires that solutes mix completely in the radial direction before they reach the outflow end of the column in the axial direction.

For this important case, the solute hydrodynamic dispersion flux has a form that is mathematically identical to the diffusion flux (Bear, 1972):

$$J_{lh} = -D_{lh}\frac{\partial C_l}{\partial z} \tag{7.12}$$

where D_{lh} (in centimeters squared per day) is the hydrodynamic dispersion coefficient. This coefficient has frequently been observed to be proportional to the pore water velocity $V = J_w/\theta$ (Anderson, 1979; Bear, 1972; Bigger and Nielsen, 1967):

$$D_{lh} = \lambda V \tag{7.13}$$

where λ (in centimeters) is the dispersivity. The size of the dispersivity depends on the scale over which the water flux and solute convection is averaged. Typical

TABLE 7.1   Values of Diffusion and Dispersion Parameters in Example 7.1

θ	V	D_{lh}	$\xi_l(\theta)$	D_l^s	D_e	D_{lh}/D_e
0.25	0.8	0.8	0.039	0.039	0.84	0.95
0.30	3.3	3.3	0.072	0.072	3.37	0.98
0.35	5.7	5.7	0.120	0.121	5.82	0.98
0.40	12.5	12.5	0.189	0.189	12.69	0.99

values of λ are 0.5–2 cm in packed laboratory columns and 5–20 cm in the field, although they can be considerably larger in regional groundwater transport (Fried, 1975).

The total flux of dissolved solute in the convection–dispersion model is thus

$$J_l = J_w C_l - D_{lh} \frac{\partial C_l}{\partial z} - D_l^s \frac{\partial C_l}{\partial z} \qquad (7.14)$$

which is commonly written as

$$J_l = -D_e \frac{\partial C_l}{\partial z} + J_w C_l \qquad (7.15)$$

where D_e is the effective diffusion–dispersion coefficient. Unless water is flowing very slowly through repacked soil, D_e is normally dominated by the hydrodynamic dispersion term (see Example 7.1).

EXAMPLE 7.1:   In a repacked sandy soil column of porosity $\phi = 0.5$, the measured dispersivity λ was 1 cm. Assuming that $D_l^w = 1$ cm day^{-1}, calculate D_e, D_{lh}, and $\xi_l(\theta)D_l^w$ at applied water fluxes of 0.2, 1.0, 2.0, and 5.0 cm day^{-1}, which create average water contents of 0.25, 0.3, 0.35, and 0.4, respectively.

Since $V = J_w/\theta$, the water velocities are 0.8, 3.33, 5.7, and 12.5 cm day^{-1}, respectively, and the corresponding tortuosity factors are [Equation (7.11)] 0.039, 0.072, 0.121, and 0.189. Thus, the final values of the various dispersion and diffusion terms are given in Table 7.1.

7.2   CONVECTION–DISPERSION EQUATION

For those solutes whose vapor phase is negligible, the solute transport equation (7.4) may be written as [combining (7.4)–(7.6) and (7.15)]

$$\frac{\partial}{\partial t} (\rho_b C_a + \theta C_l) = \frac{\partial}{\partial z} \left(D_e \frac{\partial C_l}{\partial z} \right) - \frac{\partial}{\partial z} (J_w C_l) - r_s \qquad (7.16)$$

Equation (7.16) is called the convection–dispersion equation (Bigger and Nielsen, 1967). Special forms of this equation will be studied in the next few sections of this chapter.

7.2.1 Transport of Inert, Nonadsorbing Solutes

The simplest chemical transport processes in soil are those that involve nonvolatile dissolved solutes that neither react ($r_s = 0$) nor adsorb ($C_a = 0$) to soil solids. Examples of these kinds of tracers are chloride or bromide ions flowing through soil that has only negatively charged or neutral minerals. These ions do not react chemically (except for anion exclusion) with the kinds of ions normally found in soil solution and are not attracted to clay or organic matter surfaces (see Section 1.1.6). Thus, they may act as tracers of the water flow pathways.

Imagine an experiment where water is flowing uniformly at steady state through a homogeneous soil column of length L that is at a constant water content. Equation (7.16) reduces to

$$\frac{\partial C_l}{\partial t} = D \frac{\partial^2 C_l}{\partial z^2} - V \frac{\partial C_l}{\partial z} \quad \text{where } D = D_e/\theta \tag{7.17}$$

At $t = 0$, we instantaneously switch the water inlet valve of the soil column from its initial tracer-free source over to a chloride solution at a concentration C_0, which continues to flow at the same flux rate J_w through the column (see Fig. 7.1). From $t = 0$ onward, we begin to monitor the chloride concentration at the outflow end $z = L$ of the column. The plot of outflow concentration versus time (Fig. 7.2) is called an outflow curve, or sometimes a "breakthrough" curve (representing the solute breaking through the outflow end).

The curves shown in Fig. 7.2 represent mathematical solutions to (7.17) corresponding to the initial and boundary conditions describing the experiment. There are a number of important features illustrated in this figure, which display various characteristics of convective–dispersive flow.

Breakthrough Time   The center of each of the solute fronts, drawn for different values of the dispersion coefficient D, arrive at the outflow end of the column at the same time $t_b = L/V$, called the breakthrough time. When dispersion is neglected (the $D = 0$ curve), all of the solutes move at the same identical velocity V, and the front arrives as one discontinuous jump to the final concentration C_0 at $t = t_b$. This model, in which dispersion is neglected, is called the "piston flow" model of solute movement, so named because the solute is displaced through the soil like a piston.

Effect of Dispersion   As seen in Fig. 7.2, the effect of dispersion on the breakthrough curve is to cause some early and late arrival of solute with respect to the breakthrough time. This deviation is due to diffusion and small-scale convection ahead of and behind the front moving at velocity V and becomes more pronounced as D becomes larger.

Drainage Breakthrough Curves   Instead of plotting outflow concentration as a function of time, concentration can be plotted against the cumulative water drain-

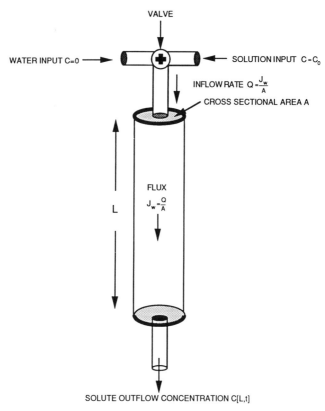

Figure 7.1   Schematic diagram of a soil column outflow experiment, where solute is added at $t = 0$.

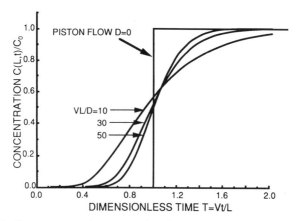

Figure 7.2   Outflow concentration versus time for a step change in solute input at $t = 0$. Curves correspond to different values of $D(V = 2 \text{ cm day}^{-1}, L = 30 \text{ cm})$.

age d_w passing through the outflow end of the column. If the water is flowing at steady state, then $d_w = J_w t$, and the outflow concentrations in Fig. 7.2 are identical when plotted against d_w except for a change in scale for the abscissa. The drainage evolved at the breakthrough time $d_{wb} = J_w t_b$ is by definition in steady state equal to the amount of water that must be added to the column to move solute from the inflow port to the outflow end:

$$d_{wb} = J_w t_b = J_w L/V = L\theta \tag{7.18}$$

which is the volume of water per unit area held in the wetted soil pores of the column during transport. For this reason $d_{wb} = L\theta$ is called a "pore volume." Thus, it requires approximately one pore volume of water to move a mobile solute through a soil column.

Equation (7.18) provides additional insight into the piston flow model ($D = 0$) and its assumptions. According to the piston flow model, the incoming solute replaces the water initially present in the soil, or, equivalently, pushes it ahead of the front like a piston. Thus, one may calculate solute transport with the piston flow model by estimating how long it will take to replace the water between the point of entry and the final location. This is illustrated in the next example.

EXAMPLE 7.2:   Calculate the amount of time required to transport nitrate (a mobile anion) from the bottom of the root zone to groundwater 10 m below it if the average water content of the soil is 0.15 and the average annual drainage rate is 0.5 m yr^{-1}.
    The total amount of water in the soil profile is $L\theta = (10)(0.15) = 1.5$ m, which would require 3 yr to add at 0.5 m yr^{-1}. Thus, $t_b = 3$ yr. Equivalently, we may calculate the nitrate velocity $V = J_w/\theta = 10/3$ m yr^{-1}, and a corresponding breakthrough time $t_b = L/V = 3$ yr.

Transport of Pulses through Soil   In some soil column outflow experiments, a narrow pulse of solute, rather than a front, might be added to the inlet end at $t = 0$. Figure 7.3 shows the corresponding outflow curves for the narrow pulse input, calculated with the mathematical solution to (7.17) and the appropriate boundary conditions (Jury and Sposito, 1985):

$$C(L, t) = \frac{C_0 L}{2\sqrt{\pi Dt^3}} \exp\left(-\frac{(L - Vt)^2}{4Dt}\right) \tag{7.19}$$

Note that as D becomes larger, the pulse becomes more and more spread out.

7.2.2   Transport of Inert, Adsorbing Chemicals

Certain chemicals, although they do not react chemically or biologically in soil, adsorb to stationary soil solids. For this case, the transport equation (7.16) may be written as (assuming homogeneous soil and steady-state water flux)

$$\frac{\rho_b}{\theta}\frac{\partial C_a}{\partial t} + \frac{\partial C_l}{\partial t} = D\frac{\partial^2 C_l}{\partial z^2} - V\frac{\partial C_l}{\partial z} \tag{7.20}$$

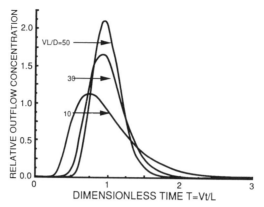

Figure 7.3 Outflow concentration versus time (7.19) for a narrow pulse input of solute at $t = 0$. Curves correspond to different values of $D(V = 2$ cm day^{-1}, $L = 30$ cm).

Adsorption Isotherms Equation (7.20) differs from (7.17) only by the addition of the first term, which represents the rate of change of mass storage in the sorbed phase. In order to solve (7.20), we must develop a relationship between the adsorbed concentration C_s and the dissolved concentration C_l. This relationship at equilibrium is called an adsorption isotherm (see Section 1.1.6). Figure 7.4 shows different kinds of isotherm shapes found for various compounds in soil.

A special form of the isotherm, called the linear adsorption isotherm, may be expressed as

$$C_a = K_d C_l \tag{7.21}$$

where K_d, the slope of the isotherm, is called the distribution coefficient.

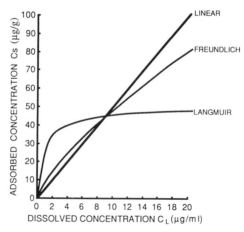

Figure 7.4 Adsorption isotherm shapes. Labels correspond to the names given to the isotherm models.

If we assume that the linear isotherm is valid at all times in soil (i.e., the dissolved and adsorbed phases are in instantaneous equilibrium), then the time derivative of C_a may be written as

$$\frac{\partial C_a}{\partial t} = K_d \frac{\partial C_l}{\partial t} \tag{7.22}$$

Using Equation (7.22), (7.20) may be written as

$$\left(1 + \frac{\rho_b K_d}{\theta}\right) \frac{\partial C_l}{\partial t} = D \frac{\partial^2 C_l}{\partial z^2} - V \frac{\partial C_l}{\partial z} = R \frac{\partial C_l}{\partial t} \tag{7.23}$$

where

$$R = 1 + \rho_b K_d / \theta \tag{7.24}$$

is the retardation factor. Finally, we may divide each side of Equation (7.23) by R, producing

$$\frac{\partial C_l}{\partial t} = D_R \frac{\partial^2 C_l}{\partial z^2} - V_R \frac{\partial C_l}{\partial z} \tag{7.25}$$

where $D_R = D/R$ and $V_R = V/R$ are the retarded dispersion coefficient and solute velocity, respectively.

Breakthrough Time Equation (7.25) now has the same mathematical form as (7.17), and therefore it has the same solution as the one for (7.17) illustrated in Figs. 7.2 and 7.3, except that the solute breakthrough time t_{bR} for the adsorbed chemical is now

$$t_{bR} = L/V_R = RL/V = Rt_b \tag{7.26}$$

Thus, for steady-state water flow, an adsorbing solute would take R times as long to reach the outlet end as a mobile solute. Similarly, since it requires $d_{wb} = L\theta = 1$ pore volume of drainage to move a mobile solute a distance L, it would require $RL\theta = R$ pore volumes to move an adsorbed solute the same distance.

> *EXAMPLE 7.3:* For the soil in Example 7.2, calculate how long it will take to move a pesticide that has a K_d of 2 cm^3 g^{-1} to the groundwater. Assume that $\rho_b = 1.5$ g cm^{-3} and $\theta = 0.15$.
>
> The retardation factor $R = 1 + \rho_b K_b / \theta = 21$. Thus, by (7.26) and Example 7.2, the breakthrough time of this chemical would be $21(3) = 63$ yr, requiring 21 pore volumes, or 31.5 m, of water to pass through the profile.

Effect of Dispersion Although the dispersion coefficient in (7.25) is reduced by a factor R, the travel time is increased, so that the amount of solute spreading as

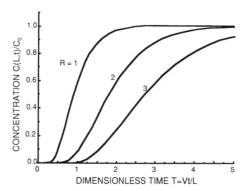

Figure 7.5 Solute outflow concentrations versus time for a step change in solute concentration at $t = 0$ for mobile ($R = 1$) and adsorbed ($R = 2, 3$) solutes ($V = 2$ cm day^{-1}, $L = 30$ cm, $D = 4$ cm^2 day^{-1}).

a function of time observed at the outflow end actually appears to be greater than that of a mobile chemical. Figure 7.5 shows plots of breakthrough curves for adsorbed and mobile solute fronts for adsorbed compounds with retardation factors $R = 2, 3$. The breakthrough times of the adsorbed compounds are longer than that of the mobile chemical t_b, and the adsorbed front is spread over a greater range of times when it arrives at the outflow end.

7.2.3 Effect of Soil Structure on Transport

Soil structure can create preferential flow channels for water and dissolved solutes, which can greatly influence the characteristics of the transport process. Figure 7.6 shows breakthrough curves for chloride pulses leached through 1-m-long columns containing undisturbed and repacked loamy sand soil irrigated at a steady water flow rate of 8 cm day^{-1}. The undisturbed columns were obtained by pushing 20-cm-diameter aluminum cylinders into the surface at a field site and carrying the excavated columns back to the laboratory. The repacked soil columns were obtained by breaking down, sieving, and repacking soil from the same site to the average bulk density of the field columns.

Several obvious differences between the two representative breakthrough curves in Fig. 7.6 are notable. Chloride arrives after less drainage in undisturbed column than in the repacked column and also has effluent appearing after the repacked column pulse has completely passed through the system. Thus, soil structure creates greater dispersion or pulse spreading than homogeneous soil.

The early arrival of solute can be attributed to preferential flow of water through the larger channels of the wetted pore space, causing water to flow more rapidly than if the water in the entire soil matrix was being displaced. This type of flow regime would be expected to be present in structured soils that have large flow channels and also wetted regions within finer pores where water is moving much more slowly (or not at all) than in the large channels. An example of this kind of

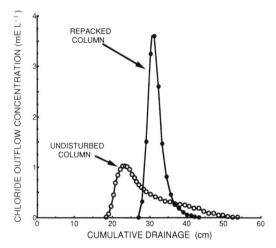

Figure 7.6 Outflow concentration versus drainage volume for long, wide, undisturbed (dashed line) and repacked (solid line) soil columns at 8 cm day^{-1} flow rate and $L = 94$ cm. (After Khan, 1988.)

flow regime would be an aggregated soil. Because of their finer pore sizes, the aggregates hold a substantial volume of water at normal soil matric potentials, but this water is relatively stagnant compared to that flowing in the large channels between aggregates. For soils of this type, some of the solute can be transported much more rapidly than one would expect based on the total volumetric water content and the piston flow model (7.18).

The late arrival of solute in the structured soil compared to the repacked soil can be attributed to removal of solute from the flow channels by transverse diffusion into stagnant regions of the soil while the pulse is passing through the column and gradual release of the solute to the flow channels by diffusion after the pulse has passed through the system.

EXAMPLE 7.4: Calculate the breakthrough time for a front of solute moving through a vertical, 100-cm soil column of saturated hydraulic conductivity $K_s = 5$ cm h^{-1} and $\theta_s = 0.5$ that has 10 cm of water ponded on it. Further assume that somewhere in the 100-cm^2 cross section of the column is a 1-mm-diameter wormhole that extends through the column and through which water flows according to Poiseuille's law (3.9), while it flows according to Darcy's law (3.13) in the soil.

Within the soil matrix, the steady water flux rate may be calculated with Darcy's law (3.13):

$$J_w = -K_s \frac{L + d}{L} = -5.5 \text{ cm h}^{-1}$$

Therefore, within the matrix the solute moves with a downward velocity equal to

$$V = J_w / \theta_s = 11 \text{ cm h}^{-1}$$

Thus, solute breakthrough in the soil occurs at a time

$$t_b = L/V = 9.091 \text{ h}$$

In the wormhole, the water volume flow rate is given by Poiseuille's law (3.9):

$$Q = \frac{\pi R^4 \rho_w g (L + d)}{8Lv} = 0.265 \text{ cm}^3 \text{ s}^{-1}$$

and the water velocity in the wormhole is

$$V_h = \frac{Q}{\pi r^2} = \frac{\rho_w g R^2}{8v} \frac{L + d}{L} = 33.7 \text{ cm s}^{-1}$$

Thus, the breakthrough time through the wormhole is

$$t_b = L/V_h = 2.97 \text{ s}$$

Clearly, structural voids can have a dominant effect on water and solute flow if they are filled with water that is flowing.

Mobile–Immobile Water Model   A model developed to include preferential flow of solute, first described by Coats and Smith (1956) and later applied to transport in soil columns by van Genuchten and co-workers (van Genuchten and Wierenga, 1976, 1977; van Genuchten et al., 1976), represents the wetted pore space with two water contents: a mobile water content θ_m through which water is flowing and an immobile water content $\theta_{im} = \theta - \theta_m$, which contains stagnant water. For this system, the solute concentration is divided into an average concentration C_m in the mobile region and a second concentration C_{im} in the immobile region. The solute is transported by a convection–dispersion process in the mobile region and also exchanges solute with the immobile region by a rate-limited diffusion process. For an inert, nonreactive solute the conservation equation (7.4) may be written as (assuming constant J_w, θ_m, θ_{im})

$$\theta_m \frac{\partial C_m}{\partial t} + \theta_{im} \frac{\partial C_{im}}{\partial t} = D_e \frac{\partial^2 C_m}{\partial z^2} - J_w \frac{\partial C_m}{\partial z} \tag{7.27}$$

where now $C_T = \theta_m C_m + \theta_{im} C_{im}$ is the total solute mass per soil volume. The rate-limited mass transfer between the mobile and stagnant regions is expressed in the model as

$$\theta_{im} \, \partial C_{im}/\partial t = \alpha(C_m - C_{im}) \tag{7.28}$$

where α is the rate coefficient. Equation (7.28) states that the rate of change of solute stored in the immobile region is proportional to the difference in concentration between the two regions. Thus, when solute is added to the soil in a pulse, it will first diffuse from the mobile to the stagnant region and then return to the mobile region after the pulse has passed through the system.

Figure 7.7 illustrates solutions to the mobile–immobile water model (MIM) (7.27) and (7.28) for different values of the model parameters. These figures show how versatile this model is for representing shapes of breakthrough curves. For example, in Fig. 7.7a the mobile water fraction θ_m/θ is varied from 0.3 to 1.0 while keeping α, J_w, and D_e constant. The effect of lowering θ_m/θ on the outflow

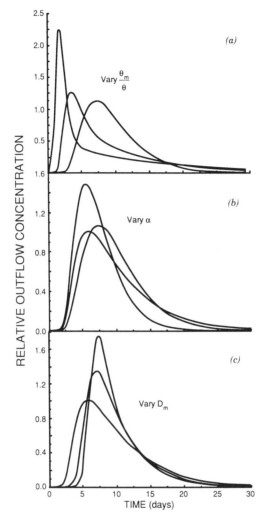

Figure 7.7 Outflow concentrations versus time for a narrow pulse input to the MIM equations (7.27) and (7.28) as a function of (a) mobile water fraction; (b) rate parameter; and (c) mobile region dispersion coefficient.

pulse is to move the breakthrough curve further to the left as more and more of the wetted pore space is excluded from an active role in transport. The limiting case $\theta_m/\theta = 1$ reduces to the simpler one-region convection–dispersion model (7.17) discussed earlier.

In Fig. 7.7b, the rate parameter α is varied over a large range. In the limit $\alpha \to \infty$, the two concentrations merge ($C_m = C_{im}$) because they mix instantaneously [see (7.28)], and the MIM again reduces to the convection–dispersion model (7.17). At the other extreme of zero mixing when $\alpha = 0$, $\partial C_{im}/\partial t = 0$, and (7.27) reduces to the convection–dispersion model (7.17) with a *total* water content θ_m, since the immobile region is blocked off completely. For this case the solute moves at a velocity $V_m = J_w/\theta_m$ and breaks through sooner than if $\alpha = \infty$. At intermediate values of α, the solute first arrives at a time intermediate between the arrival times of the two limiting cases, but now has a long effluent tail corresponding to rate-limited diffusion from the stagnant regions.

Figure 7.7c illustrates the effect of varying D_e on the outflow shape for constant α and θ_m. Here the influence of dispersion is far less pronounced than in the simple convection–dispersion model because much of the mixing and pulse spreading has been described instead by rate-limited diffusion between mobile and stagnant regions. The remaining spreading within the mobile region is less significant for this reason.

The two-region model discussed in the preceding is very useful in illustrating effects of stagnant regions on flow (Nkedi-Kizza et al., 1983). As a simulation model, however, it suffers from the drawback that its parameters cannot be measured independently in real soils but rather must be curve fitted by simultaneous optimization of all its parameters to outflow data. Recently (Rao et al., 1980) some procedures have been developed to measure some of its parameters independently in separate experiments.

7.2.4 Reactions of Chemicals in Soil

First-Order Decay   Many organic chemicals in soil are broken down by microbial or chemical reactions. Although this reaction can depend in a complex manner on temperature, pH, microbial population density, carbon content, and other factors, under optimum conditions it may be described as a first-order decay process (Hamaker, 1972). A mass of chemical $M(t)$ undergoing first-order decay losses material at a rate proportional to its mass. This loss rate is expressed mathematically as

$$\frac{dM}{dt} = -\mu M \tag{7.29}$$

where μ is the first-order decay rate constant.

EXAMPLE 7.5:   Calculate the mass remaining in the soil as a function of time for a chemical obeying (7.29). Assume that $M = M_0$ at $t = 0$.

Equation (7.29) may be integrated if all factors that depend on M are placed on the same side:

$$\frac{dM}{M} = -\mu \, dt \tag{7.30}$$

Equation (7.30) may now be integrated from $t = 0$ to t, with the result

$$\int_{M_0}^{M} \frac{dM}{M} = \ln(M/M_0) = -\mu \int_0^t dt = -\mu t \tag{7.31}$$

or

$$M(t) = M_0 \exp(-\mu t) \qquad (7.32)$$

Thus, a chemical undergoing first-order decay loses its mass at an exponentially declining rate over time.

Convection–Dispersion Equation and First-Order Decay A mobile chemical obeying the convection–dispersion equation (CDE) and undergoing first-order decay must obey (7.16) with $r_s = \mu\theta C_l$. Therefore, the transport equation may be written as (assuming steady flow)

$$\theta \frac{\partial C_l}{\partial t} = D_e \frac{\partial^2 C_l}{\partial z^2} - J_w \frac{\partial C_l}{\partial z} - \mu\theta C_l \qquad (7.33)$$

The breakthrough curve at $z = L$ for this chemical is shown in Fig. 7.8 for various values of μ. The effect of first-order decay on the outflow curve is to remove the long tail of the breakthrough curve preferentially, since this portion of the curve contains solutes that spend the longest time in the column.

Piston Flow Model and First-Order Decay The piston flow model approximation ($D = 0$) for first-order decay is quite simple. The chemical moves with a velocity $V = J_w/\theta$ and reaches a distance $z = L$ in a time $t_b = L/V = L\theta/J_w$. If the chemical was added as a front of concentration C_0 at time $t = 0$, it will have a concentration $C_0 e^{-\mu t_b} = C_0 e^{-\mu L/V}$ when it arrives at $z = L$. If it is added as a pulse of mass M_0, the mass of the pulse that passes $z = L$ will be $M_0 e^{-\mu L/V}$.

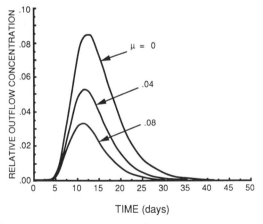

Figure 7.8 Outflow concentration versus time for a narrow pulse input of solute undergoing first-order decay with a decay constant μ ($V = 2$ cm day^{-1}, $L = 30$ cm, $D = 4$ cm^2 day^{-1}).

7.2.5 Transport of Volatile Organic Compounds through Soil

Volatile organic compounds, including many pesticides, display quite complex behavior in soil because they may move in the vapor phase as well as within solution. They adsorb to stationary soil solid material and can be degraded by biological and chemical reactions. To describe all of these processes, the chemical mass balance equation (7.4) must be combined with the full solute storage expression (7.5) and both vapor (7.7) and dissolved (7.14) solute fluxes, producing

$$\frac{\partial}{\partial t}(aC_g + \theta C_l + \rho_b C_s) = \frac{\partial}{\partial z}\left(D_g^s \frac{\partial C_g}{\partial z}\right) + \frac{\partial}{\partial z}\left(D_e \frac{\partial C_l}{\partial z}\right) - \frac{\partial}{\partial z}(J_w C_l) - r_s$$

$$(7.34)$$

This equation has three unknowns (C_a, C_l, C_g). Therefore, it cannot be solved until two additional equations are specified defining the relationship between the phases. These are sometimes called phase partitioning laws.

Phase Partitioning Laws We have already discussed equilibrium and nonequilibrium laws relating the dissolved and adsorbed phases. The relationship most often used to describe liquid–vapor partitioning is the linear, equilibrium formulation known as Henry's law.

Henry's law was originally formulated as a relationship between the solute mole fraction X in solution and the vapor pressure P_v of the vapor in equilibrium with the solution

$$P_v = k_H X \qquad (7.35)$$

where k_H (pressure per mole fraction) is Henry's constant. To convert this to a relation between C_g and C_l, we first use the ideal gas law

$$P_v V = mRT/M \qquad (7.36)$$

where V is volume, m is mass, T is temperature, R is the universal gas constant, and M is the molecular weight of the gas. Since $m/V = C_g$ by definition, we may write (7.36) as

$$P_v = C_g RT/M \qquad (7.37)$$

The mole fraction X is equal to

$$X = \frac{\text{moles of solute}}{\text{moles of solute} + \text{moles of water}} = \frac{m_s/M_s}{m_s/M_s + m_w/M_w} \qquad (7.38)$$

where the subscripts s and w refer to solute and water.

If we let $m_w = \rho_w V_w$ and $C_l = m_s/V_w$ and multiply the numerator and denominator of (7.38) by M_s/V_w, we obtain

$$X = \frac{C_l}{C_l + \rho_w M_s/M_w} \approx \frac{M_w C_l}{\rho_w M_s} \tag{7.39}$$

assuming $C_l \ll \rho_w M_s/M_w$.[1]

Thus, combining (7.35), (7.37), and (7.39), we obtain

$$C_g \frac{RT}{M_s} = k_H \frac{M_w C_l}{\rho_w M_s} \tag{7.40}$$

or

$$C_g = K_H C_l \tag{7.41}$$

where

$$K_H = M_w k_H/\rho_w RT \tag{7.42}$$

is the dimensionless form of Henry's constant.

Linear Partitioning Laws If a compound satisfies both linear, equilibrium adsorption (7.21) and the equilibrium form of Henry's law (7.41), then the total concentration C_T in (7.5) may be partitioned into phases as follows (Jury et al., 1983b):

$$\begin{aligned} C_T &= \rho_b C_a + \theta C_l + a C_g \\ &= \rho_b K_d C_l + \theta C_l + a K_H C_l \\ &= (\rho_b K_d + \theta + a K_H) C_l \equiv R_l C_l \end{aligned} \tag{7.43}$$

where

$$R_l = \rho_b K_d + \theta + a K_H \tag{7.44}$$

is the liquid phase partition coefficient. Similarly,

$$R_a = \rho_b + \theta/K_d + a K_H/K_d \tag{7.45}$$

$$R_g = \rho_b K_d/K_H + \theta/K_H + a \tag{7.46}$$

are the adsorbed phase and gas phase partition coefficients, respectively.

[1]This is the same as assuming $m_s/M_s \ll m_w/M_w$.

The phase mass fraction, which is the fraction of the total solute mass residing in a given phase, is expressed for each phase as

$$f_a = \rho_b C_a / C_T = \rho_b / R_a \tag{7.47}$$

$$f_l = \theta C_l / C_T = \theta / R_l \tag{7.48}$$

$$f_g = a C_g / C_T = a / R_g \tag{7.49}$$

By definition,

$$f_a + f_l + f_g = 1 \tag{7.50}$$

EXAMPLE 7.6: Table 7.2 presents chemical phase parameters K_d, K_H for six common pesticides in a loamy soil. Calculate the phase mass fractions for each compound assuming that $\theta = 0.25$, $a = 0.25$, and $\rho_b = 1.5$ g cm^{-3}.

From (7.44)–(7.49) and the parameters provided, Table 7.3 is generated.

The previous example illustrates several features of pesticide partitioning. When the compound is attracted to soil solids, much of the mass resides in that phase. Also, the mass fractions residing in the vapor phase are negligible for most of the

TABLE 7.2   Distribution Coefficients, Degradation Half Lives, and Henry's Constants for Six Pesticides in a Loamy Soil

Compound	K_d (cm^3 g^{-1})	$\tau_{1/2}$ (d)	K_H
Atrazine	1.6	64	2.5×10^{-7}
Bromacil	0.7	350	3.7×10^{-8}
DBCP	1.3	225	8.3×10^{-3}
DDT	2400	3837	2.0×10^{-3}
Lindane	13	266	1.3×10^{-4}
Phorate	6.6	82	3.1×10^{-4}

Source: Jury et al. (1984b).

TABLE 7.3   Calculated Phase Mass Fractions[a] for Six Compounds in Table 7.2

Compound	$100 f_a$	$100 f_l$	$100 f_g$
Atrazine	90.57	9.43	2.4×10^{-6}
Bromacil	81.20	18.80	7.0×10^{-8}
DBCP	88.55	11.35	9.0×10^{-2}
DDT	99.993	0.007	1.4×10^{-5}
Lindane	98.73	1.27	1.7×10^{-4}
Phorate	97.54	2.46	7.6×10^{-4}

[a] Percentage in each phase.

compounds. However, this does not mean that vapor phase transport is negligible, as will be seen in the next section.

Effective Liquid–Vapor Diffusion   If we ignore liquid dispersion, which might be reasonable if the convective velocity of the compound is low in soil, then using (7.7) and (7.10), the combined liquid and vapor diffusion of the compound may be written as

$$J_g + J_{ld} = -\xi_g(a)D_g^a \frac{\partial C_g}{\partial z} - \xi_l(\theta)D_l^w \frac{\partial C_l}{\partial z} \tag{7.51}$$

Assuming equilibrium linear partitioning, we may rewrite (7.51) in terms of the partition coefficients as

$$J_g + J_{ld} = -\left(\frac{\xi_g(a)D_g^a}{R_g} + \frac{\xi_l(\theta)D_l^w}{R_l} \right) \frac{\partial C_T}{\partial z} \tag{7.52}$$

Equation (7.52) may be written compactly as

$$J_g + J_{ld} = -D_E \frac{\partial C_T}{\partial z} \tag{7.53}$$

where

$$D_E = \frac{\xi_g(a)D_g^a}{R_g} + \frac{\xi_l(\theta)D_l^w}{R_l} \tag{7.54}$$

is the effective vapor and liquid diffusion coefficient (Jury et al., 1983b). Whether this coefficient represents predominantly liquid or vapor diffusion for a given compound depends on the value of its vapor and liquid partitioning coefficients and on the moisture status of the medium.

EXAMPLE 7.7:   For the six pesticides in Example 7.6, calculate and plot D_E as a function of water content, assuming that the porosity $\phi = 0.5$, the Millington–Quirk tortuosity models [(6.12) and (7.11)] are valid, and $D_l^w = 0.432$ cm^2 day^{-1}, $D_g^a = 4320$ cm^2 day^{-1} for all pesticides (Jury et al., 1983a,b).
   The effective diffusion coefficient D_E in (7.54) may be written as

$$D_E = \frac{(\phi - \theta)^{10/3} K_H D_g^a + \theta^{10/3} D_l^w}{[\rho_b K_d + \theta + (\phi - \theta)K_H]\phi^2} \tag{7.55}$$

as a function of water content. Figure 7.9 shows a plot of D_E (thick line), along with the contribution from the vapor term (thin line) and the liquid term (dotted line), for the six compounds as a function of θ.
   The total diffusion coefficient is vapor dominated for the compounds with large K_H (DBCP, DDT) and liquid dominated for the low-K_H compounds (Atrazine, Bro-

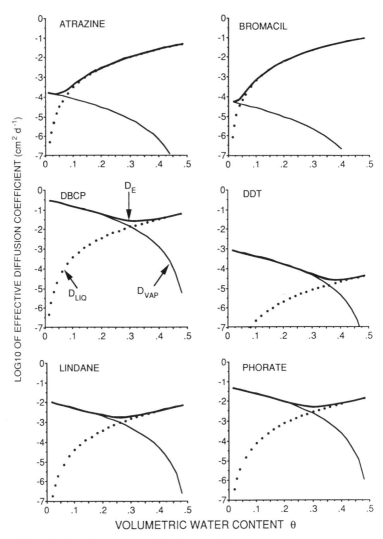

Figure 7.9   Effective liquid–vapor diffusion coefficients versus water content for six pesticides. Thin line is vapor phase contribution, dotted line is liquid phase contribution, and thick line is total contribution from both phases.

macil). Lindane and Phorate are vapor dominated at low θ and liquid dominated at high θ. As a rough rule, compounds with $K_H \gg 10^{-4}$ are vapor dominated and those with $K_H \ll 10^{-4}$ are liquid dominated at 50% water saturation (Jury et al., 1984a; Letey and Farmer, 1973).

Total Solute Flux   If we neglect solute dispersion,[2] the total solute flux is the sum of the solute convection and diffusion terms. Thus, using Equations (7.43) and

---

[2] Solute dispersion can be included if desired.

(7.53),

$$J_s = J_w C_l - D_E \frac{\partial C_T}{\partial z} = V_E C_T - D_E \frac{\partial C_T}{\partial z} \qquad (7.56)$$

where $V_E = J_w / R_L$ is the solute velocity.

If the flux equation (7.53) is plugged into the transport equation (7.34) and first-order decay is assumed, the new equation may be written as (assuming uniform soil properties and constant J_w)

$$\frac{\partial C_T}{\partial t} = D_E \frac{\partial^2 C_T}{\partial z^2} - V_E \frac{\partial C_T}{\partial z} - \mu_E C_T \qquad (7.57)$$

where μ_E is the effective degradation coefficient.

Equation (7.57) is now identical to the convection–dispersion equation (7.33) for a compound undergoing first-order decay. It may be solved exactly for many boundary conditions of interest, yielding different behavior for compounds with distinct chemical and adsorption properties (K_d, K_H, μ_E). Thus, it makes an ideal screening model for predicting the relative behavior of different chemicals in the same environment (Jury et al., 1983a,b, 1984a–c).

Volatilization of Chemicals from Soil   An important problem that may be studied by solving (7.57) with the appropriate boundary condition is that of chemical volatilization through the soil surface. Jury et al. (1983a,b) formulated the upper boundary condition for this problem by assuming that the chemical flux at the surface had to pass through an air boundary layer by vapor diffusion before entering the well-mixed region of the atmosphere.

Figure 7.10 illustrates the boundary layer region above the soil surface. It is conceptualized as an air layer of thickness d through which vapor diffuses according to Fick's law (6.10), written in the finite difference form as

$$J_{sb} = \frac{-D_g^a \left(C_g(0, t) - C_g^a \right)}{d} \qquad (7.58)$$

where $C_g(0, t)$ is the vapor concentration at the soil surface, J_{sb} is the flux through the boundary layer, and C_g^a is the concentration in the atmosphere above the air layer, which Jury et al. (1983a,b) assumed was zero. They also estimated the value of d from measured water evaporation rates and recommended $d = 0.5$ cm as an average value for a bare surface. The transport equation (7.57) was solved by these authors, together with the upper boundary condition,

$$J_s(0, t) = \left(-D_E \frac{\partial C_T}{\partial z} + V_E C_T \right)_{z=0} = J_{sb} \qquad (7.59)$$

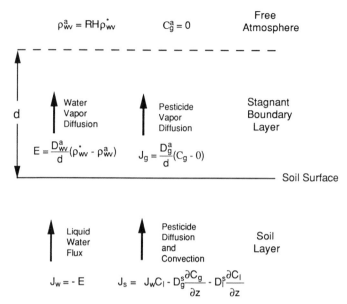

Figure 7.10 Illustration of transport through stagnant air boundary layer above soil surfaces, through which chemical and water vapor must move by diffusion. (After Jury et al., 1983b.)

and the initial condition

$$C_T(z, 0) = C_0 \quad \text{for } -L < z < 0$$

$$C_T(z, 0) = 0 \quad \text{for } z < -L \tag{7.60}$$

where L is the depth of incorporation of the compound. The model calculations were conducted under standard conditions consisting of uniform upward water flux (evaporation), zero water flux, and uniform downward water flux (leaching).

The behavior of the chemicals can be summarized in various ways, most compactly in terms of the percentage of the initial mass $M = C_0L$ that has been volatilized, has degraded, has leached below a specified layer, or remains in the soil.

Figure 7.11 summarizes the volatilization flux as a function of time for the six chemicals in Table 7.2 under the standard soil and climate conditions given in Table 7.4.

The compounds behave differently over time and react differently to the presence or absence of water flux. Atrazine and Bromacil both have insignificant volatilization fluxes when water is not flowing, but their volatilization fluxes increase with time when water is flowing upward and can eventually become significant. In contrast, the other compounds all have volatilization rates that decrease as a function of time under all water flow regimes. The volatilization rates of DBCP, Lindane, and Phorate are all significantly enhanced by evaporation whereas DDT,

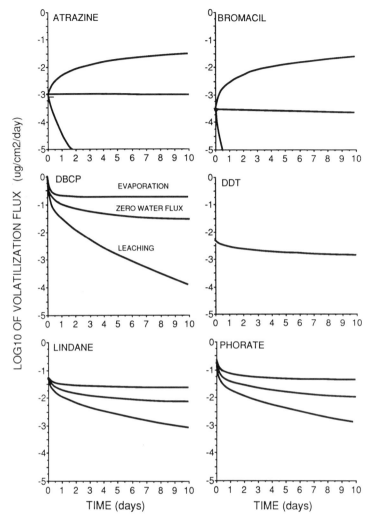

Figure 7.11 Predicted volatilization flux rate versus time for six pesticides under uniform water evaporation, uniform downward water flow, and zero water flow.

which is virtually immobile in soil in the liquid phase, has the same volatilization-time curve irrespective of the rate and direction of water flow.

Jury et al. (1984a) explained this volatilization behavior by the influence of the boundary layer. Compounds with a low K_H, such as Atrazine and Bromacil, normally have a low vapor density in soil, and the boundary layer represents a much greater resistance to flow than the soil. As a result, when water is flowing upward, the chemical accumulates at the interface and the volatilization rate increases with time as the vapor density at the surface increases.

Compounds with a large K_H, such as the other four pesticides in Fig. 7.11, have

TABLE 7.4   Soil and Environmental Conditions for Simulations in Figure 7.11

Condition	Value	Units
Bulk density	1.5	$g \, cm^{-3}$
Porosity	0.5	—
Water content	0.25	—
Evaporation rate	0.5	$cm \, day^{-1}$
Leaching rate	0.5	cm^{-1}
Depth of incorporation	10	cm
Organic carbon fraction	0.01	—
Boundary layer thickness	0.5	cm
Air diffusion coefficient	4320	$cm^2 \, day^{-1}$
Water diffusion coefficient	0.432	$cm^2 \, day^{-1}$
Time of simulation	10	days

a higher vapor density in soil, and the boundary layer offers comparatively little resistance. As a result, the solute is carried off as rapidly as it is transported to the surface and does not build up with time. The authors classified compounds into different categories based on whether $K_H \gg 2.5 \times 10^{-5}$ (category 1) or $K_H \ll 2.5 \times 10^{-5}$ (category 3) (Jury et al., 1984a). Their predictions were verified experimentally for Prometon (category 3) and Lindane (category 1) (Spencer et al., 1988). In this study, Prometon accumulated over time at the surface, reaching values in the top few millimeters of soil that were over 20 times higher than the initial concentration. In contrast, Lindane had a very low surface concentration at the end of the study.

The environmental fate summary for the six pesticides during the 10-day simulations is given in Table 7.5.

Notable in Table 7.5 are the significant losses of DBCP and Phorate under water evaporation and the high degradation of Atrazine and Phorate.

TABLE 7.5   Environmental Fate Summary for Six Pesticides in Simulations Summarized in Table 7.4

Compound	Amount Volatilized (%)		Amount Degraded (%)		Amount Remaining (%)	
	$E = 0.5$	$E = 0$	$E = 0.5$	$E = 0$	$E = 0.5$	$E = 0$
Atrazine	2.08	0.07	9.23	9.30	88.69	90.63
Bromacil	1.32	0.02	1.95	1.98	96.73	98.00
DBCP	24.28	6.68	0.60	0.77	75.12	92.55
DDT	0.22	0.21	0.18	0.18	99.60	99.61
Lindane	2.88	1.26	2.53	2.56	94.59	96.17
Phorate	5.74	2.61	7.84	7.99	86.41	89.40

7.3 TRANSFER FUNCTION MODEL OF SOLUTE TRANSPORT THROUGH SOIL

The convection–dispersion equation of solute transport (CDE) is based on a set of assumptions that are likely to be valid only under certain conditions in soil. The CDE makes the hypothesis that the dispersion process is formally equivalent to diffusion even though dispersion is a convective transport process. In order for the dispersive mixing to behave analogously to diffusion, solute transported by convection along the direction of mean transport at a rate other than the average convective velocity must have time to mix with other regions of the wetted porous medium before the outlet end is reached; otherwise, the CDE cannot describe the transport with a constant dispersion coefficient (Gelhar and Axness, 1983; Taylor, 1953).

The simple hypothetical example shown in Fig. 7.12 illustrates the criteria for the valildty of the CDE. In this situation the transport medium consists of two parallel regions with water contents θ_1 and $\theta_2 = 2\theta_1$, so that the water velocity through each region is V_1 and $V_2 = 0.5V_1$. A solute front is added to the inlet end at $t = 0$ and proceeds down the column as shown in the figure.

At an early time t_1, the solute front has progressed much farther down the column in the fast region than in the slow region, and transfer of solute from the fast to slow regions (by diffusion or transverse convection) has just begun. Later, at time t_2, the fronts are closer together because more transfer between the regions has occurred. Finally, at t_3 the mixing between the regions has had time to reach completion and there is only one front, which moves at the average velocity $3V_1/4$.[3] For times $t > t_3$, the CDE will describe transport and longitudinal mixing of solute through the system.

Because the CDE is valid only after sufficient time has elapsed to smooth out transport by convective velocity differences along the direction of motion, it cannot

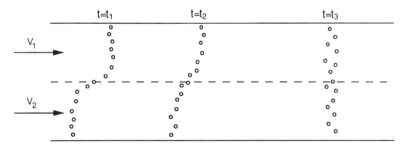

Figure 7.12 Solute front moving horizontally through a porous medium where $V = V_1$ in upper layer and $V = V_2 = V_1/2$ in lower layer. Solute mixes between the two fronts as it moves through the system.

[3]This is the average velocity calculated from the position of the pulse at a given time.

be valid at early times and requires longer to be valid in large, structured systems (such as natural fields) than in small, uniform systems (such as repacked soil columns).

For this reason, Jury (1982) developed an alternative formulation for solute transport, called the transfer function model, which does not require the restrictive assumptions of the CDE and can be applied to problems at different scales.

7.3.1 Solute Transport Volume

The macroscopic system described by the transfer function model is shown in Fig. 7.13. The soil volume, which might be a soil column or the surface zone of an agricultural field, is distinguished by an entrance surface through which solute enters and an exit surface through which it leaves. Within the volume is a fraction of the wetted pore space, called the solute transport volume fraction θ_{st}, through which solute moves. Solute may be produced or destroyed in the volume by reactions but may not leave the volume through the lateral faces. For these conditions, the fundamental transfer function equation may be written as (Jury et al., 1986)

$$Q_{out}(t) = \int_0^t g(t - t')\, Q_{in}(t')\, dt' \tag{7.61}$$

or equivalently,

$$Q_{out}(t) = \int_0^t g(\tau)\, Q_{in}(t - \tau)\, d\tau \tag{7.62}$$

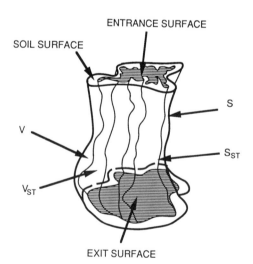

Figure 7.13 Schematic illustration of a field soil unit of volume V bounded by a surface S and enclosing a transport volume V_{st} and its bounding surface S_{st}. (After Jury et al., 1986.)

where Q_{in}, Q_{out} are the solute mass flux rates into and out of the volume, respectively, t' is the entry time, t is the exit time, and $\tau = t - t'$ is the solute lifetime in the volume. The function $g(\tau)$ is the solute lifetime probability density function, that is, $g(\tau)\,\Delta\tau$ is the probability that a solute that enters the volume at $\tau = 0$ will disappear from the volume between τ and $\tau + \Delta\tau$.

Just as in the case of the CDE model, the solute transport description will simplify considerably when special cases are considered.

7.3.2 Solute Lifetime and Travel Time Distribution Functions

No Solute Reactions   When inert solutes that do not undergo chemical or biological reactions move through the transport volume, the transfer function formulation changes in several respects. First, the input and output mass fluxes become Q_{ent} and Q_{ex}, the mass flux through the entry and exit surfaces, respectively. Second, the solute lifetime now becomes the solute travel time and $g(\tau)$ becomes the solute travel time probability density function. Thus, $g(\tau)\,\Delta\tau$ is the probability that a solute arriving through the entrance surface at $\tau = 0$ will leave through the exit surface between τ and $\tau + \Delta\tau$.

7.3.3 Transfer Function Equation

If water is flowing through the volume at an average flux rate $J_w(t)$, then the transfer function equation may be written in terms of solute concentration as

$$C_{ex}(t) = \frac{1}{J_w(t)} \int_0^t g(\tau)\, J_w(t - \tau)\, C_{ent}(t - \tau)\, d\tau \qquad (7.63)$$

Equation (7.63) is still restricted in its application in two ways. First, $J_w(t)$ must be spatially uniform within the transport volume and, second, $g(\tau)$ is a function of so many variables that it would have to be measured for a large number of different experiments to have any utility. For this reason, Jury (1982) proposed reformulating (7.63) in terms of the cumulative net applied water,

$$I(t) = \int_0^t J_w(t')\, dt' \qquad (7.64)$$

so that

$$C_{ex}(I) = C[L, I] = \int_0^I f_L(I')\, C_{ent}(I - I')\, dI' \qquad (7.65)$$

where L is the depth of the outflow surface and $f_L(I)$ is the net applied water probability density function (pdf); that is, $f_L(I)\,\Delta I$ is the probability that solute added to $z = 0$ at $I = 0$ will exit at $z = L$ between I and $I + \Delta I$ net applied water.

If this assumption is valid, then $f_L(I)$ becomes the fundamental solute transport property for the transport volume, and (7.65) applies for any time-dependent water input rate as long as the water storage changes within the volume remain small.

The assumption that solute outflow will be a unique function of the drainage or net applied water volume has been shown to be accurate for repacked soil columns subjected to different water flow rates in the laboratory (Khan, 1988). It has also been used successfully to describe area-averaged solute concentrations in a large field study (Butters et al.; 1989, Jury et al., 1982b, 1990) and has been shown to be accurate by computer simulation of solute transport in a sandy soil (Wierenga, 1977). The assumption is less accurate in structured soils, where preferential flow channels become active at high water flux rates (Dyson and White, 1986b).

7.3.4 Measurement of Transfer Function Parameters

There are two types of standard experiment that can be performed to measure $f_L(I)$: the narrow-pulse input and the solute front input. In the narrow-pulse input experiment, a narrow concentration pulse of mass M_0 is added to the inlet end of the transport volume at $I = 0$, and its average outflow concentration $C(L, I)$ at $z = L$ is measured as a function of I. The manner in which the outflow concentration is measured depends on the nature of the transport volume. For soil columns, $C(L, I)$ would correspond to the drainage effluent. For tile-drained fields, $C_{ex}(I)$ would correspond to the tile drain concentration (Jury, 1975a,b). For large fields that drain freely, $C(L, I)$ can be constructed from the average of a set of solution samplers at $z = L$ (Butters et al., 1989).

Regardless of the geometry, when a narrow pulse is added to the entrance surface,

$$f_L(I) = M_0^{-1}C_{ex}(I) \tag{7.66}$$

The mass M_0 need not be measured at the inlet end. It can be calculated from the mass recovered (Jury, 1982):

$$M_0 = \int_0^\infty C_{ex}(I') \, dI' \tag{7.67}$$

For solute front inputs, where a concentration C_0 is added to the inlet end at $I = 0$, $f_L(I)$ is equal to the slope of the outflow curve (Jury, 1982):

$$f_L(I) = C_0^{-1}dC_{ex}(I)/dI \tag{7.68}$$

This method of measuring $f_L(I)$ is not as accurate as the pulse method because the outflow data recovered in (7.68) must be differentiated.

Once $f_L(I)$ has been obtained, the transport volume fraction θ_{st} can be calculated from either the median $\tilde{I}$ or mean $\bar{I}$ water input displacement, where the mean value is calculated as

$$\bar{I} = \int_0^\infty I f_L(I)\, dI \tag{7.69}$$

Thus, $\bar{I}$ is equal to the first moment of the pdf $f_L(I)$ (Himmelblau, 1970). The median value of I is the value at the middle of the population. It is calculated from

$$\int_0^{\tilde{I}} f_L(I)\, dI = 0.5 \tag{7.70}$$

The median value of I is measured most easily by adding a solute front of concentration C_0 and determining the value of I at which the outflow concentration rises to $0.5C_0$ (White et al., 1986).

By definition,

$$\bar{\theta}_{st} = \bar{I}/L \tag{7.71}$$

and

$$\tilde{\theta}_{st} = \tilde{I}/L \tag{7.72}$$

where L is the effective thickness of the transport volume and θ_{st} corresponds roughly to the mobile water content θ_m in the MIM equations (7.27) and (7.28) or to the mobile zone of Addiscot (1977).

For some applications, the transport volume fraction θ_{st} may be the only measurement of the soil properties required to make rough predictions of solute leaching (White, 1987).

7.3.5 Model Distribution Functions

By (7.66) and (7.67), we see that the pdf $f_L(I)$ has the property

$$\int_0^\infty f_L(I)\, dI = 1 \tag{7.73}$$

Therefore, it can be represented by parametric frequency distributions that have only positive values of the independent variable I. Two such distributions are the Fickian pdf,

$$f_L(I) = \left[L/2(\pi DI^3)^{1/2} \right] \exp\left[-(L - VI)^2/4DI \right] \tag{7.74}$$

and the lognormal distribution,

$$f_L(I) = \left[1/(2\pi)^{1/2} \sigma I \right] \exp\left[-(\ln I - \mu)^2/2\sigma^2 \right] \tag{7.75}$$

where D, V, σ, μ are model parameters (Jury and Sposito, 1985; Simmons, 1982). The Fickian pdf is the solution to the CDE (7.17) for a unit-mass, narrow-pulse input of solute, with I playing the role of time. Thus, the CDE can be represented as a transfer function equation in the form of (7.65). In fact, any linear solute transport model[4] that obeys the solute conservation equation (7.4) can be represented as a transfer function equation (Jury et al., 1986). Thus, it is a more general formulation than any specific model and can be used to represent any model hypothesis that obeys mass balance and is linear in the concentration.

7.3.6 Stochastic–Convective Transfer Function Model

The foundation of the transfer function model is the pdf $f_L(t)$ or $f_L(I)$. Since it is measured by an outflow experiment in which solute moves through the entire transport volume, any model function of $f_L(I)$ representing its shape well can be used in (7.65) to predict outflow concentrations for different inputs. However, no predictions can be made about the solute concentration anywhere except at the outflow location unless additional assumptions are built into the model (Jury and Roth, 1990). The CDE (7.17) is an example of a process model that can make predictions at every location z. However, its assumptions form only one possibility for a solute transport model.

Jury (1982) proposed a new model that postulated the behavior of the solute concentration at depths $z \neq L$ in the following way: The cumulative distribution function (cdf) for solute arriving at $z = L$ is by definition equal to

$$P_L(I_0) = \int_0^{I_0} f_L(I)\, dI \tag{7.76}$$

This function is equal to the probability that a solute tracer added to the inlet at $I = 0$ will arrive at the outlet end $z = L$ before an amount I_0 of water has moved through the system. Thus, it may be determined in a solute front experiment by

$$P_L(I) = C_0^{-1} C(L, I) \tag{7.77}$$

where $C(L, I)$ is the outflow concentration at $z = L$ in response to a solute front of concentration C_0 added to the surface at $I = 0$. Jury (1982) postulated that, for $z \neq L$,

$$P_z(I_0) = P_L(I_0 L / z) \tag{7.78}$$

Equation (7.78) states that the probability that a solute tracer will reach a depth z when an amount of water $I \leq I_0$ has flowed through the system is equal to the probability that a solute tracer will arrive at $z = L$ (called the reference depth)

---

[4] A linear model is one that if C_1 and C_2 are solutions, then so is $\alpha_1 C_1 + \alpha_2 C_2$, where α_1, α_2 are constants.

when $I \leq I_0 L/z$. For example, if 5% of a solute pulse reaches $z = 30$ cm after $I = 4$ cm of water has been added, then 5% of the pulse will reach $z = 60$ cm after $I = 8$ cm of water has been added, and so on. This model may be interpreted physically as a representation of solute traveling at different velocities in isolated stream tubes whose contents do not have time to mix with adjacent stream tubes before reaching the depth where the observation is made. Thus, it may be thought of as the early time limit of solute transport (see Fig. 7.12), just as the CDE, which assumes complete mixing of adjacent stream tubes [and which does *not* satisfy (7.78)], may be thought of as the long-time limit of possible solute behavior.

The pdf $f_z(I)$ at any depth z is by definition equal to the derivative of the cdf,

$$f_z(I) = dP_z(I)/dI \tag{7.79}$$

Therefore, we may derive the pdf $f_z(I)$ for an arbitrary depth z in terms of the pdf at the reference depth $z = L$. This produces the result (Jury, 1982)

$$f_z(I) = L/z f_L(IL/z) \tag{7.80}$$

A process that obeys (7.80) is said to be stochastic–convective (Jury, 1988; Simmons, 1982). Thus, for an arbitrary z, the transfer function model (7.65) becomes

$$C(z, I) = \int_0^I L/z f_L(LI'/z) C_{\text{ent}}(I - I') \, dI' \tag{7.81}$$

in terms of the reference pdf at $z = L$. Thus, for a transport volume obeying the stochastic–convective assumption (7.80), (7.81) can be used to predict $C(z, t)$ using only the single experiment calibrating $f_L(I)$. Equation (7.81) is called the stochastic–convective transfer function model.

In contrast, the pdf for the CDE at $z \neq L$ is given by the solution of (7.17) at z for a narrow-pulse input of solute, which is simply equal to the Fickian pdf (7.74) with z substituted for L (Jury and Sposito, 1985):

$$f_z(I) = \left[z/2(\pi DI^3)^{1/2} \right] \exp \left[-(z - VI)^2/4DI \right] \tag{7.82}$$

EXAMPLE 7.8:   Verify that Equation (7.82) does not satisfy Equation (7.80).
$$L/z f_L(IL/z) = \left[L^2/2z(\pi DI^3 L^3/z^3)^{1/2} \right] \exp \left[-(L - VIL/z)^2/4DIL/z \right]$$
$$= \left[z/2(\pi DzI^3/L)^{1/2} \right] \exp \left[-(z - VI)^2/4DzI/L \right]$$
$$\neq f_z(I)$$

Convective Lognormal Transfer Function Model   The stochastic–convective assumption (7.80), when applied to the lognormal pdf (7.75), yields the simple result

$$f_z(I) = \left[1/(2\pi)^{1/2} \sigma I \right] \exp \left[-\left[\ln(IL/z) - \mu \right]^2/2\sigma^2 \right] \tag{7.83}$$

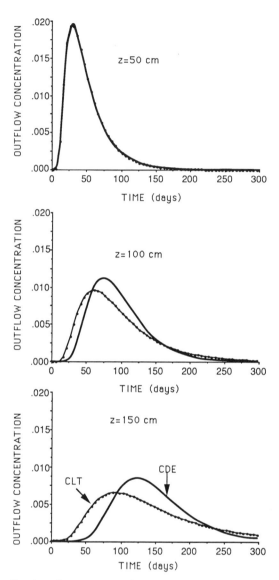

Figure 7.14 Predicted outflow concentrations of CDE (7.82) and CLT (7.83). Models are calibrated to each other at $z = 50$ cm.

The transfer function employing this pdf is called the convective lognormal transfer (CLT) function model (Jury, 1988).

Figure 7.14 plots the CLT model pdf (7.83) and the CDE pdf (7.82) for three values of z when the parameters of the two models have been adjusted for maximum agreement at $z = 50$ cm. Two important features of these curves should be noted. First, at outflow locations other than 50 cm the concentrations predicted by the models do not agree. The CLT model pulse has spread much more at depths

$z > 50$ cm than the CDE pulse. Second, at $z = 50$, the model pdfs are almost identical, showing how closely the shapes of the two mathematical functions can be made to agree. Thus, for any soil column experiment in which outflow from a single exit point is modeled, the two approaches would be judged equivalent even though they are based on completely different assumptions about the dispersion process. Consequently, a model assumption about dispersion can only be tested by observing solute movement to several different distances from the solute entry point (Khan and Jury, 1990).

7.3.7 Model Parameter Estimates

When parametric models are used to represent the frequency distribution $f_L(I)$, the model parameters must be estimated by optimizing the agreement between the model and the sample measurement of $f_L(I)$. Jury and Sposito (1985) discussed three methods for solute parameter estimation: sum of squares, method of moments, and maximum likelihood. Only the method of moments will be discussed here, because it lends itself most easily to a discussion of the differences between model hypotheses.

Method of Moments   The method-of-moments procedure for estimating model parameters from outflow experiments is used for narrow solute input pulse experiments, in which a set of concentrations $C(z, I_j), j = 1, \cdots, M$, are measured at a fixed distance z at $I_1, I_2, \cdots, I_M$ [or equivalently at $t_1, \cdots, t_N$ if $f_z(t)$ is measured]. The Nth moment T_N of I is defined theoretically as

$$T_N = \int_0^\infty I^N C(z, I) \, dI \tag{7.84}$$

where $T_0 \equiv 1$ if $C(z, I)$ has been normalized. Equation (7.84) may be evaluated in terms of model parameters if the theoretical expression for $f_z(I)$ for a given model can be integrated exactly. For example, it can be shown (Jury and Sposito, 1985; Valocchi, 1985) that for the Fickian pdf (7.82)

$$T_0 = 1 \tag{7.85}$$

$$T_1 = z/V \tag{7.86}$$

$$T_2 = z^2/V^2 + 2Dz/V^3 \tag{7.87}$$

When the data are measured, (7.84) can be evaluated numerically to yield sample measurements T_N^* of T_N. For example, using the trapezoidal rule, the integral (7.84) would be approximated as

$$T_N^* \simeq \sum_{j=1}^{N-1} \left[\tfrac{1}{2}(I_j + I_{j+1})\right]^N \left[\tfrac{1}{2}(C_j + C_{j+1})\right] \Delta I_j \tag{7.88}$$

where $C_j \equiv C(z, I_j)$ and $\Delta I_j = I_{j+1} - I_j$.

For the CDE, evaluation of the sample moments with (7.85)–(7.87) yields

$$z/V = T_1^*/T_0^* \qquad (7.89)$$

$$z^2/V^2 + 2Dz/V^3 = T_2^*/T_0^* \qquad (7.90)$$

or

$$V = z/\hat{T}_1 \qquad (7.91)$$

$$D = V^3/2z(\hat{T}_2 - \hat{T}_1^2) \qquad (7.92)$$

where $\hat{T}_1 = T_1^*/T_0^*$ and $\hat{T}_2 = T_2^*/T_0^*$ are the normalized sample moments of the distribution $f_z(I)$. Since the CDE requires that these parameters be constant at different z, evaluation of $\hat{T}_N$ at several depths can be used to test this model's validity.

The expression $\hat{T}_2 - \hat{T}_1^2$ in (7.92) has a special name. It is called the variance of the distribution $f_z(I)$ and is also equal to

$$\text{Var}_z[I] = \int_0^\infty (I - E_z[I])^2 f_z[I]\, dI \qquad (7.93)$$

where

$$E_z[I] = T_1/T_0 = \int_0^\infty I f_z[I]\, dI \qquad (7.94)$$

is called the expectation or mean value of I. By (7.93) we see that the variance is the mean-square deviation from the average value. We may use (7.91)–(7.94) to develop a fundamental definition of the dispersion of a distribution, as

$$\text{Dispersion of } f_z(I) = \frac{z^2}{2} \frac{\text{Var}_z[I]}{E_z[I]^3} \qquad (7.95)$$

This will allow us to calculate the dispersion of various solute transport models in terms of the model parameters and z. Table 7.6 (Jury and Sposito, 1985; Sposito et al., 1986) summarizes $E_z[I]$, $\text{Var}_z[I]$, and the dispersion (7.95) of the process for the three solute transport models we have studied thus far: the CDE (7.17), the MIM (7.27) and (7.28), and the CLT (7.83).

Several features are notable in Table 7.6. First, both the CDE and MIM have constant dispersion as a function of z when their parameters are constant. In contrast, the CLT has a linearly increasing dispersion with z when its parameters are constant. This explains the greater spreading observed in Fig. 7.14 compared to the CDE.

TABLE 7.6   Mean, Variance, and Dispersion of Three Transport Models.

Model	$E_z[I]$	$\text{Var}_z[I]$	Dispersion of $f_z[I]$
CDE	z/V	$2Dz/V^3$	D
MIM	$\theta z/J_w$	$\dfrac{\theta^3}{J_w^3}[2zD_M\phi_M + \dfrac{2zJ_w^2}{\theta\alpha}(1-\phi_M)]$	$D_M\phi_M + \dfrac{J_w^2(1-\phi_M)}{\theta\alpha}$
CLT	$z/L \exp(\mu + \sigma^2/2)$	$(z/L)^2 \exp(2\mu + \sigma^2)[\exp(\sigma^2) - 1]$	$\dfrac{zL}{2}\dfrac{\exp(\sigma^2)-1}{\exp(\mu+\sigma^2/2)}$

Second, by comparing the CDE and MIM, we see that part of the dispersion or spreading in the MIM is caused by the rate-limited diffusion process between the mobile and stagnant regions of the wetted pore space (Valocchi, 1985). Finally, all three models have a mean I that increases proportionally to z. Thus, the center of mass of a pulse is predicted to move at the same rate by all three models; only the spreading or dispersion differs. For this reason the piston flow model will describe mean transport equally well for all three processes, but does not describe the dispersion or spreading process.

7.3.8   Application to Experimental Studies

The first experimental application of the transfer function model of solute movement was the field study of Jury et al. (1982b), in which a narrow pulse of bromide was added to the surface of a 1.4-ha loamy sand field and was monitored by solution samplers at five depths from 30 to 180 cm at each of 14 sites during a winter of erratic rainfall. The CLT model pdf (7.83) was calibrated using the 30-cm depth-averaged concentration and was used to predict movement past the 60–180-cm depths as a function of the net applied water (rainfall evaporation) measured at the soil surface. The agreement between the model predictions and the area-averaged transport was quite good, considering that the water input rate was erratic and that no measurements of soil water properties were needed.

The model received a more substantial test on the same field in 1987, this time under steady water flow induced by bidaily sprinkler irrigation. In this study [summarized in Butters et al. (1989) and Butters and Jury (1989)], the CDE and CLT models (7.82) and (7.83) were calibrated at 30 cm and were used without further calibration to predict movement of the area-averaged pulse to depths of 450 cm. The model predictions and data for 30, 180, and 300 cm are shown in Fig. 7.15. Clearly, the CLT model predictions are in much better agreement with observation than those of the CDE through the top 3 m. Butters and Jury (1989) found that the CLT model began to overestimate the spreading of the pulse at 450 cm. However, they also showed that the final solute profile measured by soil coring to 25 m was better described by the CLT than the CDE even when the models were recalibrated at 450 cm.

The close agreement shown between the CLT model and the observed pulse movement near the surface offers convincing evidence that the dispersion process

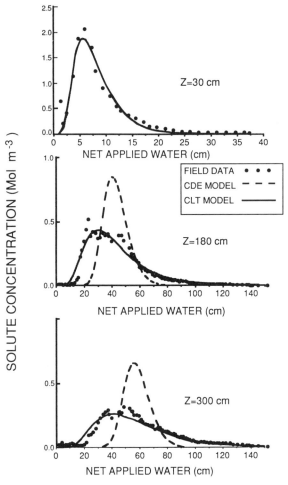

Figure 7.15 The CDE and CLT model predictions and field data at three depths using calibrations obtained by fitting at $z = 30$ cm with the method of moments. (Adapted from Butters and Jury, 1989.)

is convectively driven over the time and depth scales of this experiment. A plot of the pulse dispersion calculated by (7.95) versus depth (shown in Fig. 7.16) reveals an almost linear increase with z, indicating that convective velocity differences are not being smoothed out by lateral mixing as the solute moves downward. This increasing dispersion with distance, sometimes called the dispersion scale effect (Fried, 1975), has also been observed over several hundred days and many decimeters in a controlled groundwater experiment (Freyberg, 1986) and appears to be quite prevalent in other large-scale groundwater studies as well (Gelhar et al., 1985).

White and colleagues have used the transfer function to model solute movement

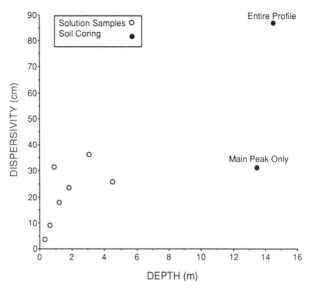

Figure 7.16   Plot of CDE dispersivity $\lambda = D/V$ versus depth calculated by sum of squares data fits at each depth. (After Butters and Jury, 1989.)

through structured clay soils (Dyson and White, 1986b; White et al., 1984, 1986), through soil drained by mole lines (White, 1987), and through layered soil (Dyson and White, 1986a). These authors use the observed movement to construct a pdf from which the median transport volume is calculated by (7.73). This parameter is subsequently used with the pdf to predict movement of other solutes through the same soil volume.

7.4   SOLUTE MANAGEMENT IN FIELD SOIL

7.4.1   Salinization of Crop Root Zones

Leaching Fraction   When poor-quality water is used for irrigation, salts can build up in the crop root zone as water is removed by plant root extraction. Management of irrigation to minimize the salt burden of the irrigation drainage water while avoiding plant stress from salinity has long been an active area of research in soil science.

Some simple calculations with the water and solute transport equations developed in this chapter will help to illustrate the principles of salinization in the root zone.

EXAMPLE 7.9:   Calculate the steady-state solute concentration as a function of depth in a crop root zone of depth L receiving high-frequency irrigation of concentration $C = C_0$ at a rate $J_w = -i_0$ in which water is extracted at a spatially uniform rate r_w in

$-L \leqslant z \leqslant 0$. Assume (i) no solute reactions, (ii) no solute dispersion, and (iii) no plant uptake of solute.

The steady-state form of the water flow equation for transport through a crop root zone with uniform water uptake was worked out in Chapter 3, Example 3.12 [see (3.83)]:

$$J_w(z) = -i_0 - ETz/L$$

When dispersion is neglected and solutes do not precipitate, dissolve, or enter plant roots, the steady-state solute flux is given by the piston flow model [(7.15) with $D_e = 0$]:

$$J_s = \text{const} = J_w(z)C(z) = -i_0 C_0 \tag{7.96}$$

Thus, combining (3.83) and (7.96), we obtain

$$C(z) = \frac{i_0 C_0}{i_0 + ET \cdot z/L} \tag{7.97}$$

We may simplify this expression by defining the leaching fraction

$$f_L = \frac{\text{drainage rate}}{\text{irrigation rate}} = \frac{i_0 - ET}{i_0} \tag{7.98}$$

After solving (7.98) for ET and substituting it into (7.97), we may write the concentration profile in terms of f_L as

$$C(z) = \frac{C_0}{1 + (1 - f_L)z/L} \tag{7.99}$$

Figure 7.17 shows a plot of C/C_0 versus z/L for various values of the leaching fraction. The maximum solute concentration occurs at the bottom of the root zone,

$$C_{\max} = C(-L) = C_0/f_L \tag{7.100}$$

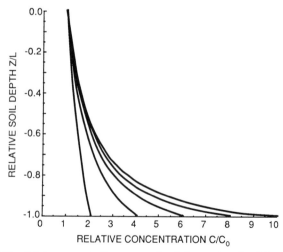

Figure 7.17 Solute concentration versus depth for various leaching fractions.

Scientists have long searched for a reliable index of the effect of soil salinity on plant growth and yield. For a fully salinized root zone such as the one described by Fig. 7.17, Raats (1975) presented the following analysis. He defined the average salinity $\overline{C}$ of the root zone as the flow-averaged concentration of the water extracted by the plant:

$$\overline{C} = \frac{\displaystyle\int_{-L}^{0} r_w(z) C(z) \, dz}{\displaystyle\int_{-L}^{0} r_w(z) \, dz} \tag{7.101}$$

where the denominator has the value ET.

In steady state, $C(z) = -C_0 i_0 / J_w(z)$ (7.96) and $r_w = -dJ_w/dz$ (3.80). Thus, (7.101) may be written as

$$\overline{C} = \frac{i_0 C_0}{ET} \int_{-L}^{0} \frac{dJ_w}{dz} \frac{dz}{J_w} = \frac{i_0 C_0}{ET} \int_{-i_0 f_L}^{-i_0} \frac{dJ_w}{J_w} = \frac{i_0 C_0}{ET} \ln\left(\frac{1}{f_L}\right) \tag{7.102}$$

Finally, using (7.98), we may rewrite (7.102) as

$$\overline{C}/C_0 = (1 - f_L) \ln(1/f_L) \tag{7.103}$$

Note that this result is independent of the shape of the water uptake distribution and depends only on f_L and C_0.

Solution Chemistry and Transport The assumptions used in the previous section to derive chemical concentrations in the root zone are best suited to extremely soluble, inert ions like chloride. However, typical irrigation waters contain many ions that contribute to the osmotic potential of the soil solution, and reactions such as precipitation–dissolution and cation exchange can affect the overall chemical concentration of the soil solution.

The least soluble salt in a typical irrigation water is calcium carbonate (lime). Calcium sulfate (gypsum) is also moderately soluble. Thus, when irrigation water that contains significant quantities of the ions that make up these salts is concentrated by plant roots, the salts precipitate out of solution and the final concentration is less than that predicted from (7.100) (Rhoades et al., 1973, 1974). Furthermore, cation exchange can alter the amount of calcium in solution and change the extent of precipitation or dissolution.

The first models constructed to take solution chemistry into account when water is concentrated in the root zone did not consider transport but conducted calculations on the solution constituents assuming thermodynamic equilibrium (Oster and McNeal, 1971; Oster and Rhoades, 1975). In these calculations, the ions underwent precipitation–dissolution and ion-pairing reactions in the presence of fixed CO_2 concentrations, and the final solution concentrations at a given point in the root zone were estimated. Since the calculations assumed steady state, the exchange complex did not influence the estimated concentrations.

The transition period prior to steady state was studied by Jury et al. (1978a,b) using the model of Oster and Rhoades to estimate equilibrium solution chemistry in finite segments of a soil profile through which water was flowing in steady state and solutes were moving in transient flow through mixing cells as described by Dutt et al. (1972). Thus, the ions would move a small distance without reacting and then equilibrate in finite segments of the profile. The authors also added cation exchange with Na^+, Ca^{2+}, Mg^{2+} to the equilibrium calculations.

Using this model, Jury and Pratt (1980) studied the influence of the solute transition period on root zone salinization and drainage salt burden for four different irrigation waters: Colorado River water (CRW), Saline power plant cooling water (SCW), municipal wastewater (MWW) and Feather River water (FRW). The composition of these waters is shown in Table 7.7.

The authors compared the estimates of the transitional model with the simple zero-reaction steady-state model (called the proportional model) and the steady-state solution chemistry model. Tables 7.8 and 7.9 show calculations for drainage concentration and drainage salt flux, respectively, as a function of f_L for a calcareous (lime-saturated) soil with a cation exchange complex of 30 me g^{-1}. These results clearly show the importance of chemical reactions on root zone salinization. For example, the saline cooling water, which is near saturation with both lime and gypsum, precipitated over 50% of the salts applied during the first eight years of irrigation, a prediction that was verified experimentally (Jury et al., 1978c).

Another finding of Jury and Pratt (1980) was that the transition period persisted for many years (and many pore volumes) because of cation exchange reactions. Figure 7.18 shows predicted ion concentrations for the MWW water at 250 cm as a function of time. The Cl$^-$ ion, which does not react, reaches steady state in about 8 years (corresponding to one pore volume). However, the other ions are still changing concentration with time after 28 years; in fact Mg^{2+} is still decreas-

TABLE 7.7   Composition of Irrigation Waters used in Simulations

Ion or Index[a]	Colorado River Water (CRW)		Saline Power Plant Cooling Water (SCW)		Municipal Wastewater (MWW)		Feather River Water (FRW)	
	mg l^{-1}	me l^{-1}	mg l^{-1}	me l^{-1}	mg l^{-1}	me l^{-1}	mg l^{-1}	me l^{-1}
NH_4^+	—	—	—	—	0.5	0.03	—	—
Ca^{2+}	78	3.9	524	26.2	87	4.35	9.0	0.45
Mg^{2+}	30	2.5	220	18.3	30	2.50	4.3	0.36
Na^+	104	4.5	1,132	49.2	140	6.09	5.8	0.25
K^+	4	0.1	—	—	14	0.36	1.6	0.04
Cl^-	92	2.6	930	26.2	160	4.51	2.8	0.08
SO_4^{2-}	283	5.9	3,010	62.7	250	5.21	7.7	0.16
HCO_3^-	146	2.4	293	4.8	192	3.15	52.5	0.86
NO_3^-	—	—	—	—	25	0.40	—	—
TDS	737	—	6,108	—	861	—	83.6	—
SAR (me l^{-1})$^{1/2}$	—	2.5	—	10.4	—	3.3	—	0.4
EC (mmho cm^{-1})	—	1.0	—	7.1	—	1.3	—	0.1

[a]TDS = total dissolved solids
SAR = sodium adsorption ratio
EC = electrical conductivity

TABLE 7.8   Predicted 150-cm Drainage Concentrations (mg l^{-1}) Using Three Models

Leaching Fraction	Drainage Water TDS for		
	Proportional Model	Steady-State Model	Dynamic Model[a]
		CRW	
0.10	7,370	6,050	4,550
0.25	2,948	2,850	2,750
0.40	1,840	1,920	1,885
		SCW	
0.10	61,080	44,500	29,647
0.25	24,432	18,714	16,850
0.40	15,070	12,452	11,830
		MWW	
0.10	8,610	6,706	5,513
0.25	3,444	3,052	2,970
0.40	2,150	2,129	2,075
		FRW	
0.10	836	852	785
0.25	334	661	655
0.40	209	616	621

[a] Data from the eighth year after irrigation commences.

ing because of exchange adsorption with Ca^{2+}, even though its final steady-state concentration will be 25 me l^{-1}. The predicted transition time for Mg^{2+} in these conditions is approximately 85 years.

7.4.2   Groundwater Contamination

More frequent and intensive monitoring of groundwater quality and improvements in the analytical detection of chemicals dissolved in water have greatly increased research on the transport of contaminants from the soil surface to groundwater.

Nitrate Pollution   Nitrates can reach groundwater from a variety of sources. Nitrogen is a primary component of organic and inorganic fertilizers (Scarsbrook, 1965) and transforms rapidly to nitrate under normal soil conditions (Alexander, 1965). Septic tanks, which serve about 30% of the households in the United States, can be major sources of nitrate pollution in groundwater, particularly if they are poorly maintained (Novotny and Chesters, 1981; Pye et al., 1983). Dairy and poultry feedlots can form substantial sources of nitrate pollution, especially if the

TABLE 7.9   Predicted Mass Emissions (t/ha yr) at 150 cm Using Three Models

Leaching Fraction	Proportional Model	Steady-State Model	Dynamic Model[a]
		CRW	
0.10	7.37	6.05	5.01
0.25	8.84	8.55	8.10
0.40	11.04	11.52	10.98
		SCW	
0.10	61.08	44.50	29.65
0.25	73.30	56.14	50.55
0.40	90.42	74.71	79.98
		MWW	
0.10	8.61	6.71	5.51
0.25	10.33	9.16	8.91
0.40	12.90	12.77	12.45
		FRW	
0.10	0.83	0.85	0.79
0.25	0.99	1.98	1.97
0.40	1.24	3.70	3.73

[a] Data from the eighth year after irrigation commences.

site is abandoned and subsequently exposed to rainfall or irrigation (Novotny and Chesters, 1981). In addition, certain soils contain nitrates deposited during geologic times that can move to groundwater when water is percolated through them (Sullivan et al., 1979; White and Moore, 1972).

The widespread appearance of nitrates in groundwater is a consequence of a number of factors. Nitrate is very mobile in soils and is easily displaced from its point of origin by water additions. It is stable in soil except when biologically transformed by denitrification, which only occurs in very wet soil or inside soil aggregates at high moisture content (Broadbent and Clark, 1985). Although nitrate is taken up rapidly by plant roots, this removal mechanism only occurs very close to the soil surface. Thus, whenever there is a source of nitrogen and an excess of water applied to the soil, nitrates have the potential to reach groundwater. Since two of the major sources of nitrogen addition to soil, irrigated agriculture and septic tanks, are also sources of excess water, it is easy to see why these operations have been associated with nitrate pollution in the past.

Although deep groundwater tables require longer to reach than shallow ones, dissolved nitrates will eventually arrive in groundwater as long as water continues to enter the soil at a rate in excess of evaporative demand. Pratt et al. (1970) sampled underneath citrus groves throughout California and found nitrate plumes

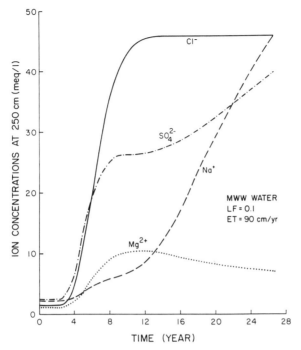

Figure 7.18 Ion concentrations at 250 cm as a function of time for MWW water applied at a leaching fraction of 0.1. (After Jury and Pratt, 1980.)

that had moved far below the root zone over a period of years but had not yet reached groundwater. In some locations, they estimated that travel times in excess of 50 years would be required to reach groundwater. This explains why groundwater under land developed over former citrus groves has continued to experience high nitrate levels for years after the high additions of nitrate fertilizer to the surface ceased (Ayers and Branson, 1973).

One of the major goals of solute transport models has been to predict the movement of contaminants such as nitrate to groundwater. A serious limitation to achieving this goal has been the absence of a procedure for quantifying the nature of the dispersion process for large-scale emission of solutes from the soil surface.

To illustrate the effect that uncertainty about the dispersion process can have on groundwater predictions, Jury (1983) compared two models, the one-dimensional CDE with a constant dispersion coefficient (7.82) and the convective–lognormal transfer function (CLT) (7.83) with a linearly increasing dispersion coefficient. As shown in Fig. 7.14, these two models virtually superimpose on each other when their parameters are optimized, but after calibration the CLT predicts much greater spreading than the CDE. Table 7.10 presents calculations by the two models of the time required to transport 1, 5, 10, 25, 50, 75, 95, and 99% of a pulse of nitrate from the surface to groundwater located at 10 m below the surface.

As the table illustrates, there is a substantial deviation between the predictions

TABLE 7.10   Predicted Arrival Times to Groundwater at $z = 10$ m for Percentage of a Pulse of Nitrate P^a

P (%)	Arrival Time (days)	
	CDE	CLT
1	266	83
5	282	121
10	292	147
25	311	203
50	332	289
75	355	415
90	377	575
95	391	699
99	419	1007

^{a}Models are calibrated at 0.3 m ($V = 3$ cm day^{-1}; $D = 15$ cm^2 day^{-1}).

of the models for early and late arrival times, which reflects the way that they represent the dispersion process. However, the prediction of the mean arrival time is similar for both models, indicating that the piston flow model will provide rough estimates of the mean transport of a solute pulse or front. The leading edge or worst-case movement differs substantially, indicating that in the absence of further experimentation, transport models making this prediction of early arrival must be regarded as uncertain within the limits of these two models.

Industrial Chemicals   Groundwater contamination from industrial chemicals such as the solvents trichloroethylene (TCE) and perchloroethylene (PCE) and petroleum hydrocarbons such as gasoline are increasingly found in groundwater as monitoring programs become more widespread. Soluble organic chemicals such as TCE and PCE behave similarly to pesticides in soil in that they undergo vapor and liquid phase movement and can adsorb to stationary soil solid materials. Thus, models such as those discussed in the previous section can be used to estimate the environmental fate of these compounds.

However, there is another class of compounds, loosely called non–aqueous phase liquids (NAPLs), that are only marginally soluble in water and exist as a separate non–aqueous phase liquid when added to soil (Dracos, 1987). The nature of the transport process for NAPLs is just beginning to be understood at present and is an active research area (Corapciouglu and Baehr, 1987). The principal manner in which NAPL transport differs from that of a soluble organic liquid is that NAPLs leave a portion of their mass trapped inside soil pores as isolated liquid globules as they move through the medium. The residual NAPL left in the soil might be as large as 20% of the wetted pore space in a saturated soil (Dracos, 1987) and will not leach out readily as water flows through the contaminated region. In the absence of artificial means for removing the compound, its only release from the stationary liquid phase thus formed is by slow dissolution into the surrounding mobile water phase. This slow release may take many years and contribute to contamination over a long period of time.

Pesticide Pollution of Groundwater   Agricultural pesticides form a special class of toxic organic compounds in that they are deliberately (rather than accidentally) added to soil and are accompanied by large and continued additions of water. Pesticide management to prevent groundwater contamination is problematical, because the compounds must be mobile enough to reach their target organism and persistent enough to eliminate it. However, persistence and mobility are not desirable properties for a toxicant to possess from an environmental perspective. Early compounds, such as DDT and Dieldrin, had very low mobility in soil and therefore did not have the potential to reach groundwater. However, they were very persistent and thus had the potential to reach the food chain by exposure through the atmosphere or migration in surface waters. To avoid this exposure route, newer pesticides have been designed to have a far more rapid breakdown in soil. For this reason they must be applied over time and have reasonably high mobility in order to reach and control the target organism.

Modern legislation in the United States and worldwide has been directed toward regulating pesticides so as to prevent groundwater contamination. Implicit in this legislation is the idea that pesticides can be screened based on their environmental fate properties at the time of their development to make preliminary decisions about their pollution potential. Jury et al. (1987) produced a simplified version of their behavior assessment model (Jury et al., 1983a) to assist in the decision-making process. Figure 7.19, taken from Jury et al. (1987), shows the results of this

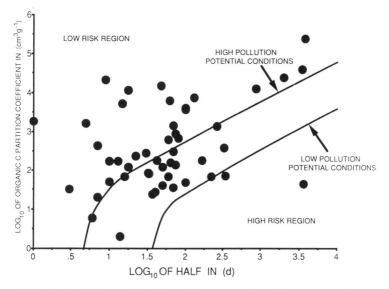

Figure 7.19   Plots of (K_{oc}, τ) for 50 pesticides, together with lines denoting relatively high and low groundwater pollution potential conditions. Compounds to the right of the line are considered to be a pollution potential groundwater contamination risk. (After Jury et al., 1987)

screening model applied to 50 pesticides in current or former use. The model calculations were constructed to screen compounds for a given set of environmental conditions so as to identify those compounds that, when applied in a pulse of mass M_0, retained more than $0.0001 M_0$ of this surface mass when they migrated below the biologically active surface zone.

In Fig. 7.19, two sets of environmental conditions are constructed. The lower line, called the low pollution potential condition, represents agricultural settings that produce a low potential for pesticides to reach groundwater (i.e., high organic carbon, low annual drainage rate), while the higher line represents high potential conditions (low organic carbon, high annual drainage rate). The compounds are plotted as a function of their organic carbon partition coefficient and their effective biological half-life and thus appear as a single point on this curve. Compounds that appear to the right of the line formed by a given condition are deemed to possess a potential groundwater contamination threat according to this criterion. Thus, the combination of high biological half-life and low organic carbon partition coefficient identify the worst potential for migration.

PROBLEMS[5]

7.1 A scientist is investigating a location where a hazardous waste spill occurred 25 years ago. In this accident, a large quantity of nitrates ($K_d = 0$) and an industrial solvent with a $K_d = 0.5$ cm^3 g^{-1} were dumped on the soil surface as a massive pulse. Rainfall and evaporation records indicate that about 1000 cm of water has infiltrated into the soil since that time. From soil coring it was determined that the average water content and bulk density of the soil profile are approximately $\theta = 0.25$ and $\rho_b = 1.5$ g cm^{-3}, respectively.

   (a) Calculate the approximate positions of the nitrate and solvent pulses today.

   (b) Assuming that the water table depth is 50 m, calculate the expected future arrival time of both pulses in the water table, assuming that the climate does not change.

7.2 Calculate the chemical transport equations analogous to (7.22) and (7.23) assuming that the chemical obeys an equilibrium Freundlich adsorption isotherm (1.13). Calculate the ratio of the effective retardation factors of a chemical with $1/N = 0.75$ at concentrations of 1 and 1000 mg l^{-1} ($\theta = 0.25$, $\rho_b = 1.5$ g cm^{-3}). What does this say about the velocity of a waste spill of high concentration? (See Rao and Davidson, 1979.)

7.3 A large undisturbed soil core is brought to the laboratory and subjected to steady-state water flow. At $t = 0$ a narrow pulse of Cl$^-$ is added to the inlet

[5]Problems preceded by a dagger are more difficult.

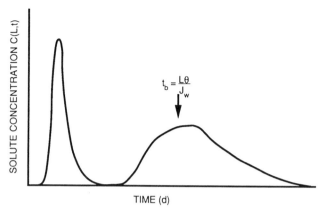

Figure 7.20 Solute outflow concentration as a function of time.

end, and the measured outflow concentration–time record looks like the curve shown in Fig. 7.20.

Give some plausible explanations for why the soil might show a break-through curve of the type in this figure.

7.4 According to the piston flow model, a chemical moving under steady-state water flux while adsorbing to equilibrium will break through at depth $z = L$ in a time

$$t_{bR} = L\theta R / J_w \quad \text{where } R = 1 + \rho_b K_d / \theta$$

If the chemical also decays by a first-order reaction, then the fraction of a chemical application of mass M_0 that is still present in the soil at time t_{bR} is

$$\frac{M(t)}{M_0} = \exp\left(-\mu t_b\right) = \exp\left(-\frac{\ln\left(2\right)t_b}{\tau_{1/2}}\right)$$

Using the properties of the six pesticides given in Table 7.2, calculate the mass fraction remaining at depth L under the conditions given in Table 7.4. Which pesticides have the greatest pollution potential? (See Rao et al., 1985.)

7.5 The residual mass fraction $\exp\left(-\mu RL\theta / J_w\right)$ in problem 7.4 is required to be less than some small number β,

$$\exp\left(\frac{\mu RL\theta}{J_w}\right) < \beta$$

Show that this relation may be written in the form

$$K_{oc} < a\tau_{1/2} - b$$

Indicate how this property may be used to select pesticide properties with low groundwater pollution potential (see Jury et al., 1987).

7.6 The travel time required for a nonadsorbing solute to reach depth L of a crop root zone receiving water input at a steady input flux rate $J_w = i_0$ is

$$t(L) = \int_o^L \frac{\theta \, dz}{J_w(z)}$$

where $V = J_w/\theta$ is the solute velocity. For the special case of uniform water uptake (3.83) and constant θ, calculate $t(L)$ as a function of i_0, ET, L, and θ. Using the definition of the leaching fraction f_L in (7.98), write the number of pore volumes,

$$P(L) = \frac{i_0 t(L)}{L\theta}$$

of applied water required to move the solute to depth L as a function of f_L alone and plot the result (see Jury et al., 1978c).

7.7 According to the piston flow model discussed in Example 7.9, the solute concentration at the bottom of the root zone is proportional to $1/f_L$ (7.100). tion (7.100)].

(a) What is the solute flux at $z = -L$ as a function of f_L?

(b) What effect does chemical precipitation have on these estimates?

(c) Discuss the recommended leaching fraction one would propose to (i) minimize the solute concentration below the root zone and (ii) minimize the solute flux below the root zone.

7.8 The solute transport equation for a mobile, nonreactive chemical according to the CDE is

$$\frac{\partial C}{\partial t} = D \frac{\partial^2 C}{\partial z^2} - V \frac{\partial C}{\partial z}$$

We wish to study the outflow at $z = -L$. Define new distance and time variables by the equations

$$y = z/L \qquad \tau = Vt/L$$

Insert these changes into the preceding solute transport equation and simplify the result. What variables does the new dimensionless equation have in it?

†7.9 By direct integration show that the Nth moment of the CLT model net applied water pdf $f_z(I)$ given by (7.83) is equal to

$$E[I^N] = \left(\frac{z}{L}\right)^N \exp\left(N\mu + N^2\sigma^2/2\right)\left[\exp\left(\sigma^2\right) - 1\right]$$

Defining the solute dispersivity as

$$\lambda = \frac{D}{V} = \left(\frac{z}{2}\right)\frac{\left(E[I^2] - E^2[I]\right)}{E^2[I]}$$

show that the dispersivity in this model has a linear scale effect (i.e., that it increases linearly with distance).

8 Methods for Analyzing Spatial Variations of Soil Properties

Much of the material presented thus far in this book has been directed toward applications where the soil is homogeneous and its properties do not vary in space. The only heterogeneous systems we have looked at were flow through stratified or layered profiles, where the soil is homogeneous in the direction normal to flow and piecewise homogeneous along the direction of flow.

However, natural field soils are quite variable, both in the vertical and horizontal directions. Moreover, variability is present at all distance scales. For example, soil porosity might vary significantly at large distance scales, where different soil series are encountered. However, it is also variable over very small distances, since it can have only two values: zero inside soil minerals and 1 between soil minerals. Thus, porosity has a meaning that is relative to the volume scale over which it is averaged (Bear, 1972). When measurements of a property like porosity are made on a scale that is small compared to the scale over which the quantity has a constant value, the measurements must be averaged to produce an estimate of the quantity. This chapter will develop methods for making estimates from a sample set of measurements of properties that vary in space or in time.

8.1 VARIABILITY OF SOIL PHYSICAL PROPERTIES

The most important characteristics of a property z that assumes a range of values z_1, z_2, ... are its average value, or mean m, and its spread about the average value. The most common index of the spread is the variance s^2, which for N samples is calculated by (Hald, 1952)

$$s^2 = \frac{1}{N-1} \sum_{j=1}^{N} (z_j - m)^2 \qquad (8.1)$$

where

$$m = \frac{1}{N} \sum_{j=1}^{N} z_j \qquad (8.2)$$

is the sample mean.

Since m is a constant, we may also express (8.1) as

$$s^2 = \frac{1}{N-1} \left(\sum_{j=1}^{N} z_j^2 - Nm^2 \right) \tag{8.3}$$

Another common index used to express the extent of variability is the sample coefficient of variation (CV):

$$CV = \frac{s}{m} \tag{8.4}$$

where $s = (s^2)^{1/2}$ is the standard deviation from the sample mean. The CV is often expressed as a percentage, by multiplying (8.4) by 100.

Table 8.1, taken from Jury (1985), summarizes sample CVs of various soil physical properties measured in the surface zone of natural field sites. Other reviews of the spatial variability of soil properties are found in Warrick and Nielsen (1980), Peck (1983), and Jury et al. (1987b,c). Several features are notable in this table. First, the so-called transport properties such as hydraulic conductivity as a rule are more variable than the retention properties such as water content, even on the same field site. Second, the CV of the transport properties on some fields is extremely large, reaching values well in excess of 100% for quantities such as the saturated hydraulic conductivity. The implications of this extensive variation are explored in the next example.

TABLE 8.1 Sample CVs Measured for Different Soil Water and Solute Properties in Unsaturated Fields

Parameter	Number of Studies	Range for CV (%)
Porosity	4	7–11
Bulk density	8	3–26
Percentage of sand or clay	5	3–55
Water content at 0.1 bar	4	4–20
Water content at 15 bars	5	14–45
pH	4	2–15
Saturated hydraulic conductivity	12	48–320
Infiltration rate	5	23–97
Solute concentration		
During transport	6	60–130
Natural	3	13–260
Solute velocity		
Unsaturated flow at constant flux	3	36–75
Ponding	2	78–194

EXAMPLE 8.1:   The saturated hydraulic conductivity of a 100-m^2 field has a mean of 100 cm day^{-1} and a CV of 50%. Assuming that the distribution is normal, calculate the values of K_s that are ± 1 and 2 standard deviations from the mean.

Since the CV is 50%, $s = 50$ cm day^{-1}. Thus, the K_s values 50, 150 are $\pm 1s$ and 0, 200 are $\pm 2s$ from the mean. If the population is really normal, then approximately 2.3% of the values of K_s should be negative, which is not physically possible. Thus, populations of variables that have only positive values (called positive definite) and that have large CV cannot be normally distributed.

The large CV calculated for some distributions often arises from a skewed population with a relatively small number of very large values. This is illustrated further in the next example.

EXAMPLE 8.2:   A soil survey team makes 20 measurements of infiltration rate on a uniform field, and a mean of 100 cm day^{-1} and a CV of 20% are calculated from these numbers. They decide to take one more reading and quit for the day. Measurement 21 happens to be taken right over a wormhole, which produces a high infiltration rate value of 500 cm day^{-1}. Calculate the new sample mean and CV for the 21 values.

The old mean was based on 20 values. Thus, the sum of all measurements is, from (8.2),

$$\sum_{j=1}^{20} z_j = 20(100) = 2000$$

The new mean, based on 21 values, is therefore

$$m = \frac{1}{21} \sum_{j=1}^{21} z_j = \frac{1}{21} (2000 + 500) = 119 \text{ cm day}^{-1}$$

Thus, the mean has increased 19% because of the new value. The old variance was $s^2 = (m \cdot \text{CV})^2 = 20 \times 20 = 400 \text{ cm}^2 \text{ day}^{-2}$. Using (8.3), we can calculate the sum of squares of the values of z_j:

$$\frac{1}{19} \left(\sum_{j=1}^{20} z_j^2 - 20 \times 100^2 \right) = 400 \quad \text{or} \quad \sum_{j=1}^{20} z_j^2 = 207{,}600$$

Thus, the new sum of squares including the final value is

$$\sum_{j=1}^{21} z_j^2 = 207{,}600 + (500)^2 = 457{,}600$$

Therefore, using (8.3), we calculate the new variance as

$$s^2 = \frac{1}{20} \left(\sum_{j=1}^{21} z_j^2 - 21(119)^2 \right) = \frac{1}{20} (457{,}600 - 297{,}381) = 8011$$

Thus, the new CV $= 89.5/119 = 0.75$.

Therefore, the CV has increased from 20 to 75% with the addition of one new large value to the sample set.

These examples point out that variability must be characterized in some detail to portray the features of the distribution as well as the average value of a set of measurements. The principal concepts of probability and statistics required to make such a characterization are given in what follows.

8.2 CONCEPTS OF PROBABILITY

Probability may be thought of as the fraction of a specific outcome from among all possible outcomes of an experiment (Himmelblau, 1970). A parameter whose values are characterized by a probability distribution is known as a random variable. Since the entire population of all possible outcomes of any event is never known to the observer, probability can only be estimated approximately by repeated observations. Certain events, such as a coin flip or a roll of a die, have only a finite number of possible outcomes, and the frequency of occurrence of a given outcome can be estimated relatively easily from a large number of trials. Other events, such as the saturated hydraulic conductivity of a large field, are not bounded a priori (except by being positive definite), and their frequency of occurrence must be inferred from the distribution of a sample set that is much smaller than the population of all possible experiments.

8.2.1 Random and Regional Variables

The application of probability to parameters whose values vary in space is different in several respects from an event that varies in time. Dice may be shaken repeatedly in separate trials, and each outcome is unknown until the trial is completed. If the throw of the dice is "random" each time (i.e., different force, angle of lift, etc.), then we expect, based on long experience, that each die face has a probability of $\frac{1}{6}$ of appearing. In contrast, properties that vary in space, such as porosity, have a single fixed value within a given small volume of space. If the property can be measured at that location without disturbing the place where the measurement was made, then its value is known with certainty at that location thereafter. Such parameters are often called regional variables (Journel and Huijbregts, 1978). To apply the concepts of probability to a regional variable, we must assume that it is a random function in space by treating measurements of the property at different locations as though they were repeated trials of the same event. Thus, for a variable to be random in space, it must have values at different locations that are characterized by a probability distribution. This distinction will become clearer after some definitions are made.

A parameter $Y(x, y, z)$ whose values vary at a given time as a function of position is called a regional variable. It may be treated as though it were a random variable Z whose value at a given point in space (x, y, z) is characterized by the probability distribution for the variable at that point. In other words, the value $Y(x, y, z)$ measured for Z at that point is regarded as a particular outcome from among a set of possibilities determined by chance. In another identical universe, perhaps an observer would have measured a different value at that point.

The problem with applying probability to spatially variable properties is that the laws of probability for a variable at a given point cannot be inferred from the single available measurement. The way around this dilemma is to assume that the value of the variable at every point in space is determined by the *same* probability law.[1] Then each measurement may be regarded as a separate sample of the random variable, and the frequency of occurrence of outcomes may be used to generate the frequency distribution function for the variable over the domain of measurement. The method by which frequency distributions are constructed is covered in the next section.

8.2.2 Frequency Distributions

The frequency distribution is the statement of the relative occurrence of a specific value from the possible outcomes of an experiment. When the number of outcomes generated is less than the total number of outcomes of the experiment, the frequency distribution generated is called a sample frequency distribution.

Cumulative Distribution Function   The cumulative distribution function (cdf) $P(Z)$ of a random variable Z is defined as

$$P[Z_0] = \text{probability that } Z \leqslant Z_0 \tag{8.5}$$

Thus, for a positive definite random variable that takes on only finite values, we can write the two limits

$$P[Z] = 0 \quad \text{if } Z < 0 \tag{8.6}$$

$$P[\infty] = 1 \tag{8.7}$$

Other values of $P[Z]$ are determined by experiment, as demonstrated in the next example.

> *EXAMPLE 8.3:*   Ten values of infiltration rate $(1, 3, 4, 2, 1, 7, 5, 13, 2, 6 \text{ cm h}^{-1})$ are measured with a double-ring infiltrometer at different locations in the field. Calculate and plot the sample estimate of $P(Z)$.
>
> By definition, each measurement counts 0.1 of the total sample probability generated from the ten measurements. Therefore, we can construct Table 8.2 of the cdf values for values of i from 0 to 13. A plot of $P(i)$ versus i is shown in Figure 8.1.

Probability Density Function   The probability density function (pdf) $f(Z)$ for a continuous random variable Z is defined as

$$\lim_{\Delta Z \to 0} f(Z_0) \, \Delta Z = \text{probability that } Z \text{ lies between}$$

$$\text{values } Z_0 \text{ and } Z_0 + \Delta Z$$

$$= P[Z_0 + \Delta Z] - P[Z_0] \tag{8.8}$$

---

[1]This is called the stationarity hypothesis (Journel and Huijbregts, 1978).

**TABLE 8.2 Sample cdf Values Calculated from
Infiltration Rate Measurements of Example 8.3**

i	$P(i)$
0	0
1	0.2
2	0.4
3	0.5
4	0.6
5	0.7
6	0.8
7	0.9
8	0.9
9	0.9
10	0.9
11	0.9
12	0.9
13	1.0

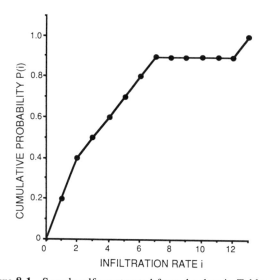

Figure 8.1 Sample cdf constructed from the data in Table 8.2.

Thus, by definition,

$$f(Z) = dP(Z)/dZ \tag{8.9}$$

Therefore, we can construct a pdf from the same information used to calculate a cdf, as shown in the next example.

EXAMPLE 8.4:   Calculate and plot $f(i)$ from Example 8.3. If we assume $P(i)$ is continuous and linearly interpolate between the values of $P(i)$ in Table 8.2 using

TABLE 8.3 Sample pdf Values for Infiltration Rate of Example 8.3

i	$f(i)$
0–2	0.2
2–7	0.1
7–12	0.0
12–13	0.1

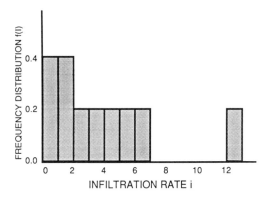

Figure 8.2 Sample pdf constructed from the data in Table 8.3.

(8.8) with $\Delta Z = 1$, then $f(i)$ is the slope of the graph in Fig. 8.1. Its values are given in Table 8.3, and the corresponding graph is in Fig. 8.2.

The probability density function $f(i)$ in Example 8.4 has a crude structure because it has been constructed from very few observations.

Moments of the pdf The pdf (8.8) represents the characteristics of the entire population of values of the random variable. If Z takes on a continuous range of values, then $f(Z)$ will have a value at every Z in the population, and

$$\int f(Z)\, dZ = 1 \tag{8.10}$$

where the integral limits span the entire range of values (i.e., 0, $\cdots$, ∞ if Z is positive definite). Equation (8.10), which is equivalent to (8.7), simply states that the sum of the probability of occurrence of each value of Z in the population is unity.

The average value of Z in the population is obtained as the sum of each value of Z multiplied by its probability of occurrence. Thus for continuous distributions, the sum is replaced by the integral

$$m = E[Z] = \int Zf(Z)\, dZ \tag{8.11}$$

where m is the population average or mean value of Z. It is also called the expectation or ensemble average $E[Z]$ of Z.

Similarly, the population average value of Z^2 (called the second moment $E[Z^2]$) is

$$E[Z^2] = \int Z^2 f(Z)\, dZ \tag{8.12}$$

and

$$\text{Var}[Z] = E\big(Z - E[Z]\big)^2 = \int \big(Z - E[Z]\big)^2 f(Z)\, dZ \tag{8.13}$$

is the population variance of Z, or the average squared deviation from the mean.

Joint Distributions Thus far we have considered only a single random variable Z. When there are two random variables Y and Z in an experiment, their joint probability of occurrence is defined by the joint cdf:

$$P[Y_0, Z_0] = \text{probability that } Y \le Y_0 \text{ and } Z \le Z_0 \tag{8.14}$$

Similarly, the joint pdf is defined as

$$f(Y, Z) = \frac{\partial^2 P}{\partial Y\, \partial Z} \tag{8.15}$$

The moments of the joint distribution are calculated as follows. The pdf's for the individual random variables (called marginal distributions) are

$$f_Y(Y) = \int f(Y, Z)\, dZ \tag{8.16}$$

$$f_Z(Z) = \int f(Y, Z)\, dY \tag{8.17}$$

from which the moments of Y and Z can be calculated by (8.11) and (8.12).

The distribution of the product of Y and Z is frequently of importance. The expectation of YZ is defined as

$$E[YZ] = \int\!\!\int YZ f(Y, Z)\, dY\, dZ \tag{8.18}$$

The deviation of this quantity from the product of the average values of Y and Z has a special name,

$$\begin{aligned}
\text{Cov}[Y, Z] &= E[YZ] - E[Y]\, E[Z] \\
&= E\big[(Y - E[Y])\, (Z - E[Z])\big] \\
&= \int\!\!\int (Y - E[Y])\, (Z - E[Z]) f(Y, Z)\, dY\, dZ
\end{aligned} \tag{8.19}$$

is called the covariance $\text{Cov}[Y, Z]$ of Y and Z. The normalized covariance, obtained by dividing (8.19) by $S_Y S_Z$, where $S_Y = (\text{Var}[Y])^{1/2}$, $S_Z = (\text{Var}[Z])^{1/2}$,

$$\rho = E\left[\frac{Y - E[Y]}{S_Y} \frac{Z - E[Z]}{S_Z}\right] \tag{8.20}$$

is called the correlation coefficient of Y and Z.

A special case of the joint distribution of two random variables occurs when the outcome of one does not affect the outcome of the other. In this case, the two variables are said to be *independent*, and

$$f[Y, Z] = f_Y[Y] f_Z[Z] \tag{8.21}$$

It is possible for the correlation coefficient in (8.20) to be zero without (8.21) being true. In this case, Y and Z are said to be uncorrelated rather than independent.

EXAMPLE 8.5: Calculate the mean and variance of the sum $Y + Z$ of two random variables Y and Z. By definition, the mean is

$$E[Y + Z] = \iint [Y + Z] f(Y, Z)\, dY\, dZ$$
$$= \iint Yf[Y, Z]\, dY\, dZ + \iint Zf(Y, Z)\, dY\, dZ$$
$$= \int Yf_Y(Y)\, dY + \int Zf_Z(Z)\, dZ$$

Therefore,

$$E[Y + Z] = E[Y] + E[Z] \tag{8.22}$$

By (8.13) and (8.22), we may write the variance as

$$\text{Var}[Y + Z] = E[(Y + Z)^2] - (E[Y + Z])^2$$
$$= E[Y^2 + 2YZ + Z^2] - (E[Y] + E[Z])^2$$
$$= E[Y^2] - (E[Y])^2 + E[Z^2] - (E[Z])^2$$
$$\quad + 2(E[YZ] - E[Y]E[Z])$$

Therefore,

$$\text{Var}(Y + Z) = \text{Var}[Y] + \text{Var}[Z] + 2\,\text{Cov}[Y, Z] \tag{8.23}$$

For the special case when Y and Z are uncorrelated, $\text{Var}[Y + Z] = \text{Var}[Y] + \text{Var}[Z]$.

Random Functions of a Parameter A random variable Z may be a function of a parameter x such as time or position. It is then denoted $Z(x)$, and the pdf of Z at x is denoted $f(Z; x)$ or $f(Z(x))$. In general, each value of the parameter has its own pdf, and $Z(x_1)$, $Z(x_2)$ must be described by the joint pdf $f(Z(x_1), Z(x_2))$. However, for reasons described in the preceding, it is impossible to measure the pdf if only one value $Z(x_i)$ is available at x_i to characterize the pdf $f(Z_i, x_i)$. To avoid this problem, it is generally assumed that the random variable is stationary, at least for its first two moments. A second-order stationary random variable $Z(x)$ has the following properties:

(i) Constant mean:

$$E[Z(x)] = m = \text{constant for all } x \qquad (8.24)$$

(ii) Covariance of $Z(x_1)$ and $Z(x_2)$ is a function only of the separation[2] $h = x_2 - x_1$ between x_1 and x_2:

$$\text{Cov}[Z(x_1), Z(x_1 + h)] = \text{Cov}[h] \qquad (8.25)$$

The condition of stationary may be envisioned by a random variable $Y(t)$ fluctuating in time (Fig. 8.3). At any point in time, the value of the variable is random but is fluctuating about the same mean value m. Moreover, the value of the variable at any instant of time t may be correlated with the values of the variable preceding it for some period of time rather than having a completely random value at each time.

These concepts are sufficient to develop the statistical theory of spatial variability used in the remainder of the chapter.

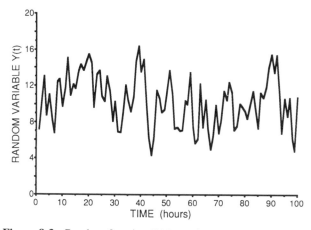

Figure 8.3 Random function $Y(t)$ varying as a function of time.

[2]The separation h is allowed to be a vector, in which case the covariance is called anisotropic.

8.3   ANALYSIS OF FREQUENCY DISTRIBUTIONS

8.3.1   Model Probability Density Functions

Probability distributions are estimated from experimental observations, which are grouped as in Examples 8.3 and 8.4 to produce a tabular representation of the cdf $P(z)$ and its derivative, the pdf $f(Z)$. In tabular or graphical form, this raw data represents a nonparametric representation of the probability distribution. An alternative approach that may be employed to generate distribution functions is to use model functions for $f(Z)$ with adjustable parameters to represent the structure of the sample measurements of the probability distribution optimally. Several commonly used model functions are described in what follows.

Normal Distribution   The pdf for the normal distribution is given by

$$f(Z) = \exp\left[-(Z-m)^2/2s^2\right]/\sqrt{(2\pi)}\, s \qquad (8.26)$$

where m and s are parameters. By direct integration using (8.11) and (8.12), it may be shown that

$$E[Z] = m \qquad (8.27)$$

$$\text{Var}[Z] = s^2 \qquad (8.28)$$

A plot of $f(Z)$ for $m = 10$ and various values of s is shown in Fig. 8.4.

The normal distribution is symmetric about its mean value m. Therefore, for large values of $CV = s/m$, part of the population assumes negative values, which

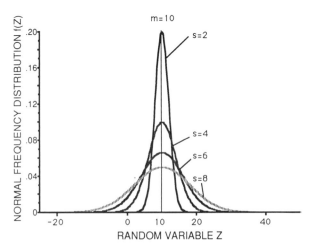

Figure 8.4 Normal distribution, showing a population with a mean of 10 and various values of the variance s^2.

makes the normal distribution a poor candidate to represent the distribution of positive definite variables with high CV (see Example 8.1).

The cumulative normal distribution $P(Z)$ is defined as

$$P(Z) = \int_{-\infty}^{Z} \frac{e^{-(z-m)^2/2s^2}}{\sqrt{2\pi}\,s}\,dz \tag{8.29}$$

which may be written as

$$P(Z) = \int_{-\infty}^{(Z-m)/s} e^{-y^2/2}\,\frac{dy}{\sqrt{2\pi}} \tag{8.30}$$

The function

$$N(u) = \frac{1}{\sqrt{2\pi}} \int_{0}^{u} e^{-y^2/2}\,dy \tag{8.31}$$

is called the normal curve of error and is tabulated in many statistics books. Since $N(-u) = -N(u)$ and $N(\infty) = 0.5$, $P(Z)$ in (8.30) may be written as

$$P(Z) = \frac{1}{2} + N\left(\frac{Z-m}{s}\right) \tag{8.32}$$

in terms of the normal curve of error. The normalized variable $u = (Z - m)/s$ is often called a u-variate. It has zero mean and unit variance. The table of the normal curve of error can be used to calculate properties of normally distributed variables, as shown in the next example.

EXAMPLE 8.6: Use Table 8.6 in the Problems to calculate the values of Z that lie at the lowest and highest 1% of a normal distribution with a mean of 10 and variance of 36.

The lower value of Z is the value Z_{min} that satisfies

$$P(Z_{min}) = 0.01 = 0.5 + N\left(\frac{Z_{min} - 10}{6}\right)$$

Therefore,

$$N\left(\frac{Z_{min} - 10}{6}\right) = -0.49 = -N\left(\frac{10 - Z_{min}}{6}\right)$$

From Table 8.6, $N(u) = 0.49$ at $u = 2.32$. Thus,

$$\frac{10 - Z_{min}}{6} = 2.32$$

$$Z_{min} = 10 - 13.92 = -3.92$$

Similarly, the maximum value of Z is calculated from

$$P(Z_{max}) = 0.99 = 0.5 + N\left(\frac{Z_{max} - 10}{6}\right)$$

$$\frac{Z_{max} - 10}{6} = 2.32$$

$$Z_{max} = 10 + 13.92 = 23.92$$

Therefore, the values between -3.92 and 23.92 comprise 98% of the population of Z.

EXAMPLE 8.7: For the distribution in Example 8.6, calculate the probability that Z is less than zero.

By definition, $P(0)$ is the probability that $Z \leq 0$. Therefore, by (8.32),

$$P(0) = 0.5 + N\left(\frac{0 - 10}{6}\right) = 0.5 - N[1.67]$$

According to the normal value of error, $N[1.67] \approx 0.4525$. Therefore,

$$P(0) = 0.5 - 0.4525 = 0.0475$$

Thus, there is a 4.75% probability that Z is negative.

Lognormal Distribution   The pdf for the lognormal distribution is given by

$$f(Z) = \exp\left\{-[\ln(Z) - \mu]^2/2\sigma^2\right\}/\sqrt{2\pi}\,\sigma Z \qquad (8.33)$$

where μ and σ are parameters. By direct integration using (8.11) and (8.12), it may be shown that (Aitcheson and Brown, 1976; Hald, 1952).

$$E[Z] = \exp(\mu + \sigma^2/2) \qquad (8.34)$$

$$\text{Var}[Z] = \exp(2\mu + \sigma^2)\left[\exp(\sigma^2) - 1\right] \qquad (8.35)$$

Therefore, by (8.4),

$$\text{CV} = \left[\exp(\sigma^2) - 1\right]^{1/2} \qquad (8.36)$$

A plot of $f(Z)$ for $E[Z] = 10$ and various values of σ is shown in Fig. 8.5. In contrast to the normal distribution (Fig. 8.4), the lognormal distribution is not symmetric and is defined only for positive Z.

The cumulative lognormal distribution $P[Z]$ is defined as

$$P[Z] = \int_0^Z e^{-[\ln(z) - \mu]^2/2\sigma^2}\,\frac{dz}{\sqrt{2\pi}\,\sigma z} \qquad (8.37)$$

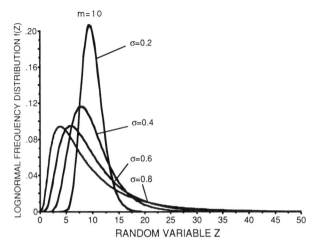

Figure 8.5 Lognormal distribution, showing a population with a mean of 10 and various values of the log variance σ^2.

which may also be written as

$$P[Z] = \int_{-\infty}^{[\ln(Z)-\mu]/\sigma} e^{-y^2/2} \frac{dy}{\sqrt{2\pi}} \tag{8.38}$$

Thus, using (8.31), we may write

$$P[Z] = \tfrac{1}{2} + N\left(\frac{\ln(Z) - \mu}{\sigma}\right) \tag{8.39}$$

in terms of the normal curve of error.

EXAMPLE 8.8: Use $N(\mu)$ in Table 8.6 to calculate the values of Z that bound the lowest and highest 1% of the population of a lognormal distribution with a mean of 10 and a variance of 36.

By definition, CV = 0.6; thus, using (8.36),

$$\exp(\sigma^2) = 1 + (CV)^2 = 1.36 => \sigma = 0.554$$

$$\exp(\mu + \sigma^2/2) = 10 => \mu = 2.149$$

Using (8.31), we may calculate Z_{min} from

$$P[Z_{min}] = 0.01 = 0.50 + N\left(\frac{\ln(Z_{min}) - \mu}{\sigma}\right)$$

from which we determine that

$$N\left(\frac{\mu - \ln(Z_{min})}{\sigma}\right) = 0.49$$

Therefore, since $N(2.32) \approx 0.49$,

$$\frac{\mu - \ln(Z_{min})}{\sigma} = 2.32$$

Therefore,

$$Z_{min} = \exp(\mu - 2.32\sigma) = 0.67$$

Similarly, $Z_{max} = \exp(\mu + 2.32\sigma) = 31$. Thus, the values between 0.67 and 31 form 98% of the population of Z.

In contrast, the normal distribution with the same mean and variance (Example 8.6) had a range of $(-3.92, 23.92)$ from the lowest to the highest 1% of the population.

Mode and Median   There are two other variable values besides the mean $m = \bar{Z}$ that can be used to characterize a pdf. They are the mode $\hat{Z}$, or most probable value, and the median $\tilde{Z}$, or the middle value of the distribution. By definition, $f(Z)$ has a maximum at $\hat{Z}$. Therefore, the mode is an extremum of $f(Z)$, which means that its derivative must vanish there (Kaplan, 1984). Thus

$$\frac{\partial f}{\partial Z} = 0 \quad \text{at } Z = \hat{Z} \tag{8.40}$$

Since $\tilde{Z}$ is the middle of the distribution, the median $\tilde{Z}$ is the value of Z that satisfies

$$P[Z] = 0.5 \quad \text{at } Z = \tilde{Z} \tag{8.41}$$

EXAMPLE 8.9:   Calculate $\hat{Z}$ and $\tilde{Z}$ for the normal distribution.
After inserting (8.26) into (8.40), we obtain

$$\frac{\partial f}{\partial Z} = -\frac{1}{s^2}(Z - m)\frac{e^{-(Z-m)^2/2s^2}}{\sqrt{2\pi}\,s} = 0$$

This function is equal to zero at $\hat{Z} = m$. By (8.32),

$$P[Z] = 0.5 \quad \text{when } N\left(\frac{Z - m}{s}\right) = 0$$

which occurs when $\tilde{Z} = m$.
Thus, for the normal distribution, $\bar{Z} = \hat{Z} = \tilde{Z}$.

EXAMPLE 8.10:   Calculate $\hat{Z}$ and $\tilde{Z}$ for the lognormal distribution. Inserting (8.33) into (8.40), we obtain

$$\frac{\partial f}{\partial Z} = -\frac{f}{Z} - \frac{f}{2\sigma^2 Z}(\ln Z - \mu) = -\frac{f}{2\sigma^2 Z}(\ln Z - \mu + \sigma^2)$$

This function is equal to zero at

$$\hat{Z} = \exp(\mu - \sigma^2) \tag{8.42}$$

By (8.39),

$$P[Z] = 0.5 \quad \text{when } N\left(\frac{\ln Z - \mu}{\sigma}\right) = 0$$

Therefore,

$$\tilde{Z} = \exp(\mu) \tag{8.43}$$

Thus, for the lognormal distribution $\hat{Z} < \tilde{Z} < \overline{Z}$. In Example 8.8, $\hat{Z} = 6.31$, $\tilde{Z} = 8.58$, and $\overline{Z} = 10$.

Most of the soil properties that have been measured in the field have been found to be either normally or lognormally distributed. The transport properties with large CV are usually better described as lognormal (Jury, 1985).

The next section discusses several methods of testing whether a sample set fits a distribution.

8.3.2 Methods for Determining Frequency Distribution

Selection of a model frequency distribution to represent the population using a sample set of finite numbers can be a difficult task, particularly if there is no information prior to the sampling about the expected form of the distribution. For the common soil properties of interest that are routinely measured in the field, we may exclude a priori any unusual distributions (e.g., bimodal) and concentrate on determining whether the distribution of the property of interest is better described as normal or lognormal. There are several established methods for doing this.

Fractile Diagram   The fractile diagram is a graphical method that can be used to provide visual information about the distribution of the property (Hald, 1952). It is based on (8.32) for the normal distribution and (8.39) for the lognormal distribution.

Imagine that a sample set of N values $Z_1, \cdots, Z_N$ of the quantity of interest has been measured. If they are arranged in order so that $Z_1 < Z_2 < \cdots < Z_N$, then the sample estimate of cumulative probability P is approximately equal to

$$P[Z < Z_j] = (J - 1)/N \quad J = 1, \cdots, N \tag{8.44}$$

If Z is normally distributed, then the theoretical value of $P[Z_j]$ is given by (8.32)

$$P[Z_j] = \tfrac{1}{2} + N\left(\frac{Z_j - m}{s}\right) \tag{8.45}$$

where m, s are calculated from the sample set using (8.1) and (8.2). If Z is normally distributed, then a plot of $(J - 1)/N$ versus $\tfrac{1}{2} + N[(Z_j - m)/s]$ should yield a straight line with zero intercept and unit slope.

Similarly, if Z is lognormally distributed, then the theoretical value of $P[Z_j]$ is given by (8.39)

$$P(Z_j) = \tfrac{1}{2} + N\left(\frac{\ln(Z_j) - \mu}{\sigma}\right) \tag{8.46}$$

where μ, σ are calculated from the log-transformed sample set using (8.1) and (8.2).

Fractile diagram analysis can be performed simply by using probability paper, which is scaled to represent $P[Z]$ on the ordinate when the unit variance, zero mean deviate $(Z - m)/s$ or $(\ln Z - \mu)/\sigma$ is plotted. Alternately, (8.45) or (8.46) can be calculated using Table 8.6[3] and plotted versus $(J - 1)/N$ on regular paper. This is illustrated in the next example.

EXAMPLE 8.11:   In the field study of Nielsen et al. (1973), the following 20 steady-state infiltration rates (in centimeters per day) were measured on ponded 6.3 × 6.3-m^2 plots over 150 ha: 5.12, 19.05, 0.54, 3.05, 11.73, 5.21, 11.98, 12.73, 1.96, 6.35, 13.26, 9.24, 40.23, 45.72, 1.16, 38.90, 12.23, 32.60, 14.20, 7.56.

Calculate m, s, μ, and σ for the sample set and evaluate whether the distribution is better described as normal or lognormal using the fractile diagram.

The sample mean and variance of these 20 values are $m = 14.64$ cm day^{-1} and $s^2 = 188.51$ cm^2 day^{-2}, while the corresponding values for the log-transformed data are $\mu = 2.18$, $\sigma^2 = 1.39$. Figure 8.6 shows the fractile diagrams of the sample

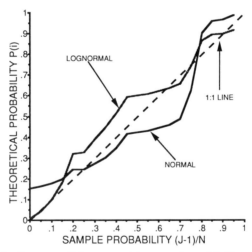

Figure 8.6   Fractile diagrams of the raw and log-transformed infiltration data in Example 8.11.

---

[3] There are also functional approximations of $N(u)$ over its useful range that allow its values to be calculated directly (see Abramowitz and Stegun, 1970).

probability (8.44) versus the theoretical probability (8.45) and (8.46) for the two assumed distributions. By inspection, the distribution appears to be better represented as lognormal than normal, although this is not a quantitative test.

Kolmogorov–Smirnov Statistic   The Kolmogorov–Smirnov (KS) statistic is a test to determine whether a sample set of N numbers belongs to a theoretical distribution. The criterion used for evaluating the sample set is the absolute maximum deviation

$$D = \max \left\{ \left| S_J - P_J \right| \right\} \quad J = 1, \cdots, N \tag{8.47}$$

where S_J is the sample cumulative probability (8.44) and P_J is the theoretical cumulative probability of the distribution being tested [Equations (8.45) and (8.46)]. Tables of KS values of D that reject the null hypothesis (that the sample set belongs to the distribution) have been tabulated for various values of N. Table 8.4 gives D values for $N = 20$ at various levels of precision.

> *EXAMPLE 8.12:* Use the KS statistic to evaluate whether the 20 infiltration values in Example 8.11 are normally or lognormally distributed.
>   The maximum deviations for the normal and lognormal models are 0.213 and 0.146, respectively. Thus, the KS test accepts both distributions at $P \leq 0.2$. The KS test would reject the normal hypothesis at $P = 0.3$.

8.3.3 Confidence Limits for the Mean

If a set of N samples are drawn from a population that is normally distributed with mean m and variance s^2, the average $\overline{X}$ of the N samples is a normally distributed random variable with the following properties (Hald, 1952):

$$\overline{E}[X] = m \tag{8.48}$$

$$\mathrm{Var}\,[\overline{X}] = s^2/N \tag{8.49}$$

Confidence limits for the mean value m of the population may be estimated from the sample measurements using the t-distribution if the population variance s^2 is not known a priori.

In this case, the probability is $P_2 - P_1$ that the population mean m is bounded by

$$\overline{X} - t_{p_1}\bar{s}/\sqrt{N} < m < \overline{X} - t_{p_1}\bar{s}/\sqrt{N} \tag{8.50}$$

TABLE 8.4   D Values of KS Statistic That Reject Null Hypothesis at Various Levels of Precision P for $N = 20$ Samples

P	0.20	0.15	0.10	0.05	0.01
D	0.231	0.246	0.264	0.294	0.352

where $\bar{s}$ is the sample variance (8.1) and t_{p_i} is the value of the t-variate at probability P_i.

Nonnormal Populations   If the population of values from which the sample is drawn is not normally distributed, then the distribution of the mean will approach a normal distribution only in the limit of large sample size. This is called the central limit theorem (Himmelblau, 1970). For small sample size, there is no exact procedure for generating confidence limits for the mean of a set of samples. In practice, (8.48) and (8.49) usually are assumed to hold for arbitrary distributions.

8.4   ANALYSIS OF SPATIAL STRUCTURE

An important aspect of sampling, which is not taken into account by such traditional statistical methods as analysis of variance, is that the samples may be correlated. Intuitively, it seems reasonable that measurements of a parameter at different places in a field are more likely to be similar to each other if they are taken close together than if they are taken far apart. Knowledge of the spatial correlation structure, or the distance over which properties are correlated with each other, is essential to the design of optimal sampling grids and interpolation methods (Warrick et al., 1986; McBratney and Webster, 1981). A function that is used to characterize the correlation structure is the variogram.

8.4.1   The Variogram

The variogram $\gamma(h)$ is a measure of the variance of a given parameter with respect to samples that are a fixed distance apart in space. Thus, for a given separation distance h,

$$\gamma(h) = \frac{1}{2N(h)} \sum_{j=1}^{N(h)} \left[Z(x_J) - Z(x_J + h) \right]^2 \tag{8.51}$$

where $Z(x_J)$ is the value of the random variable Z at $x = x_J$, $Z(x_J + h)$ is the value of Z at a distance h from x_J, and $N(h)$ is the number of pairs of points that are a distance h apart. If the random variable Z is second order stationary [i.e., it obeys (8.24) and (8.25)], then all random variables $Z(x)$ at each x have the same mean value, and the covariance Cov $(Z[x_1], Z[x_2])$ depends only on the separation distance $h = |x_2 - x_1|$ between the points. In this case $\gamma(h)$ becomes a statistical measure of the degree to which random variables $Z(x)$ at different x become alike as the distance between them decreases.

Figure 8.7 shows an ideal variogram function and labels its significant features. The function increases uniformly from a small value at $h = 0$ (called the nugget variance) to an asymptotic value (called the sill) at large h. The distance over which $\gamma(h)$ is changing is called the range and may be identified with the distance over which values of the random variable $Z(x)$ are correlated with each other.

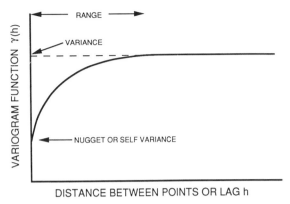

Figure 8.7 Ideal variogram of a stationary random function.

The nugget variance (the variance at zero lag) may be a source of confusion, since it implies that a property measured at the same point may have some variance with itself. The nugget variance is an extrapolated value, since in practice there is some minimum separation distance h in the sample set of measurements used to estimate $\gamma(h)$.

The sill should be equal to the population variance if the random variable $Z(x)$ is second order stationary, since (8.51) is an estimate of the variance s^2 if its sample members are uncorrelated. Warrick et al. (1986) discuss different variogram shapes and their interpretation.

Estimation of $\gamma(h)$ The estimation of $\gamma(h)$ is straightforward using (8.51), although in practice many pairs of points $N(h)$ at each lag distance h are needed to obtain an accurate representation of the spatial structure (Journel and Huijbregts, 1978; Warrick et al., 1986). Russo and Jury (1987a,b) studied the variogram estimation problem theoretically by computer simulation. They created a "field" of points with a known statistical structure and tried to deduce this structure from various sampling schemes. Their most important conclusions were as follows:

1. Two-dimensional sampling grids yield more accurate information about $\gamma(h)$ than transects (one-dimensional lines of samples) with the same number of points.

2. The minimum grid sample spacing must be less than half of the range to detect any spatial structure.

3. The sample schemes that had less than 72 pairs of points per lag commonly estimated the range of the true spatial structure with an error of 25% or greater. In practice, this means that a square grid with at least seven points on a side are required for accurate analysis.

4. If the field is not stationary but possesses a drift in addition to random spatial structure, then the spatial structure is more difficult to detect.

8.4.2 The Autocorrelation Function

The variogram $\gamma(h)$ is related to other indices of spatial structure. If the field is stationary and a population variance s^2 can be defined, then the autocorrelation function $\rho(h)$ is defined as (Warrick and Nielsen, 1980)

$$\rho(h) = 1 - \gamma(h)/s^2 \qquad (8.52)$$

Figure 8.8 shows the autocorrelation function corresponding to the idealized variogram $\gamma(h)$ in Fig. 8.7. It has a maximum value at $h = 0$ and decreases to zero at a separation distance comparable to the range of the variogram.

A quantitative measure of the range of correlation of a random spatial structure may be calculated from the autocorrelation function (8.52). This measure, called the integral scale I^*, is calculated by (Bakr et al., 1978; Lumley and Panofsky, 1964; Russo and Bresler, 1981)

$$I^* = \int_0^\infty \rho(h)\,dh \qquad (8.53a)$$

if the spatial structure was calculated from a one-dimensional row of data points, and

$$I^* = \left(\int_0^\infty h\rho(h)\,dh \right)^{1/2} \qquad (8.53b)$$

if the data points are in two dimensions. Jury (1985) and Warrick et al. (1986) present tables of measured values of I^* from various field studies. In general, I^* is not an intrinsic property of the field but depends on the scale over which it is measured.

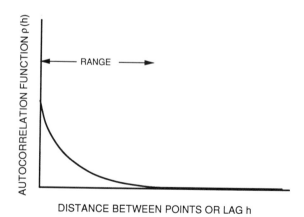

Figure 8.8 Autocorrelation function corresponding to the variogram in Fig. 8.7.

Model Variogram Functions   There are a number of variogram model functions used to represent the structure of a set of data. The most common ones, taken from Journel and Huijbregts (1978) and Warrick et al. (1986), are as follows:

Linear model:

$$\gamma(h) = \begin{cases} C_0 + h(C_1 - C_0)/a & 0 < h < a \\ C_1 & h \geqslant a \end{cases} \tag{8.54}$$

Spherical model:

$$\gamma(h) = \begin{cases} C_0 + (C_1 - C_0)\left[\dfrac{3}{2}\dfrac{h}{a} - \dfrac{1}{2}\left(\dfrac{h}{a}\right)^3\right] & 0 < h < a \\ C_1 & h > a \end{cases} \tag{8.55}$$

Gaussian model:

$$\gamma(h) = C_0 + (C_1 - C_0)\left[1 - \exp\left(-h^2/\lambda^2\right)\right] \tag{8.56}$$

Exponential model:

$$\gamma(h) = C_0 + (C_1 - C_0)\left[1 - \exp(-h/\lambda)\right] \tag{8.57}$$

Figure 8.9 shows plots of the four functions $(\gamma - C_0)/(C_1 - C_0)$ versus h. All of the functions have been given the same integral scale I^*.

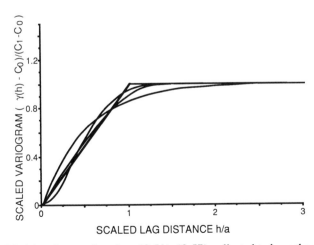

Figure 8.9   Model variogram functions (8.54)–(8.57), adjusted to have the same integral scale.

8.4.3 Kriging

The random variable Z may be estimated at a particular point x_0 in space by inter-polating values of Z around that point that are within the range of correlation. Thus, a linear estimator $Z^*(x_0)$ may be defined as (Journel and Huijbregts, 1978)

$$Z^*(x_0) = \sum_{J=1}^{N} \lambda_J Z_J \qquad (8.58)$$

where $Z_J = Z(x_J)$, $J = 1, \cdots, N$, are the N measured points around x_0 that are within the zone of correlation. Kriging derives values of λ_J that satisfy two criteria (assuming Z is second order stationary):

1. Z^* has the same mean value as each of the Z_J and
2. Z^* has the minimum possible variance about the true value $Z(x_0)$.

Derivation of the correct values of λ that satisfy these two criteria is straightforward but involves techniques beyond the scope of this book. Interested readers may consult Journel and Huijbregts (1978). Numerous computer software programs are available to calculate λ_J from a set of data.

When the assumptions of the theory are well met, (8.58) becomes an excellent method for constructing interpolated values of the variable between the measured points, from which contour lines may be drawn. In addition, the theory allows a calculation of the estimation variance, which may be used to determine where values of the function are the most uncertain. This information could be very useful if a second sampling of the same field is planned.

8.4.4 Drift

When the differences in values of a variable $Z(x)$ at different points are systematic rather than random, the field of points is said to be nonstationary. A nonstationary field cannot be analyzed by the preceding methods above because the covariance, variogram, and autocorrelation function depend on the locations of the points as well as on the space between them. In such a case, a spatial structure analysis may still be performed if the so-called deterministic or drift component of the field is first removed. Formally, one may write

$$Z(x) = m(x) + \epsilon(x) \qquad (8.59)$$

where $m(x)$ is a deterministic drift function and $\epsilon(x)$ is a zero-mean, second-order stationary random function.

Removal of the drift function is problematical. Only values of $Z(x)$ are measured, and it is not known a priori what fraction of $Z(x)$ is drift and what fraction is random. Russo and Jury (1987b) discuss various techniques for drift identification and removal. Usually, simple drift functions are postulated and the residual function $\epsilon(x)$ is analyzed to determine whether it has the attributes of a second-order stationary function. The most common method of identification of stationarity is the shape of the variogram, which should approach an asymptotic value at large lags if the process is stationary (Journel and Huijbregts, 1978).

Problems[4]

8.1 A hypothetical set of samples of infiltration rate i and percentage of clay (PC) are taken at random locations throughout a very large field. The data are summarized in Table 8.5. Construct cumulative distribution functions for the two variables using an interval of 2 cm day^{-1} for i and 0.5 for PC. Calculate the sample mean and variance of the data and the correlation coefficient.

TABLE 8.5 Infiltration Rate Measurements and Percentage of Clay at 25 Locations in Large Field Site

Sample No.	i (cm day^{-1})	Amount of Clay (%)	$\ln(i)$
1	2.0	3.5	0.000
2	5.0	2.0	1.609
3	25.0	1.1	3.219
4	3.0	1.3	1.099
5	15.0	0.8	2.708
6	4.0	2.2	1.386
7	5.0	0.9	1.609
8	7.0	1.2	1.946
9	6.0	1.0	1.792
10	5.0	3.0	1.609
11	9.0	2.8	2.197
12	0.5	5.3	-0.693
13	4.5	1.0	1.504
14	6.0	1.3	1.792
15	7.5	0.9	2.015
16	2.0	2.6	0.693
17	8.0	1.0	2.079
18	3.0	1.5	1.099
19	2.0	2.0	0.693
20	5.0	1.3	1.609
21	15.0	0.8	2.708
22	7.0	1.0	1.946
23	3.0	2.5	1.099
24	3.5	1.6	1.253
25	2.0	1.5	0.693

[4] Problems preceded by a dagger are more difficult.

8.2   Repeat problem 1 using the log-transformed data for the infiltration rate and an interval of 0.5 for ln (i). Plot i and ln (i) on probability paper and decide whether the distribution of values is better described as normal or lognormal. What does the correlation coefficient tell you about the nature of the relationship between i and PC?

8.3   The saturated hydraulic conductivity K_s was measured at 20 sites in the field study of Nielsen et al. (1973) and was found to be lognormally distributed with sample estimates of $E[\ln (K_s)] = \mu = 2.58$ and Var $[\ln (K_s)] = \sigma^2 = 1.01$.

(a) Calculate the mean, mode, and median of the K_s distribution.

(b) Calculate the cumulative probability $P(M) = $ Prob $(K_s < M)$ of obtaining a sample that is less than the mean $M = E(K_s)$. Hint: $P(M) = 0.5 + N\{[\ln (M) - \mu]/\sigma\}$.

†8.4   An agricultural field has a lognormal steady-state infiltration rate with $E[\ln (i)] = \mu = 1.5$, Var $[\ln (i)] = \sigma^2 = 1.5$.

(a) If water is ponded continuously over the entire field, calculate the steady-state mean infiltration rate.

(b) Calculate the fraction of the field that has an infiltration rate greater than the mean M. This is equal to $1 - P(M)$.

(c) Calculate the fraction of the flow that is entering through the part of the field that has an infiltration rate greater than the mean M. The flow fraction through this part of the field is

$$\text{Fraction greater than } M = \frac{\int_M^\infty if(i)\, di}{\int_0^\infty if(i)\, di} = \frac{1}{M} \int_M^\infty \frac{i}{\sqrt{2\pi\sigma i}}$$

$$\cdot \exp - \frac{[\ln(i) - \mu]^2}{2\sigma^2}\, di$$

Hint: Change the variable in the integral to $y = [\ln (i) - \mu]/\sigma$ and complete the square. The new integral can be expressed in terms of the normal curve of error.

(d) If a chemical tracer was added with the ponded water, what would the spatial distribution of the front edge of the plume look like?

8.5   A set of spatial measurements of clay content are made at 100 locations spaced at regular 1-m intervals along a line of 100 m in the field. Because of a nonuniform deposition process, the mean clay content increases linearly along the transect. Thus, an approximate random model of the percentage of clay (PC) is

$$PC(x) = ax + b + \xi(x)$$

where a and b are constants and $\xi(x)$ is a zero-mean random variable. Show that the contributions to the variogram [Equation (8.5)] from the linear drift term is a quadratic function of the spacing between points. Discuss how this variogram shape might be used to detect the presence of drift in a data set.

8.6 Theoretical analyses (e.g., Russo and Jury, 1987a,b) have shown that the variogram function (8.5) must be constructed from about 100 or more pairs of points per lag interval. Assuming that the data are taken at locations along a square grid, calculate the lag distances and the number of pairs of points per lag on a grid of 36 points in a 6×6 pattern of separation L. What would the result have been if the 36 points were laid out at equal spaces L along a line?

8.7 A chemical tracer pulse is added to the surface of a soil and is leached downward under continuous irrigation at a rate $i_0 = 1$ cm day^{-1}. The CDE and CLT models are fit to the pulse data obtained by solution samplers at the 1-m depth with parameters

$$P(t \leqslant t_0) \approx \frac{1}{2} + N\left(\frac{Vt_0 - z}{\sqrt{2Dt_0}}\right) \quad \text{(CDE model)}$$

$$P(t \leqslant t_0) = \frac{1}{2} + N\left(\frac{\ln(t_0 L/z) - \mu}{\sigma}\right) \quad \text{(CLT model)}$$

TABLE 8.6 Selected Values from Normal Curve of Error

u	$N(u)$
2.32	0.49
1.96	0.475
1.64	0.45
1.29	0.40

Using Table 8.6 calculate the amount of time t_0 predicted by each model for the first $P(t_0) = 0.01, 0.05, 0.1$ of the pulse to reach the $z = 500$ cm depth. [*Note:* $N(-u) = -N(u)$.] Since these models both describe the single data set measured at $L = 100$ cm equally well, discuss the problem of predicting arrival times of the leading edge of the plume in groundwater.

List of Symbols

LOWERCASE ARABIC

a	volumetric air content; air pressure potential head
b	overburden potential head
c	specific heat
c_a	specific heat of air
d	damping depth; stagnant boundary layer thickness
d_w	cumulative drainage volume in breakthrough experiments
d_b	drainage breakthrough volume
$f_L(I)$	net applied water pdf in transfer function model at depth L
$f_z(I)$	net applied water pdf in transfer function model at depth z
f_a	mass fraction for adsorbed phase
f_l	mass fraction for dissolved phase
f_L	leaching fraction
f_g	mass fraction for gaseous phase
g	acceleration of gravity
$g(t)$	solute lifetime or travel time pdf
h	matric potential head; lag distance
h_T	total soil water potential head
h_H	transfer coefficient for sensible heat
h_f	matric potential of wetting front in Green-Ampt model
h_v	transfer coefficient for water vapor
i	irrigation or infiltration rate
i_o, i_f	constants in Horton infiltration equation
i	unit vector in x direction
j	unit vector in y direction
k	unit vector in z direction
k_H	Henry's constant
m	mean of a random variable
m_s	mass of solute
m_w	mass of water
p	hydrostatic pressure potential head
q	O_2 consumption rate in Lemon-Wiegand model
r_g	sink term in gas conservation equation
r_H	sink term in heat conservation equation
r_s	sink term in solute conservation equation
r_w	sink term in water conservation equation

294

s	solute potential head; slope of $P_v^*(T)$ curve; standard deviation
t_c	time to end of first stage of evaporation
t_b	breakthrough time
t_{bR}	breakthrough time of chemical retarded by adsorption
t_P	t-statistic at probability P
u	zero mean, unit variance normal random variate
z	gravitational potential head
z_{soil}	elevation of the location of interest in the soil
z_0	reference elevation
z_{wt}	water table elevation

UPPERCASE ARABIC

A	area
B_λ	Planck energy density
C	heat capacity per unit volume
Cov ( )	covariance of two random variables
CV	coefficient of variation
C^*	constant in BET equation
C_1, C_2	constants in Planck radiation equation
C_a	concentration in adsorbed phase (mass/mass soil)
C_g^a	solute vapor concentration in atmosphere above soil surface
C_{ent}	solute concentration entering transport volume
C_{ex}	solute concentration exiting transport volume
C_g	concentration in gas phase (mass/volume air)
C_{im}	solute concentration in immobile water zone
C_l	concentration in dissolved phase (mass/volume solution)
C_m	solute concentration in mobile water zone
C_p	O_2 concentration in dissolved phase at root–air interface
C_{soil}	soil heat capacity
C_T	total solute concentration (mass/volume soil)
C_w	water capacity function
D	drainage rate; dispersion coefficient in CDE model; KS statistic
D_0	water diffusivity in Green–Ampt model
D_e	effective liquid diffusion–dispersion coefficient
D_E	effective liquid–vapor diffusion coefficient
D_g^a	diffusion coefficient of gas in air
D_g^s	diffusion coefficient of gas in soil
D_{lh}	hydrodynamic dispersion coefficient
D_l^s	liquid diffusion coefficient of solute in soil
D_l^w	liquid diffusion coefficient of solute in water
D_m	dispersion coefficient in mobile water zone
D_R	dispersion coefficient of chemical retarded by adsorption
D_{Tv}	thermal vapor diffusivity

D_w	water diffusivity
E	evaporation rate
$E[\ ]$	expectation or ensemble average of a random variable
E_p	potential evaporation rate
ET	evapotranspiration rate
H	hydraulic head
H_f	latent heat of fusion
H_v	latent heat of vaporization
I	electrical current; cumulative infiltration; net applied water
I^*	integral scale
$\bar{I}$	mean water displacement volume
$\hat{I}$	median water displacement volume
J_g	soil gas flux
J_H	soil heat flux
J_{Hc}	conductive soil heat flux
J_{Hl}	latent soil heat flux
J_l	solute flux in dissolved phase
J_{lc}	convective solute flux in dissolved phase
J_{ld}	diffusive solute flux in dissolved phase
J_{lh}	dispersive solute flux in dissolved phase
J_s	soil solute flux
J_{sb}	gaseous solute flux through surface boundary layer
J_w	soil water flux
$K(h)$	unsaturated hydraulic conductivity
K_0	hydraulic conductivity in Green–Ampt model
K_d	distribution coefficient in linear isotherm
K_f	Freundlich coefficient in Freundlich isotherm
K_H	dimensionless Henrys constant
K_{oc}	organic C partition coefficient
K_s	saturated hydraulic conductivity
K_T	thermal diffusivity
K_w	Dissociation constant for water
L_c	capillary bundle length
M	solute mass in soil volume
M_a	molecular weight of air
M_s	molecular weight of solute
M_w	molecular weight of water
N	constant in Freundlich isotherm equation
$N(u)$	normal curve of error
P	precipitation
$P(Z)$	cdf of a random variable Z
P_0	reference air pressure
P_a	air pressure
P_e	envelope pressure
P_l	liquid pressure

P_v	water vapor pressure
P_v^*	saturated water vapor pressure
Q	constant in Langmuir equation; volume flow rate
Q_{in}	solute input flux to transport volume
Q_{out}	solute output flux from transport volume
Q_{ent}	solute input flux to transport volume through surface
Q_{ex}	solute output flux from transport volume through exit surface
R	runoff rate; universal gas constant; retardation factor
R_a	phase partition coefficient for adsorbed phase
R_e	radius of root plus water film (Lemon–Wiegand model)
R_{earth}	long-wave thermal radiation from earth
R_{ext}	extraterrestrial radiation
R_g	phase partition coefficient for gaseous phase
RH	relative humidity
R_h	hydraulic resistance
R_l	phase partition coefficient for dissolved phase
R_N	net radiation
R_{nt}	net thermal radiation
R_s	solar radiation
R_{sky}	long-wave thermal radiation from sky
S	sensible heat flux in atmosphere
T	temperature
T_A	annual average soil temperature
T_N	Nth travel time moment
T_{ref}	reference temperature
V	volume; solute velocity in CDE equation
$Var[\ ]$	variance of a random variable
V_E	effective solute velocity
V_R	solute velocity of chemical retarded by adsorption
V_m	constant in BET equation; solute velocity in mobile zone
W	water storage
X	moles of solute in solution
Y	regional variable
Z	random variable
$Z^*(x)$	estimation of $Z(x)$ at x
$\hat{Z}$	mean of Z
$\tilde{Z}$	median of Z
$\hat{Z}$	mode of Z

LOWERCASE GREEK

α	constant in Kostiakov equation; rate coefficient in MIM model
β	constant in Horton equation
γ	contact angle; constant in Kostiakov equation; psychrometer constant
$\gamma(h)$	variogram function

ϵ	emissivity
ϕ	porosity
ϕ_m	mobile water fraction
η	coefficient of viscosity
λ	thermal conductivity; dispersivity
λ^*	thermal conductivity of moist soil neglecting latent heat transfer
λ_{eff}	thermal conductivity including latent heat transfer
μ_T	chemical potential of soil water
μ	solute degradation rate coefficient; parameter in lognormal pdf
μ_E	effective solute degradation rate coefficient
ν_m	gamma ray mass absorption coefficient of soil minerals
ν_w	gamma ray mass absorption coefficient of water
θ_g	gravimetric water content
θ_v	volumetric water content (also θ)
θ_m	volumetric water content of mobile water zone
θ_{im}	volumetric water content of immobile water zone
θ_r	residual water content
θ_s	saturated water content
θ_i	initial water content
θ_0	surface water content in infiltration models
θ_{st}	solute transport volume water content
$\bar{\theta}_{st}$	mean solute transport volume water content
$\hat{\theta}_{st}$	median solute transport volume water content
ρ	correlation coefficient
$\rho(h)$	autocorrelation function
ρ_a	air density
ρ_b	soil bulk density
ρ_m	soil mineral density
ρ_v	water vapor density
ρ_v^*	saturated water vapor density
ρ_w	density of liquid water
σ	surface tension of water; constant in Stefan–Boltzmann equation
τ	period of a thermal wave
$\tau_{1/2}$	degradation half-life
ψ_T	total soil water potential
ψ_w	wetness potential
ψ_z	gravitational potential
ψ_s	solute potential
ψ_b	overburden potential
ψ_m	matric potential
ψ_a	air pressure potential
ψ_p	hydrostatic pressure potential
ψ_{tp}	tensiometer pressure potential
ξ_g	soil gas diffusion tortuosity factor
ξ_l	soil liquid diffusion tortuosity factor

ξ' modified gas diffusion tortuosity factor for water vapor

ω ratio of gas phase thermal gradient to average thermal gradient

UPPERCASE GREEK

Σ radiant energy flux

Σ_E radiant energy flux at Earth's atmosphere

Θ reduced water content

References

Abramowitz, M. and I. A. Stegun, 1970. *Handbook of Mathematical Functions*. Dover, New York.

Addiscot, T. M., 1977. A simple computer model for leaching in structured soils. *J. Soil Sci.* **28**:554–563.

Aitcheson, J. and J. A. C. Brown, 1976. *The Lognormal Distribution*. Cambridge University Press, New York.

Alexander, M. E., 1965. Nitrification. In: J. W. Bartholomew and F. E. Clark (Eds.), *Soil Nitrogen*. Monograph 10. American Society of Agronomy, Madison, WI.

Allmaras, R. R., W. C. Burrows, and W. E. Larson, 1964. Early growth of corn as affected by soil temperature. *Soil Sci. Soc. Am. Proc.* **28**:271–275.

Anderson, M. P., 1979. Using models to simulate the movement of contaminants through groundwater flow systems. *CRC Crit. Rev. Env. Contr.* **9**:97–156.

Arfken, G., 1985. *Mathematical Methods for Physicists*, 2nd ed. Academic, New York.

Atterberg, A., 1912. Die mechanische Bodenanalyse und die Klassifikation der Mineralöboden Schwedens. *Intern. Mitt. Bodenk.* **2**:312–342.

Ayers, R. S. and R. L. Bransom, 1973. Nitrates in the upper Santa Ana River basin in relation to groundwater pollution. *Cal. Agr. Exper. Stn. Bull.* **861**:1–60.

Babcock, K. L. and R. Overstreet, 1957. The extra-thermodynamics of soil moisture. *Soil Sci.* **83**:455–464.

Bakr, A. A., L. W. Gelhar, A. L. Gutjahr, and H. R. Macmillan, 1978. Stochastic analysis of spatial variability of subsurface flow. I. Comparison of one and three dimensional flows. *Water Resour. Res.* **14**:263–272.

Baver, L. D. and G. M. Horner, 1933. Water content of soil colloids as related to their chemical composition. *Soil Sci.* **30**:329–353.

Baver, L. D. and H. F. Rhoades, 1932. Aggregate analysis as an aid to the study of soil structure relationships. *Agr. J.* **24**:920–930.

Bear, J., 1972. *Dynamics of Fluids in Porous Media*. Elsevier, New York.

Beavers, A. H. and C. E. Marshall, 1951. The cataphoresis of clay minerals and factors affecting their separation. *Soil Sci. Soc. Am. Proc.* **15**:142–145.

Bernal, J. D. and R. H. Fowler, 1933. A theory of water and ionic solution, with particular reference to hydrogen and hydroxyl ions. *J. Chem. Phys.* **1**:515–548.

Bertrand, A. R. and H. Kohnke, 1957. Subsoil conditions and their effects on oxygen supply and the growth of corn roots. *Soil Sci. Soc. Am. Proc.* **21**:135–139.

Beven, K. and P. Germann, 1982. Macropores and water flow in soils. *Water Resour. Res.* **18**:1311–1325.

Biggar, J. W. and D. R. Nielsen, 1967. Miscible displacement and leaching phenomena. *Agronomy* **11**:254–274.

Bird, R. B., W. Stewart, and E. Lightfoot, 1960. *Transport Phenomena*. Wiley, New York.

Black, T. A., W. R. Gardner, and G. W. Thurtell, 1969. The prediction of evaporation, drainage, and soil water storage for a bare soil. *Soil Sci. Soc. Am. Proc.* **33**:655–660.

Blake, G. R. and K. H. Hartge, 1986. Bulk density. pp. 363–375. In: A. Klute (Ed.), *Methods of Soil Analysis. Part 1*. Monograph 9. American Society of Agronomy, Madison, WI.

Bohn, H., B. L. McNeal, and G. Oconnor, 1979. *Soil Chemistry*. Wiley, New York.

Bolt, G. and M. G. M. Bruggenwert, 1976. *Soil Chemistry. A. Basic Elements*. Elsevier, Amsterdam.

Bolt, G. H., 1955a. Ion adsorption by clays. *Soil Sci.* **79**:267–276.

Bolt, G. H., 1955b. Analysis of the validity of Gouy Chapman theory of the electron double layer. *J. Coll. Sci.* **10**:206–218.

Bolt, G. H., 1976. Soil physics terminology. *Bull. Int. Soc. Soil Sci.* **49**:26–36.

Bolt, G. H., 1982. *Soil Chemistry. B. Physico-Chemical Models*. Elsevier, Amsterdam.

Bolt, G. H. and M. Peech, 1953. The application of Gouy theory to soil systems. *Soil Sci. Soc. Am. Proc.* **17**:210–213.

Borchardt, G. A., 1988. Montmorillinite and other smectite minerals. In: J. B. Dixon and S. B. Weed (Eds.), *Minerals in Soil Environments*, 2nd ed. Soil Science Society of America, Madison, WI.

Boulier, J. F., J. Y. Parlange, M. Vauclin, D. A. Lockington, and R. Haverkamp. 1987. Upper and lower bounds of the ponding time for near-constant surface flux. *Soil Sci. Soc. Am. J.* **51**:1424–1428.

Bouwer, H., 1966. Rapid field measurement of air entry value and hydraulic conductivity of soil as significant parameters in flow systems analysis. *Water Resour. Res.* **2**:729–738.

Bouyoucos, G. J., 1913. An investigation of soil temperature and some of the most important factors influencing it. *Mich. Agr. Exp. Stn. Tech. Bull.* **17**.

Bower, C. A. and F. B. Gschwend, 1952. Ethylene glycol retention by soils as a measure of surface and interlayer swelling. *Soil Sci. Soc. Am. Proc.* **16**:342–345.

Bowers, S. A. and R. J. Hanks, 1962. Specific heat capacity of soils and minerals as determined with a radiation calorimeter. *Soil Sci.* **94**:392–396.

Bradfield, R., 1928. Factors affecting the coagulation of colloidal clay. *Missouri Agr. Exp. Sta. Res. Bull.* **60**.

Bradley, W. F., 1940. The structural scheme of attapulgite. *Am. Mineral.* **25**:405–410.

Brindley, G. W., 1951. The kaolin minerals. Ch. 11. In: G. W. Brindley and G. Brown (Eds.), *X-Ray Identification and Crystal Structures of Clay Minerals*. Mineralogical Society of London, London.

Broadbent, F. E. and F. E. Clark, 1965. Denitrification. In: J. W. Bartholomew and F. E. Clark (Eds.), *Soil Nitrogen*, Agronomy Monograph 10. American Society Agronomy, Madison, WI.

Brooks, R. H. and A. T. Corey, 1966. Properties of porous media affecting fluid flow. *J. Irr. Dr. Div. Am. Soc. Civ. Eng.* **92**:61–88.

Brown G., 1961. *X-Ray Identification and Crystal Structures of Clay Minerals*. Mineralogical Society of London, London.

Bruce, R. R. and A. Klute, 1956. The measurement of soil water diffusivity. *Soil Sci. Soc. Am. Proc.* **20**:458–462.

Bruce, R. R. and R. J. Luxmoore, 1986. Water retention: Field methods. pp. 663–684. In: A. Klute (Ed.), *Methods of Soil Analysis. Part 1*. Monograph 9, American Society Agronomy, Madison, WI.

Brunauer, S., P. H. Emmett, and E. Teller, 1938. Adsorption of gases in multimolecular layers. *J. Am. Chem. Soc.* **60**:309–319.

Buckingham, E., 1904. Contributions to our knowledge of the aeration of soils. Bulletin 25. U.S. Department of Agriculture Bureau of Soils, Washington, DC.

Buckingham, E., 1907. Studies on the movement of soil moisture. Bulletin 38. U.S. Department of Agriculture Bureau of Soils, Washington, DC.

Burchill, S., M. H. B. Hayes, and D. J. Greenland, 1978. Adsorption. In: D. J. Greenland and M. H. B. Hayes (Eds.), *The Chemistry of Soil Processes*. Wiley, Chichester.

Burger, H. C., 1919. Das Leitvermogen verdunnter mischkristall-freier Legierungen. *Phyz. Z.* **20**:73–75.

Butters, G. L. and W. A. Jury, 1989. Field scale transport of bromide in an unsaturated soil. 2. Dispersion modeling. *Water Resour. Res.* **25**:1582–1588.

Butters, G. L. and W. A. Jury, and F. Ernst, 1989. Field scale transport of bromide in an unsaturated soil. I. Experimental methodology and results. *Water Resour. Res.* **25**:1575–1581.

Campbell, G. S. and G. W. Gee, 1986. Water potential: Miscellaneous methods. pp. 619–632. In: A. Klute (Ed.), *Methods of Soil Analysis. Part 1*. Monograph 9. American Society of Agronomy, Madison, WI.

Carslaw, H. S. and J. C. Jaeger, 1959. *Conduction of Heat in Solids*. Oxford University Press, London.

Cassel, D. K. and A. Klute, 1986. Water potential: Tensiometry. pp. 563–594. In: A. Klute (Ed.), *Methods of Soil Analysis. Part 1*. Monograph 9, American Society of Agronomy, Madison WI.

Chang, J. H., 1961. *Climate and Agriculture*. Aldine, Chicago.

Chang, J. H., 1968. Microclimate of sugar cane. *Hawaiian Planter's Rec.* **56**:195–225.

Chang, J. H., R. B. Campbell, H. W. Brodie, and L. D. Baver, 1965. Evapotranspiration research at the HSPA Experiment Station. *Proc. 12th Congr. Int. Soc. Sugarcane Tech.*, pp. 10–24.

Chapman, H. D., 1965. Cation exchange capacity. pp. 891–901. In: C. A. Black et al. (Eds.), *Methods of Soil Analysis. Part 2*. Monograph 9. American Society of Agronomy, Madison, WI.

Chepil, W. S., 1962. A compact rotary sieve and the importance of dry sieving on physical soil analysis. *Soil Soc. Am. Proc.* **26**:4–6.

Childs, E. C., 1969. *The Physical Basis of Soil Water Phenomena*. Wiley-Interscience, New York.

Childs, E. C. and M. Bybord, 1969. The vertical movement of water in stratified porous material. I. Infiltration. *Water Resour. Res.* **5**:446–459.

Childs, E. C. and N. Collis-George, 1950. The permeability of porous materials. *Proc. Roy. Soc. London. Ser. A* **201**:392–405.

Chong, S. K., R. E. Green, and L. R. Ahuja, 1982. Infiltration prediction based on estimation of Green–Ampt wetting front pressure head from measurement of soil water redistribution. *Soil Sci. Soc. Am. J.* **46**:235–238.

Chudnovskii, A. F., 1966. Plants and Light. I. Radiant energy. pp. 1–51. In: *Fundamentals of Agrophysics*, Israel Prog. for Scientific Translations, Jerusalem.

Clark, F. E. and W. D. Kemper, 1967. Microbial activity in relation to soil water and soil aeration. pp. 447–480. In: H. R. Haise and T. W. Edminster (Eds.), *Irrigation of Agricultural Lands*. Monograph 11. American Society of Agronomy, Madison, WI.

Clothier, B. E., D. R. Scotter, and A. E. Green, 1983. Diffusivity and one-dimensional absorption experiments. *Soil Sci. Soc. Am. J.* **47**:641–644.

Coats, K. H. and B. D. Smith, 1956. Dead end pore volume and dispersion in porous media. *Soc. Pet. Eng. J.* **4**:73–84.

Cole, R. C., 1939. Soil microstructure as affected by cultural treatments. *Hilgardia* **12**:429–472.

Corapciouglu, M. Y. and A. L. Baehr, 1987. A compositional multiphase model for groundwater contamination by petroleum products. I. Theoretical considerations. *Water Resour. Res.* **23**:191–200.

Currie, J. A., 1960. Gaseous diffusion in porous media. II. Dry granular materials. *Br. J. Appl. Phys.* **11**:318–324.

Currie, J. A., 1962. The importance of diffusion in providing the right conditions for plant growth. *J. Sci. Food Agr.* **13**:380–385.

Currie, J. A., 1965. Diffusion within the soil microstructure—A structural parameter for soils. *J. Soil Sci.* **16**:279–289.

Currie, J. A., 1970. Movement of gases in soil respiration. *Soc. Chem Ind. Monogr.* **37**:152–171.

Curtis, A. A. and K. K. Watson, 1984. Hysteresis-affected water movement in scale-heterogeneous profiles. *Water Resour. Res.* **20**:719–726.

Dalton, F. N. and M. Th. Van Genuchten, 1986. The time domain reflectrometry method for measuring soil water content and salinity. *Geoderma* **38**:237–250.

Darcy, H. 1856. *Les Fontaines Publiques de la Ville de Dijon*. Dalmont, Paris.

Davidson, J. M., D. R. Nielsen, and J. W. Biggar, 1963. The measurement and description of water flow through Columbia silt loam and Hesperia sandy loam. *Hilgardia* **34**:601–617.

Day, P. R., 1942. The moisture potential of soils. *Soil Sci.* **54**:391–400.

Demolon, A. and S. Hénin, 1932. Recherches sur la structure des limons et la synthese des aggregates. *Soil Res.* **3**:1–9.

De Vries, D. A. 1958. Simultaneous transfer of heat and moisture in porous media. *Trans. Am. Geophys. Un.* **39**:909–916.

De Vries, D. A. 1963. Thermal properties of soils. In: Van Wijk, W. R. (Ed.), *Physics of Plant Environment*. North-Holland, Amsterdam.

De Vries, D. A. and A. J. Peck, 1968. On the cylindrical probe method of measuring thermal conductivity with special reference to soils. *Aust. J. Phys.* **11**:255–271.

Dirksen, C., 1975. Determination of soil water diffusivity by sorptivity measurements. *Soil Sci. Soc. Am. Proc.* **39**:22–27.

Dirksen, C., 1979. Flux-controlled sorptivity measurements to determine soil hydraulic property functions. *Soil Sci. Soc. Am. J.* **43**:827–834.

Dixon, J. B. and S. B. Weed, 1988. *Minerals in Soil Environments*, 2nd. ed. Soil Science Society of America, Madison, WI.

Dobsen, M. C., F. T. Ulaby, M. T. Hallikainen, and M. A. El-Rayes, 1986. Microwave dielectric behavior of wet soil. 2. Dielectric mixing models. *IEEE Trans. Geosci. Remote Sensing* **GE-23**:35–46.

Doorenbos and Pruitt, 1976. *Crop Water Requirements*. Irrigation and Drainage Paper 24. FAO, Rome.

Douglas, L. A., 1988. Vermiculite. pp. 259–292. In: J. B. Dixon and S. B. Weed (Eds.), *Minerals in Soil Environments*, 2nd. ed. Soil Science Society of America, Madison, WI.

Dracos, T., 1987. Immiscible transport of hydrocarbons infiltrating in unconfined aquifers. In: J. H. Vandermeulen and S. E. Hrudney (Eds.), *Oil in Freshwater: Chemistry, Biology, Countermeasure Technology*. Pergamon, New York.

Drees, L. R., L. P. Wilding, N. E. Smeck, and A. L. Senaki, 1988. Silica in soils: Quartz and disordered silica polymorphs. In: J. B. Dixon and S. B. Weed (Eds.) *Minerals in Soil Environments*, 2nd. ed. Soil Science Society of America, Madison, WI.

Dutt, G. C. and K. K. Tanji, 1962. Predicting concentrations of solutes in water percolated through a column of soil. *J. Geophys. Res.* **67**:3437–3439.

Dutt, G. R., M. J. Shaffer, and W. J. Moore, 1972. Computer simulation of dynamic biophysicochemical processes in soils. Ariz. Ag. Expt. Stn. Bull 196.

Dyson, J. S. and R. E. White, 1986a. A simple approach to solute transport in layered soils. *J. Soil Sci.* **40**:525–542.

Dyson, J. S. and R. E. White, 1986b. The effect of irrigation rate on solute transport in soil during steady water flow. *J. Hydrol.* **107**:19–29.

Edelman, C. H. and J. C. L. Favagee, 1940. On the crystal structure of montmorillinite and halloysite. *Z. Krist.* **102**:417–431.

Edlefsen, N. E. and A. B. C. Anderson, 1943. Thermodynamics of soil moisture. *Hilgardia* **15**:31–298.

Fanning, D. S., V. R. Keramidas, and M. A. El-Desoky, 1988. Micas. In: J. B. Dixon and S. B. Weed, (Eds.) *Minerals in Soil Environments*, 2nd. ed. Soil Science Society of America, Madison, WI.

Farrell, D. A., E. L. Greacen, and C. G. Gurr, 1966. Vapor transfer in soil due to air turbulence. *Soil Sci.* **102**:305–313.

Fatt, I. and H. Dykstra, 1951. Relative permeability studies. *Petrol. Trans. AIME* **192**:249–255.

Fieldes, M., 1962. The nature of the active fraction of soils. pp. 62–78. In: *Trans. Jt. Mtg. Comm. IV–V*. International Society of Soil Science, New Zealand.

Fieldes, M. and R. K. Schofield, 1960. Mechanism of ion adsorption by inorganic soil colloids. *N.Z.J. Sci.* **3**:563–569.

Flegg, P. B., 1953. The effect of aggregation on diffusion of gases and vapors through soils. *J. Sci. Food Ag.* **4**:104–108.

Freeze, R. A., 1969. The mechanism of natural ground water recharge and discharge. *Water Resour. Res.* **5**:153–171.

Freyberg, D., 1986. A natural gradient experiment on solute transport in a sand aquifer 2. Spatial moments and advection and dispersion of nonreactive trancers. *Water Resour. Res.* **22**:2031–2046.

Fried, J. J., 1975. *Groundwater Pollution*. Elsevier, New York.

Fripiat, J. J., 1964. Surface chemistry and soil science. pp. 3–13. In: *Experimental Pedology*, Proc. of 11th Easter School in Agric. Sci., University of Nottingham.

Gapon, Y. N., 1933. On the theory of exchange adsorption in soils. *J. Gen. Chem.* USSR (English translation) **3**:144–160.

Gardner, W. H., 1986. Water content. pp. 493–544. In: A. Klute (Ed.), *Methods of Soil Analysis. Part 1*. Monograph 9, American Society of Agronomy, Madison, WI.

Gardner, W. R., 1958. Some steady state solutions of the unsaturated moisture flow equation with application to evaporation from a water table. *Soil Sci.* **85**:228–232.

Gardner, W. R. and M. Fireman, 1958. Laboratory studies of evaporation from soil columns in the presence of a water table. *Soil Sci.* **85**:244–249.

Gardner, W. R. and D. Kirkham, 1952. Determination of soil moisture by neutron scattering. *Soil Sci.* **73**:392–401.

Gardner, W. R., D. I. Hillel, and Y. Benyamini, 1970. Post irrigation movement of soil water. I. Redistribution. *Water Resour. Res.* **6**:851–861.

Gast, R. 1977. Surface and colloid chemistry. In: J. B. Dixon and S. B. Weed (Eds.), *Minerals and Soil Environments*. Soil Science Society of America, Madison, WI.

Gee, G. W. and J. W. Bauder, 1986. Particle-size analysis. pp. 383–412. In: A. Klute (Eds.), *Methods of Soil Analysis. Part 1*. Monograph 9, American Society of Agronomy, Madison, WI.

Geiger, R., 1965. *The Climate Near the Ground*. Harvard University Press, Cambridge, MA.

Gelhar, L. W. and C. Axness, 1983. Three dimensional stochastic analysis of macrodispersion in aquifers. *Water Resour. Res.* **19**:161–180.

Gelhar, L. W., A. Mantaglou, C. Welty, and K. R. Rehfeldt, 1985. A review of field scale physical solute transport processes in unsaturated and saturated porous media. EPRI Topical Report EA-4190. Electric Power Research Institute, Palo Alto, CA.

Germann, P., 1985. Kinematic wave approach to infiltration and drainage into and from soil macropores. *Trans. Am. Soc. Ag. Eng.* **28**:745–749.

Goates, J. R. and S. J. Bennett, 1957. Thermodynamic properties of water adsorbed on soil minerals: kaolinite. *Soil Sci.* **83**:325–330.

Green, D. W., H. Dabiri, C. F. Weinaug, and R. Prill, 1970. Numerical modeling of unsaturated ground water flow and comparison of the model to field experiment. *Water Resour. Res.* **6**:862–864.

Green, R. E., 1974. Pesticide–clay–water interactions. pp. 3–38. In: W. D. Guenzi (Ed.), *Pesticides in Soil and Water*. Soil Science Society of America, Madison, WI.

Green, R. E. and J. C. Corey, 1971. Calculation of hydraulic conductivity: A further evaluation of predictive methods. *Soil Soc. Sci. Am. Proc.* **35**:3–8.

Green, W. H. and G. A. Ampt, 1911. Studies in soil physics. I. The flow of air and water through soils. *J. Agr. Sci.* **4**:1–24.

Greene-Kelley, R. 1962. Charge densities and heats of immersion of some clay minerals. *Clay Min. Bull.* **5**:1–8.

Greenland, D. J. and J. P. Quirk, 1964. Determination of the total specific surface areas of soils by adsorption of cetyl pyridinium bromide. *J. Soil Sci.* **15**:178–191.

Greenwood, D. J. 1971. Soil aeration and plant growth. *Rep. Prog. Soil Chem.* **55**:423–431.

Grim, R. E., 1953. *Clay Mineralogy*. McGraw Book Co., New York.

Grim, R. E., 1962. *Applied Clay Mineralogy*. McGraw Book Co., New York.

Gurr, C. G., T. J. Marshall, and J. T. Hutton, 1952. Movement of water in soil due to temperature gradients. *Soil Sci.* **74:**333–345.

Hald, A., 1952. *Statistical Theory with Engineering Applications.* Wiley, New York.

Hamaker, J. W., 1972. Decomposition: Quantitative aspects. p. 253–340. In: C.A.I. Goring and M. Hamaker (Eds.) *Organic Chemicals in the Soil Environment*, Marcell Dekker, Inc. New York.

Hamaker, J. W. and J. M. Thompson, 1972. Adsorption. In: *Organic Chemicals in the Soil Environment.* Marcell Dekker, New York.

Handbook of Chemistry and Physics, 1987. Weast, R. C. (ed-in-chief). Chemical Rubber Company, Boca Raton, FL.

Hendricks, S. G., 1942. Lattice structure of clay minerals and some properties of clays. *J. Geol.* **50:**276–290.

Hénin, S., O. Robichet, and A. Jongerius, 1955. Principes pour l'évaluation de la structure du sol. *Ann. Agr.* **6:**537–557.

Hillel, D. I., 1986. Unstable flow in layered soils: A review. *Hydrol. Proc.* **1:**143–147.

Hillel, D. I. and W. R. Gardner, 1969. Steady infiltration into crust-topped profiles. *Soil Sci.* **108:**137–142.

Hillel, D. I. and W. R. Gardner, 1970. Transient infiltration into crust-topped profiles. *Soil Sci.* **109:**69–76.

Himmelblau, D. M., 1970. *Process Analysis by Statistical Methods.* Sterling Swift, Manchecka, TX.

Hofmann, V., K. Endell, and D. Wilm, 1933. Kristallstruktur und Quellung von Montmorillinit. *Z. Krist.* **86A:**304–348.

Holmes, J. W. and J. S. Colville, 1970. Forest hydrology in a Karstic region of Southern Australia. *J. Hydrol.* **10:**59–74.

Horton, R. E., 1933. The role of infiltration in the hydrologic cycle. *Trans. Am. Geophys. Un.*, 14th Ann. Mtg., pp. 446–460.

Horton, R. E., 1939. Analysis of runoff—plot experiments with varying infiltration capacity. *Trans. Am. Geophys. Un.*, 20th Ann. Mtg., Part IV., pp. 693–694.

Horton, R. E., 1940. An approach towards a physical meaning of infiltration capacity. *Soil Soc. Sci. Am. Proc.* **5:**399–417.

Houghton, H. G., 1954. On the annual heat balance of the Northern Hemisphere. *J. Met.* **11:**1–9.

Hsu, P. H., 1988. Aluminum hydroxides and oxyhydroxides. In: J. B. Dixon and S. B. Weed (Eds.), *Minerals in Soil Environments*, 2nd. ed. Soil Science Society of America, Madison, WI.

Humbert, R. P. and B. T. Shaw, 1941. Studies of clay particles with the electron microscope. *Soil Sci.* **52:**481–487.

Idso, S. B., R. J. Reginato, R. D. Jackson, B. A. Kimball, and F. S. Nakayama, 1974. The three stages of drying of a field soil. *Soil Soc. Sci. Am. Proc.* **38:**831–837.

Jackson, M. L., 1964. Chemical composition of soils. pp. 71–141 In: F. E. Bear (Ed.), *Chemistry of Soil.* Reinhold, New York.

Jackson, R. D., 1972. On the calculation of hydraulic conductivity. *Soil Soc. Sci. Am. Proc.* **36:**380–383.

Jackson, R. D. and S. A. Taylor, 1965. Heat transfer. pp. 349–360. In: C. Black (Ed.), *Methods of Soil Analysis*, Vol. 1. American Society of Agronomy, Madison, WI.

Jenny, H. 1932. Studies on the mechanism of ionic exchange in colloidal aluminum silicates. *J. Phys. Chem.* **36**:2217–2258.

Jenny, H. 1938. *Properties of Colloids.* Stanford University Press., Stanford CA.

Jenny, H. and R. F. Reitemeier, 1935. Ionic exchange in relation to the stability of colloidal systems. *J. Phys. Chem.* **39**:593–604.

Jeppson, R. W. and W. R. Nelson, 1970. Inverse formulation and finite difference solution to a partially saturated seepage from canals. *Soil Soc. Sci. Am. Proc.* **34**:9–14.

Jones, M. J. and K. K. Watson 1987. Effect of soil water hysteresis on solute movement during intermittent leaching. *Water Resour. Res.* **23**:1251–1256.

Journel, A. and Ch. J. Huijbregts, 1978. *Mining Geostatistics.* Academic, London.

Jury, W. A., 1973. Simultaneous movement of heat and water through a medium sand Ph.D. Thesis, University of Wisconsin.

Jury, W. A., 1975a. Solute travel-time estimates for tile-drained soils. I. *Theory Soil Sci. Soc. Am. Proc.* **39**:1020–1024.

Jury, W. A., 1975b. Solute travel-time estimates for tile-drained soils. II. Application to experimental studies. *Soil Sci. Soc. Am. Proc.* **39**:1024–1028.

Jury, W. A., 1982. Simulation of solute transport using a transfer function mode. *Water Resour. Res.* **18**:363–368.

Jury, W. A., 1983. Chemical transport modeling: Current approaches and unresolved problems. In: D. W. Nelson, D. E. Elrich, and K. K. Tanji (Eds.), *Chemical Mobility and Reactivity in Soil Systems.* Special Publication No. 42. American Society of Agronomy, Madison, WI.

Jury, W. A., 1985. Spatial variability of soil physical parameters in solute migration: A critical literature review. EPRI Topical Report EA 4228. Electric Power Research Institute, Palo Alto, CA.

Jury, W. A., 1988. Solute transport and dispersion PP1-17. In: W. L. Steffen and O. T. Denmead (Eds.) *Flow and Transport in the Natural Environment*, Springer Verlag, Berlin.

Jury, W. A. and J. Letey, 1979. Water vapor movement in soil: Reconciliation of theory and experiment. *Soil Sci. Soc. Am. J.* **43**:823–827.

Jury, W. A. and E. E. Miller, 1974. Measurement of the transport coefficients for coupled flow of heat and moisture in a medium sand. *Soil Sci. Soc. Am. Proc.* **38**:551–557.

Jury, W. A. and P. F. Pratt, 1980. Estimation of the salt burden of irrigation drainage waters. *J. Environ. Qual.* **9**:141–146.

Jury, W. A. and K. Roth, 1990. *Transfer Functions and Solute Transport Through Soil: Theory and Applications.* Berkhäuser, Basel.

Jury, W. A. and G. Sposito, 1985. Field calibration and validation of solute transport models for the unsaturated zone. *Soil Sci. Soc. Am. J.* **49**:1331–1341.

Jury, W. A., G. Sposito, and R. E. White, 1986. A transfer function model of solute movement through soil. I. Fundamental concepts. *Water Resour. Res.* **22**:243–247.

Jury, W. A., W. R. Gardner, C. B. Tanner, and P. Staffigna, 1976. Model for predicting movement of nitrate and water through a loamy sand. *Soil Sci.* **122**:36–43.

Jury, W. A., H. Frenkel, and L. H. Stolzy, 1978a. Transient changes in the soil water system from irrigation with saline water. I. Theory. *Soil Sci. Soc. Am. J.* **42**:579–584.

Jury, W. A., H. Frenkel, D. Devitt, and L. H. Stolzy, 1978b. Transient changes in the soil water system from irrigation with saline water. II. Analysis of experimental data. *Soil Sci. Soc. Am. J.* **42**:585–590.

Jury, W. A., H. Fluhler, H. Frenkel, D. Devitt, and L. H. Stolzy, 1978c. Use of saline water and minimal leaching for crop production. *Hilgardia* **46**:169–193.

Jury, W. A., J. Letey, and T. Collins, 1982a. Analysis of chamber methods used to measure N_2O production in the field. *Soil Sci. Soc. Am. J.* **46**:250–256.

Jury, W. A., L. H. Stolzy, and P. Shouse, 1982b. A field test of the transfer function model for predicting solute movement. *Water Resour. Res.* **18**:368–374.

Jury, W. A., W. F. Spencer, and W. J. Farmer, 1983a. Behavior assessment model for trace organics in soil. I. Description of model. *J. Environ. Qual.* **12**:558–564.

Jury, W. A., W. F. Spencer, and W. J. Farmer, 1983b. Use of models for assessing relative volatility, mobility, and persistance of pesticides and other trace organics in soil systems. pp. 1–43. In. J. Saxena (Ed.) *Hazard Assessment of Chemicals*, Vol 2, Academic Press, New York.

Jury, W. A., W. J. Farmer, and W. F. Spencer, 1984a. Behavior assessment mode for trace organics in soil. II. Chemical classification and parameter sensitivity. *J. Environ. Qual.* **13**:567–572.

Jury, W. A., W. F. Spencer, and W. J. Farmer, 1984b. Behavior assessment model for trace organics in soil. III. Application of screening model. *J. Environ. Qual.* **13**:573–579.

Jury, W. A., W. F. Spencer, and W. J. Farmer, 1984c. Behavior assessment model for trace organics in soil. IV. Review of experimental evidence. *J. Environ. Qual.* **13**:580–585.

Jury, W. A., D. D. Focht, and W. J. Farmer 1987a. Evaluation of pesticide ground water pollution potential from standard indices of soil-chemical adsorption and biodegradation. *J. Environ. Qual.* **16**:422–428.

Jury, W. A., D. Russo, G. Sposito, and H. Elabd. 1987b. The spatial variability of water and solute transport properties in unsaturated soil. I. Analysis of property variation and spatial structure. *Hilgardia* **55**:1–32.

Jury, W. A., D. Russo, and G. Sposito, 1987c. The spatial variability of water and solute transport properties in unsaturated soil. II. Analysis of scaling theory. *Hilgardia* **55**:33–57.

Jury, W. A., J. S. Dyson, and G. L. Butters, 1990. A transfer function model of field scale solute transport under transient water flow. *Soil Sci. Soc. Am. J.* **54**:327–332.

Kahn, A., 1958. The flocculation of sodium montmorillinite by electrolytes. *J. Coll. Sci.* **13**:51–60.

Kaplan, W., 1984. *Advanced Calculus*. Addison-Wesley, Reading, MA.

Keen, B. A., 1933. Experimental methods for the study of soil cultivation. Empire J. Exp. Agr. **1**:97–102.

Kelley, W. P., 1948. *Cation Exchange in Soils*. American Chemical Society Monograph 109. Reinhold, New York.

Kelley, W. P., W. H. Dore, and S. M. Brown, 1931. The nature of base-exchange material for bentonite, soils, and zeolites as revealed by chemical investigations and x-ray analysis. *Soil Sci.* **41**:259–274.

Kemper, W. D., 1961a. Movement of water as affected by free energy and pressure gradients. I. Application of classic equations for viscous and diffusive movements to the liquid phase in finely porous media. *Soil Sci. Soc. Am. Proc.* **25**:255–260.

Kemper, W. D., 1961b. Movement of water as affected by free energy and pressure gradients. II. Experimental analysis of porous systems in which free energy and pressure gradients act in opposite directions. *Soil Sci. Soc. Am. Proc.* **25**:261–265.

Kemper, W. D., 1965. Aggregate stability. pp. 511–519. In: C. A. Black et al. (Eds.), *Methods of Soil Analysis. Part 1*. Monograph 9. American Society of Agronomy, Madison, WI.

Kemper, W. D. and W. S. Chepil, 1965. Size distribution of aggregates. pp. 499–510. In: C. A. Black et al. (Eds.), *Methods of Soil Analysis. Part 1*. Monograph 9, American Society Agronomy, Madison, WI.

Kersten, M. S., 1949. *Thermal Properties of Soils*, Univ. of Minn. Inst. of Tech. Bull. 28.

Khan A. U. H., 1988. A laboratory test of the dispersion scale effect. Ph.D. Thesis. University of California, Riverside.

Khan, A. U. H. and W. A. Jury, 1990. A laboratory test of the dispersion scale effect. *J. Contam. Hydrol.* **5**:119–132.

Kissel, D. E., J. T. Ritchie, and E. Burnett, 1974. Nitrate and chloride leaching in a swelling clay soil. *J. Environ. Qual.* **3**:401–404.

Klute, A. and C. Dirksen, 1986. Hydraulic conductivity and diffusivity: Laboratory methods. In: A. Klute (Ed.), *Methods of Soil Analysis. Part 1*. Monograph 9. American Society of Agronomy, Madison, WI.

Kostiakov, A. N., 1932. On the dynamics of the coefficient of water percolation in soils and the necessity of studying it from a dynamic view for the purposes of amelioration. *Trans. 6th Congr. Int. Soc. Soil Sci.*, Russian part A: 17–21.

Kristensen, K. J. and E. R. Lemon, 1964. Soil aeration and plant-root relations. III. Physical aspects of oxygen diffusion in the liquid phase of the soil. *Agron. J.* **56**:295–301.

Kutilek, M., 1980. Constant rainfall infiltration. *J. Hydrol.* **45**:289–303.

Lang, C., 1878. Über Wärmecapacität der Bodenconstituenten. *Forsch. Gebiete Agr.-Phys.* **1**:109–147.

Langmuir, I., 1918. The adsorption of gases on plane surfaces of glass, mica, and platinum. *J. Am. Chem. Soc.* **40**:1361–1403.

Lemon, E. R., 1962. Soil aeration and plant root relations I. *Theory Agr. J.* **54**:167–170.

Lemon, E. R. and A. E. Erickson, 1952. The measurement of oxygen diffusion in soil with a platinum electrode. *Soil Sci. Soc. Am. Proc.* **16**:160–163.

Letey, J. and Farmer, W. J., 1973. Mass transfer. In: *Pesticides in Soil and Water*. American Society of Agronomy, Madison, WI.

Letey, J. and L. H. Stolzy, 1964. Measurement of oxygen diffusion rates with the platinum electrode. I. Theory and equipment. *Hilgardia* **35**:545–554.

Libardi, P. L., K. Reichardt, D. R. Nielsen, and J. W. Biggar, 1980. Some simple field methods for estimating soil hydraulic conductivity. *Soil Sci. Soc. Am. J.* **44**:3–6.

Low, P. F., 1959. The viscosity of water in clay systems. In: *Proceedings of the Eighth National Clay Conference*. Pergamon, New York.

Low, P. F., 1961. Physical chemistry of the clay–water interaction. *Adv. Agr.* **13**:269–327.

Lumley, J. L. and A. Panofsky, 1964. *The Structure of Atmospheric Turbulence*. Wiley, New York.

McBratney, A. B. and R. Webster, 1981. The design of optimal sampling schemes for local estimation and mapping of regionalized variables. *Comput. Geosci.* **7**:335–365.

McIntyre, D. S. 1971. The platinum microelectrode method for soil aeration measurements. *Adv. Agr.* **22**:235–283.

Marshall, C. E., 1935a. Mineralogical methods for the study of silts and clays. *Z. Krist. (A)* **90**:8–34.

Marshall, C. E., 1935b. Layer lattices and base exchange clays. *Z. Krist. (A)* **91**:433–449.

Marshall, C. E., 1964. *The Physical Chemistry and Mineralogy of Soils.* Wiley, New York.

Marshall, C. E. and G. Garcia, 1959. Exchange equilibria in a carboxylic resin and in attapulgite clay. *J. Phys. Chem.* **63**:1663–1668.

Marshall, C. E., R. P. Humbert, B. T. Shaw, and O. G. Caldwell, 1942. Studies of clay particles with the electron microscope: II. The fractionation of beidillite, nontrite, magnesium bentonite, and attapulgite. *Soil Sci.* **54**:149–158.

Marshall, T. J., 1958. A relation between permeability and size distribution of pores. *J. Soil Sci.* **9**:1–8.

Marshall, T. J., 1959. The diffusion of gas through porous media. *J. Soil Sci.* **10**:79–82.

Matthias, A. D., A. M. Blackmer, and J. M. Bremner, 1980. A simple chamber technique for field measurement of emissions of nitrous oxide from soils. *J. Environ. Qual.* **9**:251–256.

Mattson, S., 1929. The laws of soil colloidal behavior. *Soil Sci.* **28**:179–220.

Miller, D. E. and W. R. Gardner, 1962. Water infiltration into stratified soil. *Soil Sci. Soc. Am. Proc.* **26**:115–119.

Millington, R. J., 1959. Gas diffusion in porous media. *Science* **130**:100–102.

Millington, R. J. and J. P. Quirk, 1959. Permeability of porous media. *Nature* **183**:387–388.

Millington, R. J. and J. P. Quirk, 1961. Permeability of porous solids. *Trans. Faraday Soc.* **57**:1200–1207.

Monteith, J. L., 1958. The heat balance of soil beneath crops. *UNESCO Arid Zone Res.* **11**:123–128.

Monteith, J. L., G. Szeicz, and K. Yabuki, 1964. Crop photosynthesis and the flux of carbon dioxide below the canopy. *J. Appl. Ecol.* **6**:321–337.

Mortland, M. M. and W. D. Kemper, 1965. Specific surface methods of soil analysis. pp. 532–544. In: C. A. Black et al. (Eds.), *Methods of Soil Analysis. Part 1.* Monograph 9, American Society Agronomy, Madison, WI.

Mualem, Y., 1976a. A catalogue of the properties of unsaturated soils. Research Report 442. Technion, Israel Institute of Technology, Haifa, Israel.

Mualem, Y., 1976b. A new model for predicting the hydraulic conductivity of unsaturated porous media. *Water Resour. Res.* **12**:513–522.

Nakshabandi, G. A. and H. Kohnke, 1965. Thermal conductivity and diffusivity of soils as related to moisture tension and other physical properties. *Agr. Met.* **2**:271–279.

National Oceanic and Atmospheric Administration (NOAA), 1986. *Climatological Data-California: 1986*, Vol. 110, *Annual Summary*, NOAA, Ashville, N.C.

Nazeroff, W. W., S. R. Lewis, S. M. Doyle, B. A. Moed, and A. E. Nero, 1987. Experiments on pollutant transport from soil into residential basements by pressure-driven airflow. *Env. Sci. Tech.* **21**:459–466.

Nelson, R. A. and S. B. Hendricks, 1943. Specific surface of some clay minerals, soils, and soil colloids. *Soil Sci.* **56**:285–296.

Nero, A. E. and W. W. Nazeroff, 1985. Characterizing the source of radon indoors. pp. 23–29. In: *Radiation Protection Dosimetry*, Vol. 7, No. 1–4. Nuclear Radiation.

Nielsen, D. R., J. W. Biggar, and K. T. Erh, 1973. Spatial variability of field-measured soil–water properties. *Hilgardia* **42**:215–259.

Nimah, M. N. and R. J. Hanks, 1973. Model for estimating soil, water, plant, and atmosphere interactions. 1. Description and sensitivity. *Soil Soc. Sci. Am. Proc.* **37**:522–527.

Nkedi-Kizza, P., J. W. Biggar, M. Th. Van Genuchten, P. J. Wierenga, H.M. Selim, J. M. Davidson, and D. R. Nielsen, 1983. Modeling tritium and chloride 36 transport through an aggregated oxisol. *Water Resour. Res.* **19**:691–700.

Novotny, V. and G. Chesters, 1981. *Handbook of Nonpoint Pollution*. Van Nostrand Reinhold, New York.

Ogata, G. and L. A. Richards, 1957. Water content changes following irrigation of bare-field soil that is protected from evaporation. *Soil Sci. Soc. Am. Proc.* **21**:355–356.

Oster, J. D. and B. L. McNeal, 1971. Calculation of soil solution composition variation with water content for desaturated soils. *Soil Sci. Soc. Am. Proc.* **35**:436–442.

Oster, J. D. and J. D. Rhoades, 1975. Calculated drainage water compositions and salt burdens resulting from irrigation with river waters in the Western United States. *J. Environ. Qual.* **4**:73–79.

Parlange, J. Y. and R. E. Smith, 1976. Ponding time for variable rainfall rates. *Can. J. Soil Sci.* **56**:121–123.

Patten, H. E., 1909. Heat transference in soils. Bulletin 59. U.S. Department of Agriculture Burea of Soils, Washington, DC.

Pauling, L., 1930. The structure of micas and related minerals. *Nat. Acad. Sci. Proc.* **16**:123.

Pauling, L., 1948. *General Chemistry*. Freeman, San Francisco.

Peck, A. J., 1983. Field variability of soil physical properties. In: D. I. Hillel (Ed.), *Advances in Irrigation*, Vol. 2. Academic, New York.

Penman, H. L., 1940a. Gas and vapor movements in the soil. I. The diffusion of vapors through porous solids. *J. Agr. Sci.* **30**:437–462.

Penman, H. L., 1940b. Gas and vapor movements in the soil. II. The diffusion of carbon dioxide through porous solids. *J. Agr. Sci.* **30**:570–581.

Penman, H. L., 1948. Natural evaporation from open water, bare soil, and grass. *Proc. Roy. Soc. A* **190**:120–145.

Phene, C., 1986. Oxygen electrode measurement. In: A. Klute (Ed.), *Methods of Soil Analysis. Part 1*. Monograph 9, American Society of Agronomy, Madison, WI.

Philip, J. R., 1957a. The theory of infiltration. 1. The infiltration equation and its solution. *Soil Sci.* **83**:345–357.

Philip, J. R., 1957b. The theory of infiltration. 2. The profile at infinity. *Soil Sci.* **83**:435–448.

Philip, J. R., 1957c. The theory of infiltration. 3. Moisture profiles and relation to experiment. *Soil Sci.* **84**:163–178.

Philip, J. R., 1957d. The theory of infiltration. 4. Sorptivity and algebraic infiltration equations. *Soil Sci.* **84**:257–264.

Philip, J. R., 1957e. The theory of infiltration. 5. Influence of initial moisture content. *Soil Sci.* **84**:329–339.

Philip, J. R., 1957f. Evaporation, and moisture and heat fields in the soil. *J. Met.* **14:**354–366.

Philip, J. R., 1958a. The theory of infiltration. 6. Effect of water depth over soil. *Soil Sci.* **85:**278–286.

Philip, J. R., 1958b. The theory of infiltration. 7. *Soil Sci.* **85:**333–337.

Philip, J. R., 1968. Absorption and infiltration in two and three dimensional systems. pp. 503–525. In: *Proceedings of the UNESCO Symposium, Water in the Unsaturated Zone*, Vol. I. Wageningen.

Philip, J. R., 1969. The theory of infiltration. *Adv. Hydrosci.* **5:**215–296.

Philip, J. R. and D. A. de Vries, 1957. Moisture movement in soils under temperature gradients. *Trans. Am. Geophys. Un.* **38:**222–228.

Portland Cement Association, 1962. *PCA Soil Primer*. Portland Cement Association, Chicago.

Pratt, P. F., W. W. Jones, and V. E. Hunsaker, 1970. Nitrate in deep soil profiles in relation to fertilizer rates and leaching volumes. *J. Environ. Qual.* **1:**97–102.

Pye, V., R. Patrick, and J. Quarles, 1983. *Ground Water Quality in the United States*. University of Pennsylvania Press, Philadelphia.

Raats, P. A. C., 1973. Unstable wetting fronts in uniform and nonuniform soils. *Soil Sci. Soc. Am. Proc.* **37:**681–685.

Raats, P. A. C., 1975. Distribution of salts in the crop root zone. *J. Hydrol.* **27:**237–248.

Raats, P. A. C. and W. R. Gardner, 1975. Movement of water in the unsaturated zone near a water table. pp. 311–358. In: J. Van Schilfgaarde (Ed.) *Drainage for Agriculture*, ASA Monograph 17 American Society of Agronomy, Madison, WI.

Radoslovich, E. W., 1960. The structure of muscovite. *Acta Cryst.* **13:**919–932.

Radoslovich, E. W. and K. Norrish, 1962. The cell dimensions and symmetry of layer-lattice silicates. *Am. Mineral.* **47:**599–616.

Raney, W. A., 1949. Field measurement of oxygen diffusion through soil. *Soil Sci. Soc. Amer. Proc.* **14:**61–65.

Rao, P. S. C. and J. M. Davidson, 1979. Adsorption and movement of selected herbicides at high concentrations in soils. *Water Res.* **13:**375–380.

Rao, P. S. C., A. G. Hornsby, and R. E. Jessup, 1985. Indices for ranking the potential for pesticide contamination of groundwater. *Proc. Soil Crop Soc. Fla.* **44:**1–8.

Rao, P. S. C., D. E. Rolston, R. E. Jessup, and J. M. Davidson, 1980. Solute transport in aggregated porous media. Theoretical and experimental evaluation. *Soil Sci. Soc. Am. J.* **44:**1139–1146.

Rawlins, S. L., 1974. Principles of managing high frequency irrigation. *Soil Sci. Soc. Am. Proc.* **37:**626–629.

Rawlins, S. L. and G. S. Campbell, 1986. Water potential: Thermocouple psychrometry. pp. 597–618. In: A. Klute (Ed.), *Methods of Soil Analysis. Part 1*. Monograph 9. American Society Agronomy, Madison, WI.

Rawlins, S. L. and P. A. C. Raats, 1975. Prospects for high frequency irrigation. *Science* **188:**604–610.

Reginato, R. J., 1974. Gamma radiation measurements of bulk density changes in soil pedon following irrigation. *Soil Sci. Soc. Am. Proc.* **38:**24–29.

Rhoades, J. D., R. D. Ingvalson, J. M. Tucker, and M. Clark, 1973. Minimizing the salt burdens of irrigation drainage waters. *J. Environ. Qual.*, **3:**311–316.

Rhoades, J. D., J. D. Oster, R. D. Ingvalson, J. M. Tucker, and M. Clark, 1974. Salts in irrigation waters. *Soil Sci. Soc. Am. Proc.* **37**:770–774.

Richter, G. and W. A. Jury, 1986. A microlysimeter field study of solute transport through a structured sandy loam soil. *Soil Sci. Soc. Am. J.* **50**:863–867.

Ritchie, J. T., 1972. Model for predicting evaporation from a row crop at incomplete cover. *Water Resour. Res.* **8**:1204–1213.

Robinson, R. A. and R. H. Stokes, 1959. *Electrolyte Solutions*. Butterworth, London.

Rolston, D. E., 1986a. Gas diffusivity. In: A. Klute (Ed.), *Methods of Soil Analysis. Part 1*. Monograph 9. American Society of Agronomy, Madison, WI.

Rolston, D. E., 1986b. Gas flux. In: A. Klute (Ed.), *Methods of Soil Analysis Part 1*. Monograph 9. American Society of Agronomy, Madison, WI.

Romell, L. G., 1922. Luftväxlingen i marken som ekologisk faktor. *Medd. Statens Skogs-farsöks-anstalt* **19**:(2).

Rose, C. W., W. R. Stern, and J. E. Drummond, 1965. Determination of hydraulic conductivity as a function of depth and water content for soil *in situ*. *Aust. J. Soil Res.* **3**:1–9.

Ross, C. S. and E. V. Shannon, 1926. The minerals of bentonite and related clays and their physical properties. *J. Am. Ceram. Soc.* **9**:77.

Roth, K., R. Shulin, H. Fluhler, and W. Attinger, 1990. Calibration of time domain reflectometry for water content measurement using a composite dielectric approach. *Water Resour. Res.* **26**:2267–2274.

Rubin, J., 1968. Numerical analysis of ponded rainfall infiltration. pp. 440–450. In: *Proceedings of the UNESCO Symposium. Water in the Unsaturated Zone*, Vol. I, Wageningen.

Russell, E. J. and A. Appleyard, 1915. The atmosphere of the soil, its composition, and causes of variation. *J. Agr. Sci.* **7**:1–48.

Russell, E. W., 1938. Soil structure. *Imp. Bull. Soil Sci. Tech. Commun.* **37**.

Russell, E. W., 1973. *Soil Conditions and Plant Growth*, 10th ed. Longman, London.

Russo, D. and E. Bresler, 1980. Scaling soil hydraulic properties of a heterogeneous field. *Soil Sci. Soc. Am. J.* **44**:681–684.

Russo, D. and E. Bresler, 1981. Soil hydraulic properties as stochastic processes. I. An analysis of spatial variability. *Soil Soc. Sci. Am. J.* **45**:682–687.

Russo, D. and W. A. Jury, 1987a. A theoretical study of the estimation of the correlation length scale in spatially variable fields. 1. Stationary fields. *Water Resour. Res.* **23**:1257–1268.

Russo, D. and W. A. Jury, 1987b. A theoretical study of the estimation of the correlation length scale in spatially variable fields. 2. Nonstationary fields. *Water Resour. Res.* **23**:1269–1279.

Russo, D., W. A. Jury, and G. L. Butters, 1989. Numerical analysis of solute transport during transient irrigation. I. Effect of hysteresis and profile heterogeneity. *Water Resour. Res.* **25**:2109—2118.

Ryden, J. C., L. J. Lund, and D. D. Focht, 1978. Direct in-field measurement of nitrous oxide flux from soils. *Soil Sci. Soc. Am. J.* **42**:863–869.

Sallam, A., W. A. Jury, and J. Letey, 1984. Measurement of gas diffusion coefficient under relatively low air-filled porosity. *Soil Soc. Sci. Am. J.* **48**:3–6.

Sawhney, B. L. 1988. Interstratification on layer silicates. In: J. B. Dixon and S. B. Weed

(Eds.,), *Minerals in Soil Environments*, 2nd ed. Soil Science Society of America, Madison, WI.

Scarsbrook, G. E. 1965. Nitrogen availability. In: J. W. Bartholomew and F. E. Clark (Eds.), *Soil Nitrogen*. Monograph 10. American Society of Agronomy, Madison, WI.

Schofield, R. K. and H. R. Samson, 1954. Flocculation of kaolinite due to the attraction of oppositely-charged crystal faces. *Disc. Faraday Soc.* **18:**135-145.

Schwertmann, U. and R. M. Taylor, 1988. Iron oxides. In: J. B. Dixon and S. B. Weed (Eds.), *Minerals in Soil Environments*, 2nd ed. Soil Science Society of America, Madison, WI.

Scotter, D. R., G. W. Thurtell, and P. A. C. Raats, 1967. Dispersion resulting from sinusoidal gas flow in porous materials. *Soil Sci.* **104:**306-308.

Shainberg, I. and W. D. Kemper, 1966a. Conductance of adsorbed alkali cations in aqueous and alcoholic bentonite pastes. *Soil Sci. Soc. Am. Proc.* **30:**700-706.

Shainberg, I. and W. D. Kemper, 1966b. Hydration status of adsorbed cations. *Soil Sci. Soc. Am. Proc.* **30:**707-713.

Shainberg, I. and W. D. Kemper, 1967. Ion exchange equilibria on montmorilinite. *Soil Sci.* **103:**4-9.

Shaw, B. T. and R. T. Humbert, 1941. Electron micrographs of clay minerals. *Soil Sci. Soc. Am. Proc.* **6:**146-149.

Shortley, G. and D. Williams, 1965. *Elements of Physics*, 4th ed. Prentice-Hall, Englewood Cliffs, NJ.

Shouse, P., W. A. Jury, and L. H. Stolzy, 1982. Field measurement and modeling of cowpea water use and yield under stressed and well-watered growth conditions. *Hilgardia* **50:**1-25.

Shulin, R., M. Th. Van Genuchten, H. Flühler, and P. Ferlin 1987. An experimental study of solute transport in a stony field soil. *Water Resour. Res.* **23:**1785-1794.

Simmons, C. S., 1982. A stochastic–convective transport representation of dispersion in one dimensional porous media systems. *Water Resour. Res.* **18:**1193-1214.

Slatyer, R. O., 1967. *Plant-Water Relationships*, Academic Press, New York.

Smith, A. 1932. Seasonal subsoil temperature variations. *J. Agric. Res.* **44:**421-428.

Smith, W. O. and H. G. Byers, 1938. The thermal conductivity of dry soil of certain of the great soil groups. *Soil Sci. Soc. Am. Proc.* **3:**13-19.

Spencer, W. F., M. M. Cliath, W. A. Jury, and Lian-Zhang Zhang, 1988. Volatilization of pesticides from soil as related to their Henry's law constants. *J. Envir. Qual.* **17:**504-509.

Sposito, G., 1981. *The Thermodynamics of Soil Solutions*. Oxford University Press, Oxford.

Sposito, G., 1984. *The Surface Chemistry of Soils*. Oxford University Press, Oxford.

Sposito, G., R. E. White, P. R. Darrah, and W. A. Jury, 1986. The transfer function model of solute transport through soil. III. The convection-dispersion equation. *Water Resour. Res.* **22:**255-262.

Srinalta, S., D. R. Nielsen, and D. Kirkham, 1969. Steady flow through a two layer soil. *Water Resour. Res.* **5:**1053-1063.

Stanhill, G., 1965. Observations on the reduction of soil temperature. *Agr. Met.* **2:**197-203.

Stolzy, L. H. and J. Letey, 1964. Characterizing soil oxygen conditions with a platinum electrode. *Adv. Agr.* **16**:249–279.

Sullivan, P. J., G. Sposito, S. M. Strathouse, and C. L. Hanson, 1979. Geologic nitrogen and the occurrence of high nitrate soils in the western San Joaquin Valley. *Hilgardia* **47**:15–31.

Talsma, T. 1969. Infiltration from semi-circular furrows in the field. *Aust. J. Soil Res.* **7**:277–284.

Tanner, C. B., 1968. Evaporation of water from plants and soil. In: T. Koslowski (Ed.), *Water Deficits and Plant Growth*. Academic, New York.

Taylor, G. I., 1953. The dispersion of soluble matter flowing through a capillary tube. *Proc. Math. Soc. Lond.* **2**:196–212.

Thomas, M. D., 1928. Aqueous vapor pressure in soils. *Soil Sci.* **25**:485–493.

Tiulin, A. F., 1928. Questions on soil structure. II. Aggregate analysis as a method of determining soil structure. *Perm. Agr. Exp. Stn. Div. Ag. Chem. Rep.* **2**:77–122.

Topp, G. C., J. L. Davis, and A. P. Annan, 1980. Electromagnetic determination of soil water content: Measurement in coaxial transmission lines. *Water Resour. Res.* **16**:574–582.

Tuorilla, P., 1928. Uber Beziehungen zwischen Koagulation, elektrokinetischzen Wanderungsgeschwindigkeiten, Ionhydration und chemischer Beeinflussung. *Kolloidchem. Beihefte* **27**:44–181.

Ulrich, R., 1894. Untersuchungen über die Wärmecapacität der Bodenconstituenten. *Forsch. Gebiete Agr.-Phys.* **17**:1–31.

U.S. Department of Agriculture, 1951. pp. 207–208. In: *Soil Survey Manual*, USDA Handbook 18. USDA, Washington, DC.

U.S. Salinity Laboratory Staff, 1954. Diagnosis and improvement of saline and alkali soils. U.S. Department of Agriculture Handbook 60.

Valocchi, A. J., 1985. Validity of the local equilibrium assumption for modeling sorbing solute transport through homogeneous soils. *Water Resour. Res.* **21**:808–820.

Van Bavel, C. H. M., 1952. Gaseous diffusion and porosity in porous media. *Soil Sci.* **73**:91–104.

Van Bavel, C. H. M., N. Underwood, and R. W. Swanson, 1956. Soil moisture measurement by neutron moderation. *Soil Sci.* **82**:29–41.

Van Duin, R. H. A., 1963. The influence of soil management on the temperature wave near the surface. Technical Bulletin 29. Institute for Land and Water Management Research, Wageningen.

Vaneslow, A. P., 1932. Equilibria of the base exchange reaction of bentonites, permutites, soil colloids, and zeolites. *Soil Sci.* **33**:95–113.

Van Genuchten, M. Th., 1980. A closed form equation for predicting the hydraulic conductivity of unsaturated soils. *Soil Soc. Sci. Am. J.* **44**:892–898.

Van Genuchten, M. Th. and W. A. Jury, 1987. Progress in unsaturated flow and transport modeling. *IUGG Rev.* **25**:135–140.

Van Genuchten, M. Th. and D. R. Nielsen, 1985. On describing and predicting the hydraulic properties of unsaturated soils. *Ann. Geophys.* **3**:615–628.

Van Genuchten, M. Th. and P. J. Wierenga, 1976. Mass transfer studies in sorbing porous media. 1. Analytical solutions. *Soil Sci. Soc. Am. J.* **40**:473–480.

Van Genuchten, M. Th. and P. J. Wierenga, 1977. Mass transfer studies in sorbing porous media. II. Experimental evaluation with tritium. *Soil Sci. Soc. Am. J.* **41**:272–278.

Van Genuchten, M. Th., P. J. Wierenga, and G. A. O'Connor, 1977. Mass transfer studies in sorbing porous media. III. Experimental evaluation with 2, 4, 5-T. *Soil Sci. Soc. Am. J.* **41**:278–285.

Van Olphen, H., 1963. *Clay Colloid Chemistry.* Interscience, New York.

Van Rooyen, M. and H. F. Winterkorn, 1959. Structural and textural influences on the thermal conductivity of soils. *Highway Res. Bd. Proc.* **38**:576–621.

Van Wijk, W. R., 1963. *Physics of Plant Environment.* North-Holland, Amsterdam.

Von Schwarz, A. R., 1879. Vergleichende Versuche über die physikalischen Eigenschaften verschiedener Bodenarten. Forsch. *Gebiete Agr.-Phys.* **2**:164–169.

Wada, K., 1988. Allophane and imogolite. In: J. B. Dixon and S. B. Weed (Eds.), *Minerals in Soil Environments*, 2nd. ed. Soil Science Society of America, Madison, WI.

Warrick, A. W. and D. R. Nielsen, 1980. Spatial variability of soil physical properties in the field. pp. 319–344. In: D. I. Hillel (Ed.), *Applications of Soil Physics*, Academic, New York.

Warrick, A. W., D. E. Myers, and D. R. Nielsen, 1986. Geostatistical methods applied to soil science. pp. 53–82. In: A. Klute (Ed.), *Methods of Soil Analysis. Part 1.* Monograph 9, American Society of Agronomy, Madison, WI.

Whisler, F. D., A Klute, and D. B. Peters, 1986. Soil water diffusivity from horizontal infiltration. *Soil Sci. Soc. Am. Proc.* **32**:6–11.

White, E. C. and D. G. Moore, 1972. Nitrates in South Dakota range soils. *J. Range Mgmt.* **25**:27–32.

White, R. E., 1986. Solute transport in macropores. In: D. I. Hillel (Ed.), *Advances in Irrigation*, Vol. 2. Academic, New York.

White, R. E., 1987. A transfer function model for the prediction of nitrate leaching under field conditions. *J. Hydrol.* **92**:207–222.

White, R. E., G. W. Thomas, and M. S. Smith, 1984. Modeling water flow through undisturbed soil cores using a transfer function model derived from ^{3}HOH and Cl transport. *J. Soil Sci.* **35**:159–168.

White, R. E., J. S. Dyson, R. A. Haigh, W. A. Jury, and G. Sposito, 1986. A transfer function model of solute movement through soil. II. Experimental applications. *Water Resour. Res.* **22**:247–254.

Wiegand, C. L. and E. R. Lemon, 1958. A field study of some plant–soil relations in aeration. *Soil Sci. Soc. Am. Proc.* **22**:216–221.

Wiegner, G., 1925. Dispersität und Basenaustauch. Report of the Second Commission: 4th International Congress of Soil Science, Rome.

Wierenga, P. J., 1977. Solute distribution profiles computed with steady-state and transient flow models. *Soil Sci. Soc. Am. J.* **41**:1050–1055.

Wiersum. L. K., 1960. Some experiences in soil aeration measurements and relationships to the depth of rooting. *Neth. J. Agric. Sci.* **8**:245–252.

Wiklander, L., 1964. Cation and anion exchange phenomena. pp. 163–205. In: F. E. Bear (Ed.), *Chemistry of the Soil.* Reinhold, New York.

Willey, C. R. and C. B. Tanner, 1963. Membrane-covered electrode for measurement of oxygen concentration in soil. *Soil Sci. Soc. Am. Proc.* **27**:511–515.

Williamson, R. E. and J. van Schilfgaarde, 1965. Studies of crop response to drainage. II. Lysimeters. *Trans. ASAE* **8:**98–100, 102.

Wollney, E., 1878. Untersuchungen über den Einfluss der Exposition auf der Erwärmung des Bodens. *Forsch. Gebiete Agr. Phys.* **1:**263–294.

Wollney, E., 1883. Untersuchungen über den Einfluss der Pflanzendecke und der Beschttung auf die physikalischen Eigenschaften des Bodens. *Forsch. Gebiete Agr. Phys.* **6:**197–256.

Wooding, R. A., 1968. Steady infiltration from a shallow circular pond. *Water Resour. Res.* **4:**1259–1273.

Yakuwa, R., 1945. Über die Bodentemperaturen in dem verschiedenem Bodenarten in Hokkaido. *Geophys. Mag. Tokyo* **14:**1–12.

Yoder, R. E., 1937. The significance of soil structure in relation to the tilth problem. *Soil Sci. Soc. Am. Proc.* **2:**21–23.

INDEX